THE POLITICS OF
SPACE

THE POLITICS OF
SPACE

A SURVEY

FIRST EDITION

Editor: Eligar Sadeh

LONDON AND NEW YORK

First Edition 2011
Routledge
2 Park Square, Milton Park, Abingdon, Oxon OX14 4RN
711 Third Avenue, New York, NY 10017, USA

Routledge is an imprint of the Taylor & Francis Group, an informa business

© Routledge 2011

All rights reserved. No part of this publication may be photocopied, recorded, or otherwise reproduced, stored in a retrieval system or transmitted in any form or by any electronic or mechanical means without the prior permission of the copyright owner.

ISBN 978-1-85743-419-4 (hbk)
ISBN 978-0-203-83938-6 (ebk)

Editor Europa New Projects: Cathy Hartley

Typeset in Times New Roman 10.5/12

The publishers make no representation, express or implied, with regard to the accuracy of the information contained in this book and cannot accept any legal responsibility for any errors or omissions that may take place.

Typeset by Taylor & Francis Books

Foreword

The goal of this volume, *The Politics of Space: A Survey*, is to provide an overview of the politics of space as played out among national space efforts, and national and international organizations. The development and exploration of space are scientific and engineering pursuits. However, politics determine which space programmes and projects are pursued, and establish cost, performance and schedule parameters for these ends.

This volume consists of three parts. The first comprises essays on different aspects of space policy. The first essay provides an explanation of space politics that often interfere with rational outcomes, leading to space programmes that are not well optimized, cost too much, and are not on schedule. The chapter draws on the examples of space politics in the USA and the politics related to national security space, export controls of space technologies, and the development of space launch. The second essay elaborates on the national security space sector of the USA. Since the rise of the space age, especially beginning in the 1980s and 1990s, the USA developed the dominant military space programme in the world for the purposes of warfare. Today, the USA is critically dependent on military space assets in the modern theatre of warfare and for providing security and intelligence.

The third essay examines the progress of development in major spacefaring states and multinational organizations other than the USA. These include Russia, the United Kingdom, France, Germany, Italy, Spain, the European Space Agency, the European Union, India, the People's Republic of China, Japan and Israel. Clearly, space capabilities have proliferated world-wide since the early dominance of the USA and former Soviet Union in the 1950s and 1960s. Indicative of this trend in proliferation, the fourth essay examines the space politics and unique role of moderate space powers in international relations, with the example of Canada, and those in developmental stages, such as Brazil and the Republic of Korea.

Following this, the fifth essay focuses on commercial space actors in the USA. Since the 1960s and the rise of commercial satellite communications, these actors have influenced space politics, whereby national space efforts, in the USA and in other spacefaring states, have sought to foster commercial and business opportunities in space. The last three essays examine the role of both governmental and non-governmental organizations. These organizations are important components of the political, economic and social environment of space activities.

The second part of this volume, an A–Z glossary, surveys space-related civil, commercial and military actors. There is a plethora of such organizations across these sectors world-wide, and this survey provides an overview of their history and development. To illustrate, the largest international organization in the space arena, the International Astronautical Federation, currently consists of 198 member organizations from 53 different countries.

The third section of the volume provides figures, tables and documentation on key issues of space politics. Important figures include: models of space political processes; issues related to national security space management, challenges and environment; geopolitics of space; and the dynamics of relations between the governmental and commercial space sectors. Tables of significance highlight programmatic attributes of national security space, comparisons in the development of space programmes among states worldwide, and the functions and roles of international organizations that deal with space. Documentation is focused on the key outcome of the political process, encompassing excerpts from a number of important space law documents.

This volume is intended for education and reference. It is useful for students of space policy, academic scholars and policy analysts at think tanks and research institutes, senior management and policy officials from international and governmental agencies and departments relevant to space issues, military officers and operators in relevant units, commands and in staff colleges and service academies, senior management and policy officials from major aerospace corporations and scientists and engineers interested or involved in space policy issues.

Eligar Sadeh
August 2010

Contents

Illustrations and Documentation	viii
Acknowledgements	x
The Contributors	xi
Abbreviations	xiv
ESSAYS	1
Politics of Space	3
ELIGAR SADEH	
National Security Space	29
PETER L. HAYS	
Developed Space Programmes	58
LAURENCE NARDON	
Moderate Space Powers	77
WADE L. HUNTLEY	
Commercial Space Actors	104
DAVID D. CHEN AND MOLLY K. MACAULEY	
International Organizations in Civil Space Affairs	120
HENRY R. HERTZFELD	
Non-governmental Space Organizations	143
JAMES A. VEDDA	
Public Sector Actors	161
ROGER B. HANDBERG	
A–Z GLOSSARY OF SPACE ORGANIZATION	175
ELIGAR SADEH AND CRIS SADEH	

Illustrations and Documentation

FIGURES

1.1	Space Politics and Policy	311
1.2	Space Policy Implementation	312
1.3	National Security Space Community	312
1.4	Proposed National Security Space Authority	314
2.1	Investment in National Security Space Programmes	315
2.2	Challenges of the Global Security Environment	316
4.1	Geopolitics of Space	318
5.1	Risk to Commercial Return Relationship	319

TABLES

2.1	Space-Enabled Reconnaissance Strike Complex	323
2.2	Force Enhancement Space Systems	324
2.3	Attributes of Military Space Doctrines	325
4.1	Moderate State Civil Satellites	326
4.2	Moderate State Military Satellites	329
4.3	Moderate State Space Programmes	330
6.1	Functions of Selected Bodies in the United Nations	332
6.2	United Nations Treaties and Resolutions of Space Law	333
6.3	Regional Space Organizations	334
6.4	Global Space Organizations	336
8.1	Government Sector Interventions in Space	338

DOCUMENTATION

6.1	International Cooperation in the Peaceful Uses of Outer Space	341
6.2	European Space Agency	342
6.3	European Space Policy	343
6.4	Asia-Pacific Space Cooperation Organization	344
6.5	International Telecommunication Union	344
6.6	World Meteorological Organization	345
6.7	Committee on Earth Observation Satellites	345

6.8	Global Earth Observation System of Systems	346
6.9	Space and Major Disasters	348
6.10	Inter-Agency Space Debris Coordination Committee	348
6.11	Wassenaar Arrangement	349

Acknowledgements

The genesis of this book began back in 2006 when Routledge and I recognized a need for establishing a basic reference text on the politics of space. To this end, I took the initiative to make that a reality, which is this volume on *The Politics of Space*.

I am grateful for the support provided by Routledge for this effort and, in particular, for the guidance and patience of Cathy Hartley, Editor Europa New Projects at Routledge.

I would like to thank as well the contributing authors to this volume, who include: Peter L. Hays, Laurence Nardon, Wade L. Huntley, David C. Chen, Molly K. Macauley, Henry R. Hertzfeld, James A. Vedda and Roger B. Handberg. The expertise and insights into space politics provided by the contributing authors are reflected here in the pages of this volume. Lastly, I would like to thank the assistance of Cristian H. Sadeh for help in completing the glossary part of this volume.

Eligar Sadeh
August 2010

The Contributors

Eligar Sadeh, PhD is President of Astroconsulting International, which addresses challenges facing Security, Civil, and Commercial space by providing Organizational Consulting, Research of Best Practices, Professional Education, and Workforce Development. Eligar also serves as a Research Associate with the Center for Space Studies at the University of Colorado, Colorado Springs, and as Editor of the academic journal *Astropolitics* published by Taylor and Francis, Routledge. Eligar held professorships in Space and Defense Studies at the University of North Dakota and at the United States Air Force Academy, and worked for Lockheed Martin as a Space Systems Engineer. He served as well as a Research Associate with the Eisenhower Center for Space and Defense Studies at the United States Air Force Academy and with the Space Policy Institute at George Washington University. Eligar has a number of publications, including *Space Strategy for the 21st Century* (Routledge, Taylor and Francis, under development for 2011), *Politics of Space: A Survey* (Routledge, Taylor and Francis, 2010), and *Space Politics and Policy: An Evolutionary Perspective* (Kluwer Academic Publishers, 2003).

David C. Chen, MA is a Principal Research Analyst at Centra Technology, with a regional focus on the Asia-Pacific region. He participated in the Organisation for Economic Cooperation and Development Space Futures Programme by modelling and forecasting future economic scenarios vis-à-vis space industries. He has also authored papers and presented on Chinese military modernization at the annual PLA conference at Carlisle Barracks and the Naval War College. He serves on the Editorial Board of *Space Policy* and is a reviewer for *China Security Quarterly*. A native Mandarin speaker, he has studied at Nanjing University, China, and in Taipei, Taiwan.

Roger B. Handberg, PhD is Chair and Professor of the Department of Political Science at the University of Central Florida. His research has focused on space policy, ballistic missile defense, science and technology policy, energy policy, and defense and security policy He has an extensive space-related publication record, which includes: *Reinventing NASA and the Quest for Outer Space*; *Ballistic Missile Defense and the Future of American Security*; *New World Vistas: Militarization of Space*; *The Dormant Frontier: Changing the Space Paradigm for the 21st Century*; *The Future of the Space Industry*; and *International Space Commerce: Building from Scratch*.

THE CONTRIBUTORS

Peter L. Hays, PhD works for SAIC supporting the Department of Defense and the Eisenhower Center for Space and Defense Studies, and teaches at George Washington University. He helps develop space policy initiatives including the National Defense University Spacepower Theory Study. He served internships at the White House Office of Science and Technology Policy and National Space Council, and taught space policy courses at the US Air Force Academy, School of Advanced Airpower Studies, and the National Defense University. Major book publications include *Spacepower for a New Millennium* and *United States Military Space*.

Henry R. Hertzfeld, PhD is a Research Professor of Space Policy and International Affairs at the Space Policy Institute of George Washington University. He is an expert in the economic, legal and policy issues of space and advanced technological development. He has served as a Senior Economist and Policy Analyst at both NASA and the National Science Foundation, and is a consultant to both US and international agencies and organizations. He is the co-editor of *Space Economics*, as well as many articles on the economic and legal issues concerning space and technology.

Wade L. Huntley, PhD is Senior Lecturer in the National Security Affairs Department at the Naval Postgraduate School in Monterey, California. His publications include four edited volumes and over fifty peer-reviewed articles, book chapters and scholarly essays on nuclear proliferation, arms control, East and South Asian regional security, U.S. policy and international relations theory. Recent space-related works include "Planning the Unplannable: Scenarios on the Future of Space," *Space Policy* 26:1 (February 2010); and "Smaller State Perspectives on the Future of Space Governance," *Astropolitics* 5:3 (Fall 2007). Previously, Dr. Huntley was Director of the Simons Centre for Disarmament and Non-Proliferation Research at the University of British Columbia in Vancouver, Canada; Associate Professor at the Hiroshima Peace Institute in Hiroshima, Japan; and Director of the Global Peace and Security Program at the Nautilus Institute in Berkeley, California. He received his doctorate in political science from the University of California at Berkeley in 1993 and has taught at several universities.

Molly K. Macauley, PhD is a Research Director and Senior Fellow at Resources for the Future. Her research expertise includes the economics of new technologies, the value of information, space economics and policy, and the use of economic incentives in environmental regulation and other policy design. She has frequently testified before Congress, and serves on national-level committees and panels, including the Space Studies Board, the Climate Working Group of NOAA, and the Earth Science Applications Analysis Group of NASA. She also served as a lead author on a project under the US Climate Change Science Program. She has published widely and serves as a visiting professor in the Department of Economics at Johns Hopkins University.

THE CONTRIBUTORS

Laurence Nardon, PhD is a senior research fellow at the *Institut français des relations internationales*, located in Paris, France. She directs the Space Policy programme for the Institute. Her field of expertise is space policy, looking at the military, commercial and exploratory aspects of space programmes in Europe, the USA, Asia and Russia. Previously, she was a research analyst at *Aérospatiale Espace et Défense*, then she was a Fulbright scholar at the Space Policy Institute (G. Washington University) in 2003, at the *Ecole des Hautes Etudes en Sciences Sociales*, and a Senior Visiting Fellow at the Center for Strategic and International Studies in Washington, DC.

James A. Vedda, PhD is Senior Policy Analyst at the Aerospace Corporation Center for Space Policy & Strategy, where he has been performing research and analyses on security, civil and commercial space issues. Previously he was assigned to the Office of the Secretary of Defense working on space policy and homeland defense issues. Before that, he was an associate professor in the Department of Space Studies at the University of North Dakota, where he taught undergraduate and graduate courses on civil, commercial and military space policy. He is the author of *Choice, Not Fate: Shaping a Sustainable Future in the Space Age*. His writing has appeared also in journals – *Space Policy, Space News, Astropolitics, Space Times, Ad Astra, Space Business News*, and *Quest*.

CONTRIBUTOR TO THE GLOSSARY

Cris Sadeh is CEO of Astroconsulting International. She holds a business degree and specializes in space management, as well as South American space programmes. She serves as a Research Associate with the US Air Force Academy, and develops cultural programmes for Astroconsulting and the US Air Force.

RESEARCH ASSISTANTS

Nathalie Rangel Iovino holds a MA degree from Fordham University and a BA from the Federal University of Goiana in Brazil. She has worked in journalism, as well as the film industry in Brazil. Her work included research on the history and politics of Brazil.

Joni Wilson, MA works as a professional executive and administrative assistant. She authored, abstracted, edited, copyedited, proofread and formatted a variety of published items, including journals, articles, books, aerospace encyclopedia, aerospace journal and abstracter for ABC-CLIO scholarly journals. She has also served as an educational specialists and public relation assistant, and has been involved in compiling, editing and writing adult texts, newsletters and journals.

Abbreviations

AAAF	Association Aéronautique et Astronautique de France
AAC	Ångström Aerospace Corporation Microtec
AAS	American Astronautical Society
AATE	Asociación Argentina de Tecnología Espacial
ABAE	Agencia Bolivariana para Actividades Espaciales
ABMA	Army Ballistic Missile Agency
ACE	Astronaute Club Européen
ADD	Agency for Defence Development
ADEOS	Advanced Earth Observing Satellite
AEB	Agência Espacial Brasileira
AEHF	Advanced Extremely High Frequency
AEXA	Agencia Espacial Mexicana
AFB	Air Force Base
AFRL	Air Force Research Laboratory
AFRL/SD	Air Force Research Laboratory Space Vehicles Directorate
AFSATCOM	Air Force Satellite Communications
AFSCN	Air Force Satellite Control Network
AFSPC	Air Force Space Command
AGI	Analytical Graphics, Inc.
AGS	Americom Government Services
AIA	Aerospace Industries Association
AIAA	American Institute of Aeronautics and Astronautics
AIDA	Associazione Italiana per l'Aerospazio
AIDAA	Associazione Italiana di Aeronautica e Astronautica
AIPAS	Associazione Italiana per L'Aerospazio
AIRSS	alternative infrared satellite system
ALOMAR	Arctic Lidar Observatory for Middle Atmosphere Research
ALOS	Advanced Land Observation Satellite
ALS	Advanced Launch System
ANASA	Azerbaijan National Aerospace Agency
ANGKASA	Agensi Angkasa Negara
APIC	Asia-Pacific International Space Year Conference
APL	The Johns Hopkins University Applied Physics Laboratory
AP-MCSTA	Asia-Pacific Multilateral Cooperation in Space Technology and Applications
APRSAF	Asia-Pacific Regional Space Agency Forum

ABBREVIATIONS

APSCO	Asia-Pacific Space Cooperation Organization
Arabsat	Arab Satellite Communications Organization
ARCA	Asociaţia Română pentru Cosmonautică şi Aeronautică
ARI	Aerospace Research Institute
ARPA	Advanced Research Projects Agency
ARR	Andøya Rocket Range
ASAL	Agence Spatiale Algérienne
ASAT	anti-satellite
ASC/CSA	Agence Spatiale Canadienne/Canadian Space Agency
ASD	AeroSpace and Defence Industries Association of Europe
ASERA	Australian Space Engineering Research Association
ASI	Agenzia Spaziale Italiana
ASI	Astronautical Society of India
ASI-ENAV	Italian National Satellite Navigation Programme for Civil Aviation
ASOpS	Advanced Space Operations School
ASR	Agenţia Spaţială Română
ASRI	Australian Space Research Institute
AST	Office of Commercial Space Transportation
A*STAR	Agency for Science, Technology & Research
ASTC	Astrotech Corporation
ÅSTC	Ångström Space Technology Center
ASTOS	Association of Specialist Technical Organisations for Space
AT&L	US Undersecretary of Defense for Acquisition, Technology and Logistics
AT&T	American Telegraph and Telephone
ATEx	Advanced Tether Experiment
ATK	Alliant Techsystems
ATR	Avions de Transport Regional
ATSB	Astronautic Technology Sdn Bhd
ATUCOM	Association Tunisienne de la Communication et des Sciences Spatiales
ATV	Automated Transfer Vehicle
BAS	Bulgarian Aerospace Agency
BASIC	Broad Area Space-based Imagery Collector
BASMATES	High Altitude Commercial Solar Balloon for Scientific Equipment
BDLI	Bundesverband der Deutscher Luft-und Raumfarhtindustrie
BIRA	Belgisch Instituut voor Ruimte-Aëronomie
BIS	British Interplanetary Society
BLP	Bangladesh Landsat Programme
BMD	Ballistic Missile Defense
bn	billion
BNSC	British National Space Centre
CALT	Chinese Academy of Launch Vehicle Technology

CAMET	China Aerospace Machinery and Electronics Corporation
CAS	Chinese Academy of Sciences
CASC	China Aerospace Corporation
CASI	Canadian Aeronautics and Space Institute
CAST	Chinese Academy of Space Technology
CASTLE	Clusters in Aerospace and Satellite Navigation Technologies Linked to Entrepreneurial Innovation
CAV	Common Aero Vehicle
CBERS	China-Brazil Earth Resources Satellite
CCD	computer-controlled display
CCDev	Commercial Crew Development
CCE	Comisión Colombiana del Espacio
CCL	Commerce Control List
CCS	Counter Communications System
CDI	Center for Defense Information
CDTI	Centro para el Desarrollo Technológico Industrial
CEAS	Council of European Aerospace Societies
CEOS	Committee on Earth Observation Satellites
CEV	Centro Espacial Venezolano (now ABAE)
CFE	Commercial and Foreign Entities
CGP	Controlled Goods Program
CGS	Carlo Gavazzi Space
CGWIC	China Great Wall Industry Corporation
CIA	Central Intelligence Agency
CIDA-E	Centro de Investigación y Difusión Aeronautico Espacial
CIRA	Centro Italiano Ricerche Aerospaziali
CITT	Centre for Innovation and Transfer of Technology
CLA	Alcântara Launch Centre
CLBI	Barreira do Inferno Launch Centre
CLTC	China Satellite Launch and Control General
cm	centimetre(s)
CMAR	CSIRO Marine and Atmospheric Research
CNES	Centre National d'Études Spatiales
CNSA	China National Space Administration
CNTC	Centre National de la Cartographie et de la Télédétection
CoCoM	Coordinating Committee for Multilateral Export Controls
comsat	communications satellite
CONAE	Comisión Nacional de Actividades Espaciales
CONEE	Comisión Nacional del Espacio Exterior
CONIDA	Comisión Nacional de Investigación y Desarrollo Aeroespacial
COPUOS	United Nations Committee on the Peaceful Uses of Outer Space
COSMO-SKYMed	Constellation of Small Satellites for the Mediterranean Basin Observation

ABBREVIATIONS

COSPAR	Committee on Space Research
COSPAS-SARSAT	Space System for the Search of Vessels in Distress Search and Rescue Satellite-Aided Tracking
COSTIND	Commission for Science, Technology and Industry for National Defense
COTS	Commercial Orbital Transportation Services
CRISP	Centre for Remote Imaging, Sensing and Processing
CRTS	Centre Royal de Télédétection Spatiale
CSIRO	Commonwealth Scientific and Industrial Research Organisation
CSIS	Center for Strategic and International Studies
CSO	Czech Space Office
CSOC	Consolidated Satellite Operations Center
CSSAR	Center for Space Science and Applied Research
CSSI	Center for Space Standards and Innovation
CTBT	Comprehensive Nuclear Test Ban Treaty
CZ	Long March
DARPA	Defense Advanced Research Projects Agency
DBS	direct broadcast satellites
DCS	defensive counterspace
DCTA	Departamento de Ciência e Tecnologia Aeroespacial
DEPANRI	National Aviation and the Republic of Indonesia
DFH	Dongfanghong
DGLR	Deutsches Gesellschaft für Luft-und Raumfahrt Lilienthal-Oberth
DHS	Department of Homeland Security
DIA	Defense Intelligence Agency
DISEC	Disarmament and International Security
DLR	Deutsches Zentrum für Luft und Raumfahrt
DMC	Disaster Monitoring Constellation
DMSP	Defense Meteorological Satellite Program
DNI	Director of National Intelligence
DNSC	Danish National Space Centre
DoC	US Department of Commerce
DoD	US Department of Defense
DoE	US Department of Energy
DoS	US Department of State
DoT	US Department of Transportation
DRDC	Defence Research Development Canada
DSCS	Defense Satellite Communications System
DSP	Defense Support Program
DSRI	Danish Space Research Institute
DTRA	Defense Threat Reduction Agency
DTU Space	Institut for Rumforskning og-teknologi Danmarks Tekniske Universitet

EAA	Export Administration Act
EAC	European Astronaut Centre
EADS	European Aeronautic Defence and Space Company
EAR	Export Administration Regulations
ECOFIN	Economic and Financial
ECOSOC	Economic and Social Council
ECSL	European Centre for Space Law
EELV	Evolved Expendable Launch Vehicle
EGNOS	European Geostationary Navigation Overlay System
EHF	extremely high frequency
EIAST	Emirates Institute for Advanced Science and Technology
ELDO	European Launcher Development Organisation
ELISA	Electronic Intelligence Satellite
ELV	expendable launch vehicle
ENVISAT	European Space Agency Environmental Satellite
EO	electro-optical
EOEP	Earth Observation Envelope Programme
ERA	European Research Area
EROS	Earth Resources Observation Satellites
ERS	European Remote Sensing
ERTS	Earth Resources Technology Satellite
ESA	European Space Agency
ESAC	European Space Astronomy Centre
ESOC	European Space Operations Centre
ESPI	European Space Policy Institute
Esrange	European Space and Sounding Rocket Range
ESRO	European Space Research Organization
ESTEC	European Space Research and Technology Centre
EU	European Union
EUCASS	European Conference for AeroSpace Sciences
EUMETSAT	European Organisation for the Exploitation of Meteorological Satellites
EUTELSAT	European Telecommunications Satellite Organization
EXA	Agencia Espacial Civil Ecuatoriana
FAA	Federal Aviation Administration
FAA-AST	Federal Aviation Administration Office of Commercial Space Transportation
FALCON	Force Application and Launch from Continental USA
FCC	Federal Communications Commission
FCPJ	Florida Coalition for Peace and Justice
FCS	Future Combat Systems
FEDIL	Luxembourg Federation of Employers
FFG	Austrian Research Promotion Agency
FFRDC	Federally funded research and development center
FIA	Future Imagery Architecture

ABBREVIATIONS

FLEVO	Facility for Liquid Experimentation and Verification in Orbit
FLTSATCOM	Fleet Satellite Communications
FY	fiscal year
FYDP	Future Years Defense Program
G8	Group of Eight
GAIA	Graphical Astronomy and Image Analysis Tool
GAO	Government Accountability Office
GBS	Global Broadcast Service
GEO Orbit	Geosynchronous Earth orbit
GEO	Group on Earth Observations
GEOINT	Geospatial intelligence
GEOSS	Global Earth Observation System of Systems
GIFAS	Groupement des Industries Françaises Aéronautiques et Spatiales
GIG	global information grid
GIS	geographic information systems
GISTDA	Geo-Informatics and Space Technology Development Agency
GLADIS	Global AIS & Data-X International Satellite Constellation
GLAE	Groupement Luxembourgeois de l'aéronautique et de l'espace
GLONASS	Global Orbital Navigation Satellite System
GMDSS	Global Maritime Distress Safety System
GMES	Global Monitoring for Environment and Security
GN	Global Network against Weapons and Nuclear Power in Space
GNC	Guidance, navigation and control
GNP	gross national product
GNSS	Global Navigation Satellite System
GOCE	Gravity Field and Steady-State Ocean Circulation Explorer
GOCNAE	Organizing Group of the National Commission for Space Activities
GORS	General Organization of Remote Sensing
GOS	Global Observing System
GPS	Global Positioning System
GPS/IMU	Global Positioning System/Inertial Measurement Unit
GRACE	Gravity Recovery and Climate Experiment
GSC	Guiana Space Centre
GSICS	Global Space-Based Inter-Calibration System
GSLV	Geostationary Satellite Launch Vehicle
HAS	Croatian Space Agency
HEO	highly elliptical Earth orbit
HICDS	Highly Integrated Control and Data Systems
HRSC	High Resolution Stereo Camera
HRTI	high-resolution terrain information
HSA	Hrvatska Svemirska Agencija
HTSSE	High Temperature Superconductivity Space Experiment

HTV	H-II Transfer Vehicle
IAA	International Academy of Astronautics
IAASS	International Association for the Advancement of Space Safety
IACG	Inter-Agency Consultative Group for Space Sciences
IADC	Inter-Agency Space Debris Coordination Committee
IAEA	International Atomic Energy Agency
IAF	International Astronautical Federation
IAI	Israel Aerospace Industries
IASB	Institut d'Aéronomie Spatiale de Belgique
IASL	Institute of Air and Space Law
IASS	International Association for the Advancement of Space Safety
IC	Intelligence Community
ICAO	International Civil Aviation Organization
ICBM	intercontinental ballistic missile
ICC2	Interferometer Constellation Control
ICG	International Committee for Global Navigation Satellite Systems
ICSU	International Council for Science
ICT	Information and Communication Technology
IFHE	Institut Français d'Histoire de l'Espace
IFR	Internationaler Förderkreis für Raumfahrt
IGDDS	Integrated Global Data Dissemination Service
IGOS	Integrated Global Observing Strategy
IGS	Information Gathering System
IISL	International Institute of Space Law
ILS	International Launch Services
IMINT	Imaging Intelligence satellites
INGO	international non-governmental organization
Inmarsat	International Maritime Satellite Organization
INOVATOR	In Orbit Verification of ÅAC Technologies on Rubin
INPE	Instituto Nacional de Pesquisas Espaciais
INSAR	Radar interferometry
INTA	Instituto Nacional de Técnica Aeroespacial
Intelsat	International Telecommunications Satellite Organization
IO	international organization
IOSA	Integrated Overhead SIGINT Architecture
IRS	Indian Remote Sensing
ISA	Israel Space Agency
ISARS	Institute for Space Applications and Remote Sensing
ISAS	Institute of Space and Astronautical Science
ISC	Iran Space Council
ISIS	International Satellites for Ionospheric Studies
ISO	Infrared Space Observatory

ISP	Infrared Surveillance Program
ISR	intelligence, surveillance and reconnaissance
ISRO	Indian Space Research Organization
ISS	International Space Station
IST	Institute of Space Technology
ISU	International Space University
ITAR	International Traffic in Arms Regulations
ITSO	International Telecommunications Satellite Organization
ITT	International Telephone & Telegraph
ITU	International Telecommunication Union
ITWAA	Integrated Tactical Warning and Attack Assessment
IUS	Inertial Upper Stage
JAXA	Japan Aerospace Exploration Agency
JBIS	Journal of the British Interplanetary Society
JCS	Joint Chiefs of Staff
JEM	Japanese Experimental Module
JFCC	Joint Functional Combatant Command
JFCOM	Joint Forces Command
JPL	Jet Propulsion Laboratory
JROC	Joint Requirements Oversight Council
JRS	Japanese Rocket Society
JSASS	Japan Society for Aeronautics and Space Sciences
JSpOC	Joint Space Operations Center
KAMD	Korea air and missile defence
KARI	Korea Aerospace Research Institute
KASI	Korea Astronomy and Space Science Institute
KazKosmos	National Space Agency of the Republic of Kazakhstan
KBTM	Design Bureau of Transport Machinery
KCST	Korean Committee of Space Technology
kg	kilogram(s)
KH	Key Hole
KHI	Kawasaki Heavy Industries
km	kilometre(s)
KML	keyhole markup language
KOMPSAT	Korea Multi-Purpose Satellite
KSAT	Kongsberg Satellite Services, AS
KSLV	Korea Space Launch Vehicle
KVN	Korean VLBI Network
Landsat	Land Remote Sensing Satellite
LAPAN	Lembaga Penerbangan dan Antariksa Nasional
LDCM	Landsat Data Continuity Mission
LEO	low Earth orbit
LISA	Laser Interferometer Space Antenna
LKE	Lockheed Khrunichev Energia
LKEI	Lockheed Khrunichev Energia International

ABBREVIATIONS

LMC	Lockheed Martin Corporation
LOLA	Liaison Optique Laser Aéroportée
LORAN	Long Range Navigation
LPI	Lunar and Planetary Institute
LUT	Local User Terminal
m	metre(s)
m.	million
M3MSat	Maritime Monitoring and Messaging Microsatellite
MANT	Magyar Asztronautikai Társaság
MASI	Ministry of Aerospace Industry
MBDA	Matra BAE Dynamics Alenia
Mbps	megabits (106 bits) per second
MCL	Munitions Control List
MDA	MacDonald, Dettwiler and Associates
MECB	Missão Espacial Completa Brasileira
MEO	medium Earth orbit
MEXT	Ministry of Education, Culture, Sports, Science and Technology
MHI	Mitsubishi Heavy Industries
MIIT	Ministry of Industry and Information Technology
MILSATCOM	military satellite communications
Milstar	Military Strategic and Tactical Relay satellite
MIOSAT	Missione Ottica su Microsatellite
MLRS	Multiple launcher rockets system
MNC	multinational company
MOSAIC	Micro Satellite Applications in Collaboration
MOST	Ministry of Science and Technology
MOSTI	Malaysian Ministry of Science, Technology and Innovation
MOU	Memorandum of Understanding
MPG	Max-Planck Society
MRI	Royal Meteorological Institute
MTCR	Missile Technology Control Regime
MUOS	Mobile User Objective System
MUSIS	Multinational Space-based Imaging System
NACA	National Advisory Committee on Aeronautics
NADIA	Navigation for Disability Applications
NAL	National Aerospace Laboratory of Japan
NARSS	National Authority for Remote Sensing and Space Sciences
NASA	National Aeronautics and Space Administration
NASDA	National Space Development Agency of Japan
NASP	National Aero-Space Plane
NASRDA	National Space Research and Development Agency of Nigeria
NATO	North Atlantic Treaty Organization
NAVSEG	Navigation Satellite Executive Group

NEO	near Earth object
NEOSSat	near Earth object surveillance satellite
NESDIS	National Environmental Satellite, Data and Information Service
NGA	National Geospatial-Intelligence Agency
NGO	non-governmental organization
NII	Networks and Information Integration
NIMA	National Imagery and Mapping Agency
NISO	Netherlands Industrial Space Organisation
NIVR	Nederlands Instituut voor Vliegtuigontwikkeling en Ruimtevaart
NKSA	North Korea National Space Agency
NLR	National Aerospace Laboratory
NLS	National Launch System
NNRMS	National Natural Resource Management System
NOA	National Observatory of Athens
NOAA	National Oceanic and Atmospheric Administration
NOAA/ NESDIS	NOAA Satellite and Information Services
NORAD	North American Aerospace Defense Command
NPO	Scientific-Production Association
NPO Lavochkin	Lavochkin Research and Production Association
NPOESS	National Polar-Orbiting Operational Environmental Satellite System
NPP	NPOESS Preparatory Project
NPT	Nuclear Non-Proliferation Treaty
NRCT	National Research Council of Thailand
NRL	Naval Research Laboratory
NRO	National Reconnaissance Office
NRSA	National Remote Sensing Agency
NRSC	National Remote Sensing Center
NSA	National Security Agency
NSAU	National Space Agency of Ukraine
NSC	Norwegian Space Centre
NSF	National Science Foundation
NSO	Netherlands Space Office
NSPO	National Space Organization of Taiwan
NSPU	National Space Programme of Ukraine
NSS	National Security Space
NSSA	National Security Space Architect
NSSI	National Security Space Integration
NSSO	National Security Space Institute
NTMV	National Technical Means of Verification
OBMS	Onboard manipulator system

ABBREVIATIONS

OCS	offensive counterspace
ODNI	Office of the Director of National Intelligence
OECD	Organisation for Economic Cooperation and Development
OHB	Orbital High-technology Bremen
OKB-1	Special Design Bureau-1
ONERA	Office National d'Etudes et de Recherches Aérospatiales
OOSA	Office of Outer Space Affairs
ORBIT Act	Open-market Reorganization for the Betterment of International Telecommunications Act
ORS	Operationally Responsive Space
OSD	Office of the Secretary of Defense
OSTP	Office of Science and Technology Policy
PA&E	Programme Analysis and Evaluation
PACS	Precision Agile Control System
PAROS	Prevention of an Arms Race in Outer Space
PERSEUS	Projet Etudiant de Recherche Spatial Européen Universitaire et Scientifique
PETA	People for the Ethical Treatment of Animals
PGMs	precision guided munitions
PMSCS	Polar Military Satellite Communications System
PNAE	National Space Activities Programme
PNT	Position-Navigation-Timing
PO Yuzmash	Yuzmash State Enterprise
POES	Polar-Orbiting Operational Environmental Satellites
POL	Point-of-Load
PRISMA	Probing Rotation and Interior of Stars: Microvariability and Activity
PRL	Physical Research Laboratory
PRORA	Aerospace Research
PSLV	Polar Satellite Launch Vehicle
PWR	Pratt & Whitney Rocketdyne
QDR	Quadrennial Defense Review
QZSS	Quasi-Zenith Satellite System
R&D	research and development
RADOM	Radiation Dose Monitor
RAIDRS	Rapid Attack Identification, Detection and Reporting System
RCA	Radio Corporation of America
RCM	Radarsat Constellation Mission
RFP	request for proposals
RLV	reusable launch vehicle
ROKVISS	Robotic Components Verification on the ISS
ROSA	Romanian Space Agency
Roscosmos	Russian Space Agency
RSC	Rocket and Space Corporation
RSC Energia	S.P. Korolev Rocket and Space Corporation Energia

ABBREVIATIONS

RTC	Central R&D Institute of Robotics and Technical Cybernetics
SA	Selective Availability
SAR	synthetic aperture radar
SASTIND	State Administration for Science, Technology and Industry for National Defence
SBIRS	Space-Based Infrared System
SBSS	Space-Based Surveillance System
SBV/MSX	Space-Based Visible sensor on the Midcourse Space Experiment
SCIAMACHY	Scanning Imaging Absorption Spectrometer for Atmospheric Cartography
SDO Yuzhnoye	Yuzhnoye State Design Office
SDR	Software Defined Radio
SECDEF	Secretary of Defense
SECOR	Sequential Correlation of Range
SEOSAR	Spanish Earth Observation Synthetic Aperture Radar
SEOSAT	Spanish Earth Observation Satellite
SES	Society of European Satellites
SETI	Search for Extraterrestrial Intelligence
SFF	Space Frontier Foundation
SGAC	Space Generation Advisory Council
SHF	super-high frequency
SIDC	Space Innovation Development Center
SIGINT	Signal Intelligence satellites
SLI	Space Launch Initiative
SLV	space launch vehicle
smallsat	small satellite
SMC	Space and Missile Systems Center
SMDC	Space and Missiles Defense Command
SMOS	Soil Moisture and Ocean Salinity
SMTI	surface moving target indications
SNSB	Swedish National Space Board
SOCHUM	Social, Humanitarian and Cultural
SOCRATES	Satellite Orbital Conjunction Reports Assessing Threatening Encounters in Space
SOPSC	Space Operations School
Space Grant	National Space Grant College and Fellowship Program
SpaceX	Space Exploration Technologies Corporation
SPARRSO	Space Research and Remote Sensing Organization
SPAWAR	Space and Naval Warfare Systems Command
SPECPOL	Special Political and Decolonization
SPOT	Satellite Pour l'Observation de la Terre
SR	Space Radar
SRC	Space Research Centre

ABBREVIATIONS

SRI	Space Research Institute
SRON	Stichting Ruimte Onderzoek Nederland
SRS	Satellite Remote Sensing
SS/L	Space Systems/Loral
SSA	Space Situational Awareness
SSB	Space Studies Board of the National Academies
SSC	Swedish Space Corporation Group
SSI	Space Studies Institute
SSN	Space Surveillance Network
SSOC	Sensor System Operations Centre
SSTL	Surrey Satellite Technology Ltd
STEM	science, technology, engineering and mathematics
STEP	Satellite Test of the Equivalence Principle
STIL-BAS	Solar-Terrestrial Influences Institute
STK	Satellite Tool Kit
STP	Space Test Program
STRATCOM	Strategic Command
STS	Space Tactics School
STSS	Space Tracking and Surveillance System
SUPARCO	Pakistan Space and Upper Atmosphere Commission
SvalSat	Svalbard Station
SWF	Secure World Foundation
SWIET	Southwest Institute of Electronics Technology
TacSat	Tactical Satellite
TAS	Thruster Actuation System
TAUVEX	Tel-Aviv University Ultra Violet Experiment
TCO	Transformational Communications Office
THEMIS	Time History of Events and Macroscale Interactions during Substorms
THEOS	Thailand Earth Observation Satellite
TiPS	Tether Physics & Survivability Experiment
TRSC	Thailand Remote Sensing Centre
TRSP	Thailand Remote Sensing Programme
TSAT	Transformational Satellite Communications System
TsKBEM	Central Design Bureau of Experimental Machine Building
TsNIIMash	Central Scientific Research Institute of Machine Building
TSPR	total systems performance responsibility
TSS	Tromsø Station
TsSKB-Progress	State Research and Production Space Rocket Center
TT&C	telemetry, tracking and control
TÜBİTAK	Scientific and Technological Research Council of Turkey
TÜBİTAK UZAY	TÜBİTAK Space Technologies Research Institute
UAL	Upper Atmosphere Lab

UAVs	unmanned aerial vehicles
UFO	ultra-high frequency follow-on
UHF	ultra-high frequency
UK	United Kingdom
UN	United Nations
UNESCO	United Nations Educational, Scientific and Cultural Organization
UNIDROIT	International Institute for the Unification of Private Law
UNISPACE	United Nations Exploration and Peaceful Uses of Outer Space
UNOOSA	United Nations Office for Outer Space Affairs
UN-SPIDER	United Nations Platform for Space-based Information for Disaster Management and Emergency Response
US(A)	United States (of America)
USA	United Space Alliance
USAF	US Air Force
USD	Undersecretary of Defense
USERS	Unmanned Space Experiment Recovery System
USGS	US Geological Survey
USMC	US Marine Corps
USN	US Navy
USRA	Universities Space Research Association
USSPACECOM	US Space Command
USSR	Union of Soviet Socialist Republics
USSTRATCOM	US Strategic Command
UV	Ultraviolet
UZAY	Middle East Technical University
VAST	Vietnamese Academy of Science and Technology
VEHRA-SH or VSH	Véhicule Hypersonique Réutilisable Aéroporté Suborbital Habité
VLS	Veículo Lançador de Satélite
vMFP	virtual major force programme
VSAT	Very Small Aperture Satellite Terminals
WGS	Wideband Global System
WMD	weapon of mass destruction
WMO	World Meteorological Organization
WRC	World Radiocommunication Conferences
XMM	X-ray Multi-Mirror Mission
ZARM	Zentrum für Angewandte Raumfahrttechnologie und Mikrogravitation
ZY	ZiYuan

Essays

Politics of Space

ELIGAR SADEH

Civil, commercial and military space policy must meet practical demands and requirements in the present, at acceptable cost. Having said this, the process of space policy interferes with these rational outcomes. The actors associated with space policy-making serve individual and organizational interests and goals, rather than being guided by an objective, rational standard. The process leads to space programmes and projects that are not well optimized, cost too much and are not on schedule.

Herein, space policy-making is introduced and explained. The case of national security space (NSS) policy-making is then examined through the example of the USA. Then space policy-making in the USA is assessed in the cases of export controls of commercial satellites and in the development of space launch systems. Space policy has to accommodate a broad range of perceptions and interests, from practical issues of national defence, commerce and technology, to less quantifiable characteristics, such as the contribution of space exploration and development for societal benefit and to the achievement of humanity as a spacefaring species.

SPACE POLICY-MAKING

There are three key stages to the making of space policy, which take place in the following order: (1) the setting of goals by the national leadership on the agenda for policy-making; (2) the formulation of appropriate means by administrative and executive agencies to achieve those goals; and (3) the allocation of resources by government to implement policy (see Illustrations and Documentation: Figure 1.1 Space Politics and Policy). Yet the initial impulse for space policy will often come not from national leadership, but from advocacy coalitions, which can include political leaders, individuals from administrative and executive agencies, private interest groups, academics and the space industry. Aside from these advocacy coalitions, there are institutional players who are regular sources of policy initiatives. Most notable, in the USA, are the civil and military space programmes administered by the National Aeronautics and Space Administration (NASA) and the US Air Force (USAF).

In recent years, this list has expanded to include non-governmental actors, especially commercial enterprises, including both established industrial organizations like the major space contractors and entrepreneurial enterprises utilizing private capital. Examples of the influence of these actors are found in the development of telecommunications; global positioning, navigation and timing systems; remote sensing; and space launch vehicles to support civil, commercial and military activities.

Public policy usually reflects an aggregation of interests, both public and private. The process of aggregation involves discovering players whose perceived interests coincide with those of the policy initiator, then modifying the proposal as necessary to garner the widest possible support. For example, an initiative to use commercial space assets to perform services for the military might call on the support of the private companies that own those assets, US Congressional delegations who represent the districts in which those companies do business, interest groups lobbying for efficiency in government and ideological supporters of the free market. The breadth of support depends on the feasibility of the proposal and on the coincidence between the proposal and existing national goals.

The US Space Shuttle programme illustrates these aspects of space policy-making. The programme was originally proposed to sustain human space flight after the Apollo programme and, therefore, justify the continued existence of NASA. The Shuttle concept evolved to meet the requirements of an expanded coalition, involving the commercial and military sectors. In the process, claims for the programme gradually expanded, so that eventually it was portrayed as a single-launch vehicle for all civil, commercial and military purposes. The Shuttle coalition became so powerful that the US President adopted the programme as the centrepiece of a national space programme. In this case, an expanding coalition of interest groups succeeded in both establishing the feasibility of this programme and setting the national agenda.

The result of this coalition building was an 'over-optimization' of the system; i.e. functions were added to the Space Shuttle's mission, which made it less capable of optimizing any single goal, whether civil, commercial or military. By the time this became apparent, the strength of the Shuttle coalition was so pervasive that attempts to coalesce support behind alternative systems failed. Only after the Space Shuttle Columbia accident were policy-makers able to plan for termination of the program as well as adopt approaches for a replacement human space flight system.

The Space Shuttle example shows how the interplay of differing agendas among the major actors in space can make rational policy-making more difficult. To move agendas from conception to policy, by way of executive decision, administrative rule making and authorization and appropriation of resources, requires these government actors to enlist the support of advocacy coalitions. These coalitions can be motivated by ideological or political conviction, or by private interest, and can include engineers, scientists, military officers, national policy-makers, space business leaders and space enthusiasts in grassroots organizations. In the US case, the coalitions fall into four general categories: space business, national defence, human destiny and space science.[1] The characteristics of these are defined below.

Politics of Space

The space business coalition focuses on material benefits and comprises those interests who regard the space environment primarily as a venue for making profit. It includes the commercial space sector and their political supporters in the US Departments of Transportation and Commerce, and parts of NASA. This group is also closely aligned with the US House Space and Aeronautics Subcommittee and the US Senate Science and Space Subcommittee. Parts of this coalition are committed to free market capitalism and regard government involvement as a constraint on the commercial development of space.

The national defence coalition is driven by national security concerns. This coalition constitutes the relevant parts of the Department of Defense (DoD) dealing with space, private defence contractors with whom the DoD works, such as Boeing, Lockheed Martin and Northrop Grumman and the US Department of State (DoS). It receives additional support from the US House and US Senate Armed Services Committees and prominent conservative and pro-military organizations and individuals.

The human destiny coalition envisions humanity as a spacefaring species and calls for the settlement of the Moon and Mars. It is focused primarily upon purposive benefits to society and is made up of NASA, NASA-associated human space flight proponents, like Wernher von Braun and Carl Sagan in the past, and a number of professional and grassroot organizations, such as the American Institute of Aeronautics and Astronautics, the American Astronautical Society, the International Astronautical Federation, the National Space Society, the Planetary Society and the Mars Society. Support for this coalition's view is also scattered throughout US Government executive agencies, universities and amongst the public.

The space science coalition includes research scientists in industry, academia and government and their allies in the US Congress and in federal agencies. Professional associations representing scientific disciplines are politically involved as well. The relationship between this coalition and others, like the human destiny coalition, is often tense. Even though federal funding for all areas of science and engineering increased after the launch of Sputnik and the rise of the space age, many in the US scientific community view NASA and the human space flight programme with suspicion. This coalition seeks a mix of material and purposive benefits. In some cases, like the Hubble Space Telescope, this group persuaded the US Congress to fund large-scale space science programmes. It also, with less policy success, publicly criticized other NASA projects, like the International Space Station (ISS).

The Space Shuttle campaign expanded to include all of these groups, to the ultimate disadvantage of the programme. The policy initiative that led to the ISS also sought support from all these groups, and has struggled to maintain that support through difficulties in implementation. Although the ISS lost the support of the space science and military sectors, the programme was sustained when US President Clinton linked it to post-Cold War foreign policy goals, through collaboration with Russia on the programme.

Following the agenda part of policy-making is policy formulation. By its very nature, formulation reflects many views, perspectives and interests and involves political accommodation leading to goal modification. The political process of formulation represents a compromise between what national space organizations may want and might regard as most effective, efficient or feasible and what national organizations perceive as the appropriate response to political forces. The political forces are shaped by historical and political conditions and, as such, policy formulation is framed by the extent to which the chosen policies are congruent with the prevailing national interests and goals.

The dynamics of agenda and formulation described above entail a process of input, communications channels, conversion structures and outcomes and depend upon the overall domestic and international environments. In the daily decision-making process associated with policy, like systems acquisition and budgeting, among other actions, the domestic environment influences the space policy process through a number of mechanisms, including US strategic culture, the state of technology and civil, commercial and NSS objectives. International regimes, such as the International Telecommunication Union (ITU), which manages radio frequency spectrum and geostationary orbital slot allocations, the Missile Technology Control Regime (MTCR), the US International Traffic in Arms Regulations (ITAR), which oversees the transfer of technology associated with space launch vehicles and satellites, and international space law, i.e. the Outer Space Treaty Regime, influence the international environment in which policy-making occurs. Feedback from the actions undertaken, and the strategies and policies created by and resulting from the process, influence future decisions and actions.

The policy process accounts for the complicated and multidimensional inputs into decision-making that lead to policy formulation. As an example, even at the input stage – the earliest stage of the process that equates with the agenda – demands for a particular policy or approach may be moderated by the budget and by national interests. Similarly, expectations and demands from the international environment may significantly contrast with needs as defined by domestic events. This complicated situation is compounded as inputs are communicated through a number of channels and acted on by the executive and legislative branches of government and the bureaucracy. These actors serve as conversion structures, which receive the varied and frequently conflicting system inputs and convert them into the formulated decisions of government.[2]

While this conversion sounds straightforward, the actual process is far more difficult and is characterized by coalition building, bargaining, compromise and goal accommodation through formal and informal mechanisms. Interest groups, as represented by government actors and advocacy coalitions, often hold competing and conflicting views. Advocacy coalitions in the US will seek to influence different US Congressional members; the US DoD and the US DoS may seek different resolutions to the situation based on organizational

Politics of Space

interests and goals; and members of the Office of the US President, such as the Office of Science and Technology Policy (OSTP), may experience increased external scrutiny when deciding on appropriate actions to pursue. Further complicating this process are the perceptions of interests and goals brought to the process by each of the participants, something else that must be considered when seeking to understand the policy process.

As inputs enter the political process from multiple sources, they are acted on through both formal and informal processes. The interagency, budget and acquisition processes represent some of the formal approaches to resolving policy issues. Organization and agency cultures and values may converge around 'widely shared values and images of international reality' and 'rules of the game' as framed by laws, regulations and bureaucratic structures through which decisions are made.[3] Conversion structures frame the decision-making process and outputs of policy formulation. Although these outputs appear rational, the process leading to them definitely does not conform to the rational decision-making model. The 'pulling and hauling' among the various political actors, which characterizes policy-making, results in formulated outputs that are far from rational and more incremental.[4] Policy-making of this nature characterizes the cases of NSS (NSS is discussed in detail in Chapter 2), commercial satellite export controls and space launch in the US, which are discussed here.

The formulated outputs in the form of policies, strategies and programmes must be acted upon through implementation. Policy implementation involves the development of the enabling space technologies and their application to the actual building of the hardware and systems to support space-related projects and programmes.[5] Technical ability and know-how directed at high-performance and high-reliability outcomes are crucial variables affecting implementation.

The technical skills of the implementers, like national space agencies and commercial industry, are influenced by political and associated organizational issues involved with developing and administering complex space technologies (see Illustrations and Documentation: Figure 1.2 Space Policy Implementation). Added to this, organizations are left to grapple with imprecise language and the interpretation of political intent, further complicating implementation. The complexities associated with policy implementation often lead to reformulation of the policy and then reimplementation, as efforts are made to optimize outcomes.

NATIONAL SECURITY SPACE

For the past decade, the United States Government has attempted to reassess its NSS policy. This began with the Space Commission report of 2001, which addressed the issue and noted that the USA is highly vulnerable to 'surprises in space' equivalent to a 'Space Pearl Harbor'.[6] According to the Space Commission, this vulnerability results from US dependency on space assets, the limited attention given to analyzing and assessing threats to these assets and the lack of responsibility and accountability resulting from the limited co-ordination and oversight of the disparate space activities and agencies.

The Space Commission report documented the political reasons for these concerns. One reason was regarding the numerous government organizations involved in NSS activities and, therefore, those involved in policy decision-making (see Illustrations and Documentation: Figure 1.3a and 1.3b). The decision process is further complicated by additional actors, some of whom were noted earlier as making up the international environment affecting decision-making, while others are within the domestic environment, such as the various advocacy coalitions and industrial players involved in and influencing the decision process.

The Space Commission report proceeded to make a number of recommendations intended to 'achieve greater responsibility and accountability' in space activities.[7] Given the Space Commission's charter of reviewing space policy management and organization, these recommendations focused on the need for enhanced national leadership for space policy at the level of the US President and executive branch of government. One key recommendation dealt with strengthening the role of the Secretary of the Air Force as the 'DOD Executive Agent for Space' for the US military, with the responsibility to be organized, trained and equipped for prompt and sustained offensive and defensive space operations.[8]

In response to this recommendation, the USAF did make several organizational changes, including enhanced leadership of space through changes in how Air Force Space Command administers the USAF space programme.[9] The Space Commission also put forward the requirement to identify, train and employ a cadre of space professionals. As a result of criticism in this area, Air Force Space Command initiated a Space Professional Development Program. The National Security Space Institute (NSSI) was created with the goal of developing credentialed space professionals.

The many challenges associated with policy, organization and management of NSS, particularly in the areas of decision-making, acquisitions and the development and initiation of new programmes, are not completely resolved, despite the Space Commission report.[10] A key reason for this lies with the ongoing competition and conflict that exists among the numerous government organizations dealing with space. One example of this concerns the multiple attempts that have all failed to craft a national security space strategy due to lack of common assumptions and vision between DOD and the United States intelligence community on how to protect and preserve the space domain and ensure access to space. The challenges are only more acute than they were a decade ago; the US dependence on space assets has grown, with reduced advantages in the space domain, and greater foreign access to space and use of space assets.

The challenges of NSS policy were addressed more recently in a 2008 Congressionally mandated report of the Institute for Defense Analysis entitled *Leadership, Management, and Organization for National Security Space: Report to Congress of the Independent Assessment Panel on the Organization and Management of National Security Space.*[11] This report concluded that space capabilities underpin US leadership and that US leadership is in jeopardy as the global access to technology is proliferating and leveling among space powers (see Chapters 3 and 4 on developed and developing space programmes, respectively) and, hence, US adversaries are gaining space-related capabilities. Without changes in the overall management and organization of NSS, space will cease to be a competitive advantage for the USA in addressing national security.

The report highlights the importance of national space strategy considerations and political leadership. There exists no overarching national space strategy to achieve the goals of national space policy. The lack of a compelling strategy for space makes more difficult the task of sustaining national and international consensus to secure the space medium for sustainable access to, and peaceful uses of, space and freedom from space-based threats. This is a problem that has become more pressing with the growing ambitions of spacefaring states, asymmetric threats from states and non-state terrorist actors, and threats to space assets from orbital debris, space weather and near Earth objects (NEOs). Without a national space strategy, the NSS arena is fragmented among numerous organizational entities, which constrain effective acquisitions for sustaining and recapitalizing NSS systems and link space system development to specific bureaucratic and programmatic ends, rather than to a set of strategic capabilities and national outcomes the USA desires for national security.

The fragmented nature of NSS presents the issue that there is no clear political leadership. There is no decision-making mechanism to arrive at a coherent and unified NSS budget and to prioritize space programmes across the US military. The fragmentation of NSS policy complicates implementation and fosters inefficiency in acquisitions and new space system development. The report recommends that the executive branch of the US Government establish a NSS political authority to remedy this situation (see Illustrations and Documentation: Figure 1.4).

EXPORT CONTROLS OF COMMERCIAL SATELLITES

The case of export controls of US commercial satellites exemplifies a policy-making process beset by bureaucratic politics, leading to policy outcomes that are not rational, i.e., the desired outcome of national security is not met and commerce in the satellite sector is harmed. The constraints on rational policy-making are a result of competition, conflict and protectionism, a bureaucratic 'pulling and hauling', among the relevant actors in this case, including the US President and Congress, the US Departments of State and Commerce, and the DoD. It is this pulling and hauling that results in policies for licensing the export of commercial satellites that are far from orderly, stable and predictable.[12]

The crux of the political issue revolves around bureaucratic control and jurisdiction over the licensing process for export of commercial satellites. As commercial satellites represent a dual-use (military and commercial) space technology, bureaucratic pulling and hauling exists between the framing of export controls as a matter of national security versus a matter of business and commerce.[13] The national security advocates, among them the President, Congress, the DoS, and the DoD, view commercial satellites and the associated technologies as items to be controlled for export within the same legal regime that controls export and trafficking of arms. The DoS, through the Office of Defense Trade Controls Policy, is the bureaucratic entity that governs ITAR and the associated Munitions Control List (MCL). The DoD, through the Defense Threat Reduction Agency (DTRA), assists the DoS in implementing its regulatory authority.

The commercial space advocates, among them the President and Congress, especially from 1988 to 1998, the Department of Commerce, and aerospace and defence industries, view commercial satellites as an indicator of US leadership with a strong market share in the global commercial satellite sector. The way to regulate export of these satellites is through the legal regime that governs dual-use technologies used commercially, namely the Export Administration Regulations (EAR) administered by the Department of Commerce's Bureau of Industry and Security. The Department of Commerce governs exports through the Commerce Control List (CCL), and from 1992 to 1999 this regime applied directly to the export of commercial satellites.

This commercial approach enabled the People's Republic of China to compete within the US market for the launch of commercial satellites. From 1992 to 1996, the Chinese Long March rocket failed to launch commercial satellites manufactured by US companies Hughes Space and Communications, which was purchased by Boeing in 2000, and Space Systems Loral. As required by the insurance companies covering these companies' assets, investigations into the launch failures were concluded and submitted to the Department of Commerce for approval. The Department authorized Hughes and Loral to communicate the technical reports to the Chinese launch officials. The transfer of the reports sparked political controversy over the statutory authority of the Department of Commerce to allow such a transfer without proper review and oversight by the DoS. Specifically, the controversy focused on the export of knowledge on the reliability of space launch vehicle technology and, more generally, was linked to the issue of ballistic missiles and US-Chinese relations.

Congress investigated this issue through the *Report of the Select Committee on US National Security and Military/Commercial Concerns with the People's Republic of China* (known as the *Cox Report*),[14] and determined that Hughes and Loral transferred to China, in violation of US export control laws, the Arms Export Control Act of 1976 and the ITAR regime, missile design information and knowledge that improved the reliability of the Chinese Long March rocket useful for civil and military purposes.[15] The Congressional response led to the National Defense Authorization Act for Fiscal Year 1999 that directed sole export control responsibility to the DoS. This led to the application of the ITAR regime to commercial satellites. The DoS's jurisdiction over commercial satellites began in 1999 and continues to the present (June 2010). According to the majority of space leaders, the application of ITAR to commercial space technologies is a misapplication of the regime and is one of the top space policy issues requiring Congressional reformulation.[16]

International and Domestic Environments

The national security and commercial space advocates' respective policy preferences are influenced by the international and domestic environments. Both the international and domestic environments date back to the Cold War and the issue of how to control dual-use technologies. The concern, then and now, is that such technologies can be used for the development of arms that, in turn, can lead to proliferation of ballistic missiles, and nuclear, biological and chemical weaponry. Dual-use technologies with these potential applications are viewed by national security advocates as sensitive items to be controlled.

One aspect of control lies with the statutory authority within the USA for dual-use technologies. This authority lies with the Export Administration Act (EAA) of 1979, in which Congress delegated to the executive branch the legal authority to regulate foreign commerce by controlling and licensing exports. The EAA is the domestic environment from which the Commerce Department's EAR regime emerged. Of note, the EAA expired in September 1990; reauthorization of EAA took place for short periods with the last incremental extension expiring in August of 2001. Since then, no new Congressional legislation has been passed to either reauthorize or rewrite EAA, and the regime functions on the basis of Presidential authority under emergency power provisions.

Within the context of the global war on terrorism and the resulting emphasis on national security, at times to the detriment of commercial interests, the Congressional failure to act on the EAA further strengthens and maintains the DoS's regulation of commercial satellites through the ITAR regime. Furthermore, the origins of the EAA are Cold War related and originate from the Export Control Act of 1949. Even though the EAA represents a lessening of restrictive export control in comparison to the Export Control Act and subsequent amendments to that Act, the legal regime is a relic of Cold War international politics and national security rivalries.[17] The EAA has not been sufficiently adapted as an export control regime for the post-Cold War international environment of non-traditional security concerns, development and proliferation of space technologies and the emergence of global space commerce, including the advent of multinational commercial alliance in the space arena (see Chapter 5 on Commercial Space Actors).

A second aspect of control deals with the Arms Export Control Act of 1976, the basis for the ITAR export control regime. This regime was established during the Cold War and has not undergone any statutory changes. Further, neither the DoS nor the DoD made any changes to the implementation modalities of any of these Cold War regimes.[18] Although the President and Congress have noted the need to review the arms export control regime to streamline the processing of export licence applications, the DoS and the DoD have not acted on these recommendations. The issue of delay and the cost of bureaucratic compliance in the granting of export licences is one of the key concerns of commercial space advocates; these concerns translate into an economic issue for the commercial satellite sector. The economic issue also posits a barrier to entry for new space commercial companies, often referred to as new space, which are attempting to enter into existing markets, such as space launch services, or to develop new markets, such as space tourism.

The third aspect of the control issue exists at international level. In 1949 a multilateral export control regime, called the Coordinating Committee for Multilateral Export Controls (CoCoM), involving North Atlantic Treaty Organization (NATO) allies, was established. This regime mirrored US domestic controls as established with the Export Control Act of 1949. CoCoM advanced restrictive export controls on sensitive dual-use technologies at the multilateral level. The regime was dissolved in 1994 and replaced in 1996 by the Wassenaar Arrangement on Export Controls for Conventional Arms and Dual-Use Goods and Technologies. The Wassenaar Arrangement, as compared to CoCoM, lessened export controls of dual-use technologies at an international level and is more loosely organized, with more limited institutional structures. It relies on: consensus by state members, resulting in a 'lowest common denominator' approach for multilateral export control; minimal reporting requirements preventing pre-export consultations among state members; and a lack of authority among state members to block transactions of other state members.[19] In addition, the Wassenaar Arrangement did not apply multilateral control over commercial satellite technology or expertise.

Less control at international level influenced the US domestic environment and raised concerns that the EAA regime is not best suited to achieve national security objectives when dealing with the export of dual-use technologies. Internationally, the USA argued for the strengthening of export control laws in other states, particularly in China and Russia. Given the expiration of EAA and the liberalization of US domestic export controls in the 1990s, the international credibility of the USA was diminished. Over the past decade, the lessening of controls of the international legal regime is a factor that favours the NSS advocates' position and their preference for ITAR as the regime to control and license exports of commercial satellites.

Communications Channels

Communications channels were identified as a function of the relevant bureaucratic strategic cultures. In this case study, the strategic cultures of the

Politics of Space

national security advocates versus the space commerce advocates frame the political debates and arguments. Commercial space advocates frame the export control issue through the lens of foreign availability and controllability of technology. The contention is that the proliferation of technology cannot be effectively controlled and US dominance of space technology cannot continue. The globalization of space commerce points to the fact that unilateral controls will not stop foreign states from acquiring technology and, in the end, US dominance in space commerce is diminished, while foreign businesses win new markets and gain incentives to enter into new markets.[20] All this is complicated by the fact that as space commerce is increasingly global, many components in the commercial satellite sector are manufactured worldwide and are considered commercial commodities. ITAR is not designed to deal with the global nature of the industry and the outcome is one where foreign commercial satellite developers seek to reduce dependence on US satellite components because of delays associated with the ITAR export licensing process. The emerging trend is one where US satellite manufacturing companies, which must adhere to ITAR restrictions, are at a growing disadvantage as the inventory of 'ITAR-free', or not US-manufactured components, satellites expands abroad.[21]

Space commerce advocates see a link between national security and robust export control industries and, hence, favour an export control regime that is streamlined, less complex and, therefore, not an impediment to exports. As an example, the US Department of Commerce presumes that the issuing of an export licence is routine unless good cause can be shown otherwise.[22] Space commerce advocates argue that national security is undermined when exports are impeded, resulting in the loss of US market share and, simultaneously, the limitation of US satellite components through export controls leads to greater foreign research and development (R&D) investment in this area. In turn, these foreign R&D investments can be leveraged to achieve parity, even surpass, the US technological lead. Space commerce advocates frame commercial satellite technology as possessing no inherent strategic or military relevance, a view shared with the state members of the Wassenaar Arrangement, with the exception of the USA.[23]

In contrast, the national security advocates maintain there is a need to control commercial space exports as sensitive military technologies. This control prevents the proliferation of technologies that could be used by hostile states against the USA or its allies, secures the DoD's reliance on the commercial sector for R&D as a result of declining defence budgets in the 1990s and sustains the US military use of commercial space assets for operations, including commercial satellites for telecommunications and remote sensing purposes. For these advocates, national security is framed in ideological and warfare terms – limiting the diffusion of technology advances US foreign policy interests and enhances national security. This framing of export control as a national security issue compelled Congress to place commercial satellites and associated technologies within the authority of the ITAR regime.[24] The

Chinese Long March rocket incident discussed earlier and the global war on terrorism served to strengthen this world view and, thus, weaken political attempts to reformulate the export control regime.

Conversion and Outputs

The political process of conversion – how the President, Congress and bureaucracies interact – is exemplified by bureaucratic politics in this case of commercial satellite exports. It is the nature of bureaucratic politics that result in policy outputs that are not rational. A rational policy-making process suggests outputs that serve the desired communications channels of at least one group of advocates. In this case, the policy outputs, albeit unintended, do not ideally realize the policy preferences of either the national security or commercial space advocates. On one hand, ITAR damages national security by placing legal and bureaucratic restrictions on the US military use of commercial space assets that rely on a robust satellite industry. This includes risks to the military use of: commercial satellites for operational support; advanced satellite technologies developed in the commercial sector; and foreign suppliers for satellite components and services needed for operations. On the other hand, export control of commercial satellites through ITAR has made the US space and satellite component industry less competitive internationally and has contributed to a weakening of the US space industrial base and market position.[25]

How did the issue of export controls of commercial satellites result in undesirable policy outputs? The answer to this question lies in the way the relevant political actors serve as conversion structures for the political process. Prior to 1992, export control of commercial satellites fell within the purview of the ITAR regime, but beginning in 1988 President Reagan began to loosen export restrictions on commercial satellites to keep US industry competitive in global markets and to advance national space policy for the development of the commercial space sector. The following Bush and Clinton Administrations shared these policy preferences. Bush and Clinton used their political and legal authority to waive trade sanctions with China put in place through Congressional legislation following the Tiananmen Square massacre. The sanctions waived included those on commercial satellites for export to launch on the Chinese Long March. The policy conflict illustrated here, between the President and Congress, set the stage for the Chinese Long March incident and the resulting Congressional legislation that reversed the lessening of export controls begun under Reagan.

The theme of policy conflict persisted as Bush made use of Presidential authority to extend EAA and vetoed Congressional legislation that would have amended and extended the full EAA on a permanent basis.[26] In this legislation, the US Congress took more of a national security position on the export of dual-use items, in conflict with Bush's post-Cold War commercial view for the increased role of economic competitiveness. Bush sustained this view by removing all items from the MCL that were on the CoCoM dual-use list. This, in turn, led to split jurisdiction on export controls between the Departments of State and Commerce from 1992 to 1996. An interagency review process initiated by President Bush determined which of the dual-use items listed on the MCL could be transferred to the CCL. Under the Commerce Department's licensing process, these transfers made it easier to export commercial satellites for foreign launches. Less advanced commercial satellites, already proliferated throughout the commercial satellite sector, were exported as commercial goods under the EAR regime. Throughout the story of commercial satellite export controls the Departments of State and Commerce have both sought influence and authority, and split jurisdiction was viewed by the actors as a political compromise to resolve this dispute.[27] Nonetheless, the differences in strategic cultures of each bureaucracy sustained the struggle for influence over export controls.

As a result of split jurisdiction, the technical parameters for determining whether commercial satellites should be treated as munitions or dual-use commercial goods became unworkable.[28] One of the issues that emerged was that the export regulatory bureaucracies at the Departments of Commerce, State and Defense lacked the requisite technical expertise to determine which technologies should be controlled as munitions and which technologies could be exported as commercial commodities.[29] This was exacerbated by the fact that regulatory monitors were asked to implement 'impossible tasks' – apply overlapping, self-contradictory and rigid sets of rules and track all hardware for export without explicit political guidance on what to protect for reasons of national security and what to consider commercial commodities. Consequently, split jurisdiction was abandoned as a policy preference by the actors. In 1996 until 1999, Congress assigned the Department of Commerce primary jurisdiction.

The political process underlying the transfer to Department of Commerce jurisdiction was characterized by bureaucratic politics and conflicts. The transfer was met with counter moves by the DoS export officials determined to exert their full authority to the extent permissible by law. Both export control bureaucracies sought regulatory authority and their self-interest to do so became a goal in and of itself. The bureaucratic politics concept that 'where you sit defines who you are' applies directly in this case; regulators of the DoS and Commerce were explicitly tied to the strategic cultural perspectives of their organizations. The DoS made it increasingly difficult and costly for satellite companies to export if even a single component remained subject to their control through ITAR. During this period, the DoS pressured Congress for a greater role in the export control regime of commercial satellite technologies.

Congressional reaction to the Chinese Long March rocket incident and the sustained efforts of national security advocates led to Congressional legislation that resulted in sole DoS jurisdiction in 1999. In many ways this move was reactive, rather than rationally considered. One indication of this is that the export violations committed by Hughes, Loral and Boeing did not damage US national security in any material way; the expertise transferred to China only marginally benefited Chinese missile programmes by improving launch reliability.[30] In fact, many of the breaches were little more than technical violations of the DoS's export control regulations dealing with services that could 'in theory' be applied to national security purposes.[31]

In this situation, the policy output of State Department jurisdiction is suboptimal; instead of seeking compromise, the DoS countered the policy preferences of the commercial space advocates. Given the drive for bureaucratic self-preservation, the DoS took the Congressional mandate for sole jurisdiction and unilaterally implemented its approach in administrative rule-making to realize its national security world view.

This outcome raises a number of issues. First, regarding the intentions of the *Cox Report* recommendations, which prompted Congress to give the DoS commercial satellite export licensing authority. On this issue, it is not clear whether the *Cox Report* recommendations intended to control the export of technology from solely a national security standpoint, or to control the technology to satisfy both national security and commercial policy preferences. This ambiguity provided the DoS with the opportunity to advance their national security perspective.

More revealing is that the DoS is implementing the law in ways that are not necessarily what Congress intended, yet Congress itself fails to act on this problem. To illustrate, the *Cox Report* called for: Congressional reauthorization of EAA; continuous updating of the export control regime; and streamlining of the licensing procedures to provide greater transparency, predictability and certainty. In all these areas, neither the DoS nor Congress have taken any substantive action. Not only did the DoS act unilaterally to do other than what was recommended by the *Cox Report*, but Congress failed in its basic oversight role to hold the DoS accountable to Congressional policy intentions and preferences. This political dynamic stalled the efforts of reform advocates.

Albeit there is pending legislation in Congress to reform export controls, the advocates are in the minority. The proposed Congressional Satellite Trade and Security Act of 2001 sought a return to Department of Commerce jurisdiction, but the measure failed to advance and through the 111th Congress.[32] Other barriers to reform include export risks and organizational constraints on expediting the DoS's process for exporting commercial satellites.[33] These barriers stem from the fact that technical expertise at the Departments of State and Defense is lacking. Even though some incremental advances in addressing these barriers have taken place, as recommended by the *Cox Report*, determining risk is in many ways unworkable and the control of satellite exports through the national security lens does not readily lend itself to streamlining the licensing process.

Other examples of the policy dynamic of the DoS countering the Department of Commerce include: unilateral reversal of the Department of

Commerce's approach which exempted many items from requiring a licence;[34] the extension of ITAR controls to US allies for commercial satellites;[35] and regulations that required the return of hardware to its state of origin for repair.[36] In addition, the DoS issued retroactive regulations for the Technology Assistance Agreements governing technology transfers for satellites that had been licensed by the Department of Commerce. The DoS's retroactive approach created a situation where new technology transfer licences had to be issued for satellites already in orbit. The DoS acted to reverse President Reagan's decision that exempted fundamental research and technical information from an export licence.[37] This affects NASA, universities and industry R&D, as they now require licences for any collaboration with foreign nationals.

Export control policies and laws applicable to commercial satellites erode the US space industrial base as well as space commercial leadership and competitiveness. The political process began with the incremental political liberalization of export controls in response to the changing international post-Cold War environment, and the rapid increase in global space commerce. This led to Department of Commerce jurisdiction.[38] All the while, the process was driven by bureaucratic politics between the Departments of Commerce and State. In the context of the numerous security concerns facing the USA over the past decade, the national political preference was that commercial interests should never take precedence over national security interests. As a result, the DoS has succeeded in policy implementation under its own worldview, a costly situation for the US space industrial base and for commercial space.

SPACE LAUNCH SYSTEMS

Space policy-making in regard to the case of space launch system development in the USA is characterized by interactions among various groups of actors. These groups interact with each other on the basis of political, organizational and technical decisions. The relevant groups include advocacy coalitions, political decision-makers, the military, NASA, space industry, international actors and technologists. The first part of the case study discussion deals with advocacy groups for access to space. Then, the role of the political decision-makers who formulate policy is examined. Formulation of policy is followed by implementation involving the development of enabling space technologies for access to space.

In this case, the technological development of space launch systems is socially constructed by the competition and conflict among the relevant groups. The social construction of access to space has created political, organizational and technological momentum that is very difficult to change. Despite this momentum, the end of the Cold War, the commercialization of space launch services and the military need for assured access provide rationales for new political and technical approaches for access to space.

Advocacy Coalitions

Advocacy coalitions concerning space launch are represented by a number of actors, including rocket pioneers, engineers, scientists, military officers, national policy-makers, space business leaders and enthusiasts. These actors can be grouped by the advocacy coalitions identified in the first part of this chapter: space business, national defence, human destiny and space science.

Space business groups first emerged with the early entrepreneurs who attempted to profit by the development of rocketry. This advocacy coalition takes the position that access to space is best achieved through privatization efforts, which realize efficient, reliable, cost-effective and ideally routine space launches. Today, this group is best represented by the new space sector, which seeks to privatize space launch.

National defence groups find their intellectual origins in the development of artillery rocketry in both Germany and Russia.[39] Of primary interest to this coalition is assured access to space. This involves the concepts of operationally responsive launch and launch on-demand capabilities, advocacy of a mixed fleet for access to space to provide redundancy, operational control of launchers and technological innovation for different military missions. Due to the influence of the national defence coalition, chemical rocketry based on multi-staged ballistic missiles was viewed as more legitimate and feasible by the political decision-makers than other types of launch vehicles proposed by the various advocacy coalitions for access to space.[40]

The human destiny coalition pursues human access to space in co-operation with the government. In many ways, it is this coalition that sets the political, organizational and technological conditions that allowed travel to space to be translated into reality by the advocates of access to space. Important to the ideas of this coalition is the role of the 'rocket pioneers' exemplified in the works of Goddard, Oberth, Tsiolkovsky, Sänger, Korolev and von Braun.[41] The ideas of innovative improvements to chemical rocketry and alternative approaches, such as nuclear rockets, air-breathing 'fly into orbit' vehicles, magnetic lifters and space elevators emanate from this coalition.

From the beginning of serious rocketry in the USA, space science groups have been involved. Examples of this include the co-operation between scientists and the military services during the US Navy's Viking programme, the US Army's various V-2 testing efforts, and the discovery of the Van Allen radiation belts by the first US satellite.[42] The space science advocacy coalition relies on expendable civil and commercial launchers for access to space. This coalition often opposes the development of costly, complex launch systems used for human space flight, exemplified by the Space Shuttle. The rationale for this position is that human space flight diverts resources away from other, more scientifically justified projects.

Political Decision-Makers

Political decision-makers are those who are involved in the policy process whereby acceptable courses of action for dealing with specific problems are

Politics of Space

identified and enacted into political and programmatic decisions. The evolution of decision-making with regard to space launch systems is one of coalition building involving a plurality of actors, including advocacy coalitions, policy-makers, space bureaucracies and commercial enterprises.

There are a number of political decisions relevant to access space. These decisions include the development of liquid and solid chemical fuels for propulsion and expendable launch vehicles (ELVs), and attempts (which have failed to date, 2010) to develop reusable launch vehicles (RLVs). The formulation of concepts for access to space is influenced by political and economic realities and by the need for consensus among the different coalitions of actors to achieve political support.[43] The political and economic imperatives and the diversity of stakeholder demands on the proposed launchers presents difficulties in attaining effective optimization of access to space systems for a particular role, whether for civil scientific uses, operational and commercial applications or national security purposes.[44]

Political decision-makers chose to pursue ballistic missile technology, rather than other ideas that were available, such as winged, air-breathing or track launchers. The rise of the 'power state' that could mobilize vast societal resources for large-scale endeavours and the development of the intercontinental ballistic missile (ICBM) in the context of the Cold War led to the emergence of the ballistic missile as the preferred means for propelling cargo and humans into space.[45] This preference is reinforced by the military's need for assured access to space and the commercialization of space launch.

The evolution of the commercialization of space launch entails the failed 'one-launch policy', which was directed at making use of the Space Shuttle as a civil, commercial and military launcher, to political decisions that involved the Departments of Commerce and Transportation in commercial space transportation. The involvement of Commerce and Transportation occurred as a result of government-led international economic competition, particularly due to the development of the European and French Ariane launch vehicle, and the US military's reconsideration of its earlier decision to depend on the Space Shuttle due to their concerns regarding assured access.

The Ariane commercial launcher, the US military's interest in reviving ELVs to achieve assured access, President Reagan's free-market ideology relating to space commerce in the 1980s, and the culminating event that precipitated the demise of the Space Shuttle as a commercial and military launcher, the Challenger accident in 1986, engendered a reformulation of space launch policy by decision-makers. The political decision on commercial space transportation is one example of this reformulation. This decision put into place a licensing regime for space launchers along with an agenda to move to governmental purchases of launch services, where the government procures the services of industry for access to space, instead of buying and operating space launch vehicles. In fact, this approach to space launch has been endorsed by the Obama Administration.

Important to the analysis of the decision-makers are the political factors that influence the making of space policy. The roles played by the US President and Congress, and the interactions between them, at times cooperative and at other times conflictual, comprise one set of factors. Related to the role of Congress is the controlling and influencing of the space bureaucracy. In particular, the relations between Congress and the NASA and space industry bureaucracies have affected the nature of access to space. Congress is also influenced by the nature of regional politics that generate Congressional acquiescence to constituency interests and concerns. Examples of this are the preference for maintaining the technological status quo in launcher vehicle technology development to ensure continued economic benefits to Congressional districts, and the emergence of commercial spaceports in a number of US states.

A second set of factors relates to foreign policy and national security concerns. Access to space technologies directly impacts these concerns, as illustrated by ballistic missile proliferation issues. The US President also makes use of space as a means to achieve foreign policy objectives, such as Apollo for the Cold War and commercial launch agreements between the USA and Russia that started in the 1990s.

The interactions among the political decision-makers structure the policy process. How the process is structured directly impacts the implementers – the organizations and technologists – of space launch systems. The structure is either congruent with what the implementers need, such as political support in terms of budgetary resources and organizational capabilities, to develop the programmes and technology, or it is incongruent with what they need. Often, there is a compromise between what the implementers want and what the decision-makers formulate.

Military

The key areas of implementation for the US military include: assuring access to space; co-operating with and using both NASA and the space industry to promote military goals; ensuring domestic production of military sensitive technologies; preventing technology transfer to foreign countries; maintaining political accountability with the executive and legislative branches of the US Government; and managing bureaucratic conflicts among the various military services and other federal agencies.

National security, particularly the need to guarantee access to space for force enhancement and reconnaissance systems, is an essential factor in the military's development of launch vehicle technology.[46] This development was largely based on the conversion of ICBM technology, which served as the technical basis for launch vehicles developed by NASA and the space industry. National security interests to assure continued access to space led to the development of improved launch capacity with the Evolved ELV (EELV) programmes of the Atlas V and Delta IV. In addition, the military's forays into space control and force application has driven an interest in hypersonic vehicles, including, for example, Dyna-Soar, the Space Shuttle, the National Aero-Space Plane (NASP) and the X-40 Space Maneuver Vehicle.[47]

NASA

In NASA's case, there are a number of relevant factors that affect realization of access to space. One factor deals with the human space flight imperative versus scientific groups within NASA and their preference for robotic scientific probes. Another factor is the degree of organizational balance between competence, which is defined as the technical culture that seeks technological innovation and is accountable to the standards of technology development, and that of control, which is defined as the bureaucratic management culture that is accountable to organizational and political demands.[48] The balance between competence and control shapes the development stories of NASA's primary launch vehicles, the Saturn launch system that supported the Apollo programme and Skylab, and the Space Shuttle, which is planned for retirement by 2011.[49]

The relations between NASA, the space industry and the military to achieve access to space, i.e. NASA use of commercial and military launchers, are an essential part of the story. Their relationship with the advocacy coalitions, particularly those directed at space commerce with the space business coalition and the human destiny coalition, has influenced NASA. This influence is evident through the various space launch initiatives that NASA has pursued, such as NASP and the Shuttle replacement ideas of the X-33/Venture Star, Orbital Space Plane and Project Constellation.

Given that NASA is a governmental agency, political accountability has placed demands that considerably shaped NASA's formulation and implementation of access to space. This is compounded by the fact that NASA has focused on organizational and bureaucratic survival in a post-Apollo era of diminished political and budgetary support. This leads to a compromise between what NASA seeks in terms of access to space based on technical and organizational capacities, and what is attainable given political factors. To illustrate, all the Shuttle replacement programs mentioned earlier were cancelled prior to full development.

NASA's relations and interactions with the US Congress and President, and other federal agencies, such as the DoD and the Department of Commerce, affected this compromise, as illustrated by the Space Shuttle. In the end, the Space Shuttle was inefficient to meet any one particular need, whether science sought by NASA, assured access sought by the military or reliable, efficient operations sought by space business.

Space Industry

For the space industry group, access to space entails commercialization and privatization of space launch. This begins historically with the failed attempts to create a commercial launch industry, including the examples of the Percheron, Conestoga and a privately owned shuttle. The Space Shuttle programme is the first concrete attempt at commercialization of space launch in the USA. Even though the Space Shuttle was justified on economic grounds, it is, in fact, a compromise of military and civil human space flight interests, which is subject to non-market factors. This compromise leads to limits on the use of the Space Shuttle as a viable commercial launcher.

Following the end of commercial launches on the Space Shuttle after the Challenger accident, the space industry group, spurred by a market pull, made various efforts to improve the technology of access to space. The market pull was a result of the rise of space commercialization characterized by the development of the market for regional and private communication satellite networks. By the end of the Cold War, Russian and Chinese launch competition, in addition to the Ariane, further pushed an agenda for space launch commercialization in the USA. This resulted in specific policies and laws for commercialization and investments in various attempts at technology development with commercial interests foremost in mind.[50]

Weak market conditions – greater capacity than demand – in conjunction with the US Government's economic rationales for the commercialization and privatization of space launch, resulted in active government involvement in developing and subsidizing launchers. The US Government's role in commercialization of space launch, which at times has been enabling and at other times constraining, is shaped by the relevant political, economic and technical factors, as well as by the particular civil, commercial and military coalitions that deal with the issue of commercialization of space launch.

United States and International Actors

Foreign state actors have impacted access to space in a number of areas. For example, foreign launch providers increase the number of systems vying for commercial and government markets. This lessens the prospects for enhancing commercialization of space launch, as capacity is in excess of market demand. The development paths of foreign launch providers, primarily in Europe with the Ariane and in Russia and China, influences the US space launch industry. Important outcomes in this respect are the multinational corporate alliances, whereby technologies for commercial launch are integrated, such as between the USA and Russia with Atlas V and Sea Launch.

Technologists

The politics of access to space would be incomplete without considering the technologists. The role of the technologists is based on the efforts of the various rocket developers and the technical approaches that they pursued. This includes: the conversion of ICBMs into ELVs, such as the Scout, Atlas, Delta and Titan; upper stage systems, like the Centaur and Agena; the Saturn launch system; and hypersonic vehicle and reusable vehicle programmes, such as X-15, Dyna-Soar and the Space Shuttle.

With the development of the Space Shuttle, innovations from the technical R&D groups reached a technological plateau. Since then, there have been only incremental improvements in space launchers, with no major breakthroughs in the technology of access to space. This technical steady state is due to the view of particular coalitions of technologists rooted in the civil, commercial and military sectors that the ballistic missile technology is an acceptable means of access to space and it is not worth the cost of improving it. Even though the commercially built configurations of the legacy Atlas, Delta and Titan, and the development of new launch systems, like the Atlas V, Delta IV and the Pegasus air-launched vehicle, suggest some degree of innovation beyond the technical steady state, attempts by technologists to advance technical innovations in a revolutionary manner have failed.

Failures exist in developing low-cost access to space, reusability and in alternatives to the Space Shuttle. Prominent examples of failure include: NASP; the Advanced Launch System (ALS) and National Launch System (NLS); the various attempts to develop second-generation RLVs through the Space Launch Initiative (SLI), from the X-series programmes to the X-33/Venture Star; and the attempts to produce an RLV that were put forward by a number of privately funded commercial start-ups.[51]

In developing space launch systems, the separation of R&D groups from project groups is an important issue.[52] As a result of this separation, technologists find it difficult to 'push' their new technologies into the actual launcher projects, and project managers frequently do not know where to find appropriate technologies to 'pull' into their systems. At NASA, this has been reflected by the separation of NASA's R&D organizations and budgets from the project groups; in the military, this is distinguished by the separation of the budgets and organizations for R&D from the budgets for development and testing. These separations have been a factor in the conservative and incremental evolution of space launch technologies since the 1970s.

Given the number of players in the access to space arena, the interests of those players must be met to ensure political support and sustainability for innovation. The more the approach to access to space is tied to the interests of one group, the greater the chances are that competing interests will constrain the political feasibly of that approach. At the same time, meeting the interests of the various actors can result in an over-optimization of systems, which increases complexity and undermines other ends, like low cost and reliability. The role of the commercial sector is also important to innovation. To date, the model of contracting out favours the technical steady state. The reformulation of contracting to allow for governmental procurements of services and developments with private start-ups in space launch, such as the NASA Commercial Crew and Cargo programme, is a trend that could favour technological innovation.

CONCLUSIONS

The remaining question is whether space policy is in some way different, placing it outside the scope of the policy-making processes described at the beginning of this chapter. The use of space assets for scientific, commercial and military purposes means that there is a unique convergence, or divergence, of interests and uses in the space medium. Others add that this medium is particularly expensive, with budgets controlled by multiple organizations and agencies such as the DoD, and individual services like USAF and NASA. Finally, there are those who note the 'black' (or classified) versus 'white' (or unclassified) nature of space, referring to the conflict between the highly restricted NSS environment and the open, commercial and scientific one.

While all of these arguments are true, they do not negate the applicability of policy-making processes discussed herein. In many ways, these arguments further illustrate the complex environment, interests and policy preferences that policy-making entails. Space policy-making, whether in the civil, commercial or military sector, involves a multitude of advocacy coalitions, government actors and agencies and commercial corporations competing over resources and objectives and control of space programmes and projects.

NOTES

1 Christopher J. Bosso and W. D. Kay, 'Advocacy Coalitions and Space Policy', in Eligar Sadeh, ed., *Space Politics and Policy: An Evolutionary Perspective* (The Netherlands: Springer Kluwer Academic Publishers, 2003).
2 Schuyler Foerster and Edward N. Wright, 'The Twin Faces of Defense Policy: International and Domestic', in Schuyler Foerster and Edward N. Wright, eds, *American Defense Policy*, 6th edition (Baltimore, MD: The Johns Hopkins University Press, 1990).
3 Morton H. Halperin and Arnold Kanter, 'The Bureaucratic Perspective: A Preliminary Framework', in Morton H. Halperin and Arnold Kanter, eds, *Readings in American Foreign Policy: A Bureaucratic Perspective* (Boston, MA: Little, Brown, 1973).
4 Incremental policy-making does not solve an identified national problem and is more apt to represent a continuing government commitment. See Charles E. Lindblom's classic essays, 'The Science of Muddling Through', *Public Administration Behavior*, 19: 2 (1959): 79–88; and 'Still Muddling ... Not Yet Through', *Public Administration Review*, (November/December 1979): 517–26.
5 The task of implementation is rocket science. It is terribly complicated. Launchers explode and spacecraft disappear. No one wants failure. Good enough is not good enough for technology within which thousands of components must work in tandem for a mission to succeed. See Howard E. McCurdy, 'Bureaucracy and the Space Program', in Eligar Sadeh, ed., *Space Politics and Policy: An Evolutionary Perspective* (The Netherlands: Springer Kluwer Academic Publishers, 2003).
6 *Report of the Commission to Assess United States National Security Space Management and Organization*, 11 January 2001.
7 *Report of the Commission to Assess United States National Security Space Management and Organization*.
8 Ibid.
9 In 2001, upon the recommendation of the Space Commission, the Space and Missile Systems Center joined the command. It previously belonged to Air Force

Materiel Command. Air Force Space Command is currently the only Air Force command to have its acquisition arm within the command. In 2002, also on a recommendation from the Space Commission, Air Force Space Command was assigned its own four-star General as commander after previously sharing a commander with US Space Command and the North American Aerospace Defense Command (NORAD).
10 United States General Accounting Office, *Report to Congressional Committees; Defense Space Activities: Organizational Changes Initiated, but Further Management Actions Needed* (GAO-03-379, 2003).
11 The Independent Assessment Panel was chartered by the US Congress to review and assess the US Department of Defense management and organization of National Security in Space and make appropriate recommendations to strengthen the US position. "The panel members are unanimous in our conviction that significant improvements in National Security Space leadership, management and organization are imperative to maintain US space preeminence and avert the loss of the US competitive national security advantage. NSS inadequacies are unacceptable today and are likely to grow, but political leadership can reverse this trend."
12 Joan Johnson-Freese, 'Alice in Licenseland: US satellite export controls since 1990', *Space Policy* 16: 3 (2000): 195–204.
13 Commercial satellites are clearly intended for commercial use and applications, but do represent applications and technologies that could be used for military purposes and military satellite development.
14 US House Report 105-851 (Washington, DC: US Government Printing Office, 1999).
15 Boeing and Loral were fined by the US Federal Government for the export violations; both companies paid the fines in 2002. Boeing was also charged with similar export violations concerning Sea Launch – a joint venture with Russian, Ukrainian and Norwegian companies – during this same period.
16 *The Space Report: The Guide to Global Space Activities* (Colorado Springs, CO: Space Foundation, 2008); and *Space 2030: Exploring the Future of Space Applications* (Paris, France: Organisation for Economic Co-operation and Development, 2004).
17 Ian F. Fergusson, *The Export Administration Act: Evolution, Provisions, and Debate* (United States Congressional Research Service, The Library of Congress, updated 5 May 2005).
18 *Defense Trade, Arms Export Control System in the Post-9/11 Environment* (United States Government Accountability Report, 2005).
19 Ian F. Fergusson, *The Export Administration Act: Evolution, Provisions, and Debate* (United States Congressional Research Service, The Library of Congress, updated 5 May 2005).
20 Export controls on space commerce create risk through uncertainties, result in losses of markets because of impacts on space industry's ability to serve international markets, and prevent efficient industry restructuring to the forces of globalization. *Space 2030: Exploring the Future of Space Applications* (Paris, France: Organisation for Economic Co-operation and Development, 2004).
21 In Europe, Alcatel Alenia Space and the European Aeronautic Defence and Space Company have both made it company policy to build 'ITAR-free' commercial satellites.
22 Joan Johnson-Freese, 'Alice in Licenseland: US satellite export controls since 1990', *Space Policy* 16: 3 (2000): 195–204.
23 Wassenaar Arrangement state members in addition to the USA, include: Argentina, Australia, Austria, Belgium, Bulgaria, Canada, Croatia, Czech Republic, Denmark, Estonia, Finland, France, Germany, Greece, Hungary, Ireland, Italy, Japan, Republic of Korea, Latvia, Lithuania, Luxembourg, Malta, Netherlands, New Zealand, Norway, Poland, Portugal, Romania, Russian Federation, Slovakia,

Slovenia, South Africa, Spain, Sweden, Switzerland, Turkey, Ukraine and the United Kingdom.
24 "It is the sense of the US Congress that business interests must not be placed above national security interests." Strom Thurmond National Defense Authorization Act for Fiscal Year 1999.
25 See *State of the Satellite Industry Report* (Washington, DC: Futron Corporation, 2006); Robert D. Lamb, *Satellites, Security, and Scandal: Understanding the Politics of Export Controls* (College Park, University of Maryland: Center for International and Security Studies at Maryland, 2005); *Space 2030: Exploring the Future of Space Applications* (Paris, France: Organisation for Economic Co-operation and Development, 2004); and *State of the Space Industry* (Washington, DC: International Space Business Council, 2000).
26 The Congressional bill vetoed by President Bush was the Omnibus Export Amendments Act of 1990.
27 Marcia S. Smith, *Space Launch Vehicles: Government Activities, Commercial Competition, and Satellite Exports* (United States Congressional Research Service, The Library of Congress, updated 1 January 2006).
28 George Abbey and Neal Lane, *United States Space Policy, Challenges and Opportunities* (American Academy of Arts and Sciences, 2005).
29 *Preserving America's Strength in Satellite Technology, A Report of the CSIS Satellite Commission* (Washington, DC: Center for Strategic and International Studies, 2002).
30 *Report of the Select Committee on US National Security and Military/Commercial Concerns with the Peoples' Republic of China*, US House Report 105–851 (Washington, DC: US Government Printing Office, 1999).
31 Robert D. Lamb, *Satellites, Security, and Scandal: Understanding the Politics of Export Controls* (College Park, University of Maryland: Center for International and Security Studies at Maryland, 2005).
32 In addition to the Satellite Trade and Security Act of 2001, Congressional sponsors have proposed amendments to the Export Administration Act and other separate bills, which would return export licensing authority for commercial satellites to Commerce.
33 The inability to accurately measure risk to national security is one of the most serious problems for the system of export controls.
34 Commerce exempted basic items, like screws and knobs for example, from export control.
35 The Strom Thurmond National Defense Authorization Act for Fiscal Year 1999 included language that MCL will not necessarily apply to the "export of a satellite or related items for launch in, or by nationals of, a state that is a member of NATO, or that is a major non-NATO ally of the United States." In implementing ITAR, the Department of State interpreted this exception to apply only to the mandated monitoring activities. Further, the expanded definitions of satellite-related components, and the additions of defence technical services and space insurance business meetings as new areas needing export licences, led to the bureaucratic micro-regulation of the US commercial satellite industry in response to accusations initially related to China.
36 The ITAR regime as applied can also have extraterritorial elements. A Technology Assistance Agreement may specify the states of which a foreign national may be a dual citizen and still have access to the transferred data or hardware. See Eric Choi and Sorin Niculescu, 'The impact of US export controls on the Canadian space industry', *Space Policy* 22: 1 (2006): 29–34.
37 In 1985 President Reagan issued an ITAR exemption for fundamental research conducted at US universities. National Security Decision Directive 189, 21 September 1985.

Politics of Space

38 Lewis R. Franklin, 'A Critique of the Cox Report Allegations of PRC Acquisition of Sensitive US Missile and Space Technology', in M. M. May, ed., *The Cox Committee Report: An Assessment* (Stanford, CA: Stanford University, Center for International Security and Cooperation, 1999).
39 T. A. Heppenheimer, *Countdown: A History of Space Flight* (New York: Wiley, 1997).
40 Curtis Peebles, *High Frontier: The United States Air Force and the Military Space Program* (Washington, DC: Air Force History and Museums Program, 1997).
41 John Bankston, *Robert Goddard and the Liquid Rocket Engine* (Bear, DE: Mitchell Lane Publishers, 2001); Milton Lehman, *Robert H. Goddard: Pioneer of Space Research* (New York: De Capo Press, 1988); Dennis Piszkiewicz, *Wernher von Braun: The Man Who Sold the Moon* (Westport, CT: Praeger Publishers, 1998); and Ernst Stuhlinger and Frederick I. Ordway III, *Wernher von Braun: Crusader for Space* (Malabar, FL: Krieger, 1996).
42 William E. Burrows, *This New Ocean: The Story of the First Space Age* (New York: Random House, 1998).
43 Andrew J. Butrica, *Single Stage to Orbit: Politics, Space Technology, and the Quest for Reusable Rocketry* (Baltimore, MD: The Johns Hopkins University Press, 2003).
44 T. A. Heppenheimer, *History of the Space Shuttle: Volume Two, Development of the Space Shuttle, 1972–1981*, (Washington, DC: Smithsonian Institution Press, 2002); and T. A. Heppenheimer, *The Space Shuttle Decision: NASA's Search for a Reusable Space Vehicle* (Washington, DC: NASA SP-4221, 1999).
45 William E. Burrows, *This New Ocean: The Story of the First Space Age* (New York: Random House, 1998); and Walter McDougall, *The Heavens and the Earth: A Political History of the Space Age* (Baltimore, MD: The Johns Hopkins University Press, 1997).
46 Curtis Peebles, *High Frontier: The United States Air Force and the Military Space Program* (Washington, DC: Air Force History and Museums Program, 1997).
47 Andrew J. Butrica, *Single Stage to Orbit: Politics, Space Technology, and the Quest for Reusable Rocketry* (Baltimore, MD: The Johns Hopkins University Press, 2003); Jay Miller, *The X-Planes: X-1 to X-31*, 2nd edition, (New York: Aerofax, 1998); and Milton O. Thompson, *At the Edge of Space: The X-15 Flight Program* (Washington, DC: Smithsonian Institute Press, 1992).
48 Richard L. Chapman, *Project Management in NASA, the System and the Men* (Washington, DC: NASA SP-324, 1973); Francis T. Hoban, *Where Do You Go After You've Been to the Moon* (Malabar, FL: Krieger Publishing Company, 1997); Arnold Levine, *Managing NASA in the Apollo Era* (Washington, DC: NASA SP 4102, 1982); Eligar Sadeh, ed., *Space Politics and Policy: An Evolutionary Perspective* (The Netherlands: Kluwer Academic Publishers, 2003); and Philip K. Tompkins, *Organizational Communication Imperatives: Lessons of the Space Program* (Los Angeles, CA: Roxbury, 1992).
49 Roger E. Bilstein, *Stages to Saturn: A Technological History of the Apollo/Saturn Launch Vehicles* (Washington, DC: NASA SP-4206, 1980); and Roger D. Launius and Dennis R. Jenkins, eds, *To Reach the High Frontier: A History of US Launch Vehicles* (Lexington, KY: The University Press of Kentucky, 2002).
50 Joan Lisa Bromberg, *NASA and the Space Industry* (Baltimore, MD: The Johns Hopkins University Press, 1999); and John L. McLucas, *Space Commerce* (Cambridge, MA: Harvard University Press, 1991).
51 Andrew J. Butrica, *Single Stage to Orbit: Politics, Space Technology, and the Quest for Reusable Rocketry* (Baltimore, MD: The Johns Hopkins University Press, 2003); Jay Miller, *The X-Planes: X-1 to X-31*, 2nd edition, (New York: Aerofax, 1998); Dale R. Reed, *Wingless Flight: The Lifting Body Story* (Washington, DC:

NASA SP-4220, 1997); and Milton O. Thompson, *At the Edge of Space: The X-15 Flight Program* (Washington, DC: Smithsonian Institute Press, 1992).
52 Stephen B. Johnson, *The Secret of Apollo: Systems Management in American and European Space Programs* (Baltimore, MD: The Johns Hopkins University Press, 2002); and Stephen B. Johnson, *The United States Air Force and the Culture of Innovation 1945–1965* (Washington, DC: USAF History and Museums Program, 2002).

National Security Space

Peter L. Hays*

This chapter discusses the example of the national security space (NSS) sector in the USA. The rapid and decisive conventional military victories during Operations Desert Storm in the Persian Gulf in 1991, Allied Force Overhead Serbia in 1999, Enduring Freedom in Afghanistan in 2001, and Iraqi Freedom in 2003 were a result of an increasing percentage of precision guided munitions (PGMs) and improving communications connectivity, which are key components of a new US way of war that is empowered by a space-enabled global reconnaissance, precision strike complex (see Illustrations and Documentation: Table 2.1 Space-Enabled Reconnaissance Strike Complex).

A primary goal of the US Department of Defense (DoD) since the end of the Cold War has been to achieve full spectrum dominance of the battle space by continuing and accelerating military transformation. DoD is developing lighter and more easily deployable forces, which are better able to leverage network-enabled operations and strike more precisely from greater distances against adversaries that may range from emerging military peers to insurgents and terrorists.

Space capabilities often provide the best and sometimes the only way to pursue these ambitious transformational goals. There are, however, many difficult and fundamental issues related to space and defence policy, including: the nature of the fundamental contributions of space to enabling the information revolution and the new US way of war; changes caused by growth in commercial space, the number of major space actors and global proliferation of counterspace capabilities; and the role and efficacy of space capabilities in structuring options for military intervention and in dissuading and deterring competition from potential adversaries in the changed geopolitical environment following the end of the Cold War, the 11 September 2001 (9/11) terrorist attacks on the USA, and Operation Iraqi Freedom.

These complex factors contribute to uncertainty about how space capabilities can best advance US national security, the most useful organizational structures to manage and transform space activities themselves, and the utility of investments in space capabilities versus other enabling military capabilities. Moreover, the USA faces significant challenges in its current plans to modernize, improve and replace almost all major military space systems, because most of these systems are essential for future transformed forces, but their acquisition has been marked by cost overruns and deployment delays. It is unclear whether the USA will be able to find and follow the best path forward for space strategy, implement the best management and organizational structures for space activities, and sustain the political will needed to continue funding the nearly simultaneous modernizations in space assets currently planned. It is also uncertain whether these new and improved space capabilities can be developed and integrated on cost and on time, and whether these future systems will deliver on their promise of accelerating transformational capabilities and effects.

The NSS sector in the USA consists of DoD activities, conducted primarily by the US Air Force (USAF) to enhance national security, National Reconnaissance Office (NRO) programmes to collect intelligence data from space, and civil, commercial and international space activities that support US national security. The NSS sector is divided into separate sectors known as the military or defence space sector, and the intelligence space sector. It is also useful to consider all of the space professionals, training and education, infrastructure, industry and policies that support the NSS sector.

Following implementation of one of the recommendations of the January 2001 Commission to Assess National Security Space Management and Organization (Space Commission) report, the DoD now uses an accounting procedure known as the virtual major force programme (vMFP) to track NSS spending, but it is still difficult to follow the NSS budget.[1] According to figures provided to the Congressional Research Service, total (classified and unclassified) DoD space spending amounted to US $19.4bn in fiscal year (FY) 2003, $20bn in FY 04, $19.8bn in FY 05 and $22.5bn was requested for FY 06.[2] Under the vMFP baseline for NSS procurement and research and development (R&D), unclassified military space acquisition spending grew from $4.9bn to $6.9bn, or more than 40%, between FY 05 and FY 06, rose almost 12% to $7.7bn in FY 07, and then climbed another 13% to $8.7bn for FY 08.[3]

Overall trends in planned major military space acquisition to 2024 (see Part III: Figure 2.1. Investment in National Security Space Programmes) illustrates the risk of cost growth whereby space acquisition expenditure will peak at $14.4bn in 2010 or almost double present funding if current programmes follow the historic trend of an average 69% rise in costs for space research, development, engineering and testing, as well as an average growth of 19% in space procurement costs.[4] Clearly, the path ahead for currently planned NSS improvements and modernizations will be very difficult, if not unsustainable. This problem, along with a number of other daunting short-term challenges, is discussed after an overview of conceptual frameworks for analysis, a review of major NSS actors and management structures, and current major space acquisition programmes and budgets.

ANALYSIS OF NATIONAL SECURITY SPACE

Three major analytical frameworks shape most discussions about NSS capabilities: space activity sectors; military space mission areas; and military space doctrines. There are four space activity sectors, which include civil, commercial, military and intelligence. Many traditional space activities fall neatly into one of these sectors, although the growing number of dual-use (civil-military or commercial-military) space systems, digital convergence, and growth in the commercial space sector make it increasingly difficult to delineate among the sectors.[5]

There are also four military space mission areas: space support, force enhancement, space control and force application.[6] Currently, force enhancement is the most important military space mission area; due to growth in the number and effectiveness of space systems, many analysts believe these capabilities now produce effects that have moved beyond force enhancement and currently enable a wider range of military missions to be undertaken or even contemplated (see Illustrations and Documentation: Table 2.2 Force Enhancement Space Systems). Finally, there are four major military space doctrines: sanctuary, survivability, control and high ground.[7] The attributes associated with these doctrines entail primary value and functions, employment strategies, conflict missions and desired organizational structures (see Illustrations and Documentation: Table 2.3 Attributes of Military Space Doctrines).

ACTORS AND MANAGEMENT STRUCTURES OF NATIONAL SECURITY SPACE

Over the past decade, the NSS enterprise has undergone considerable turmoil in terms of its major actors and organizational structure (see Illustrations and Documentation: Figure 1.3 National Security Space Community). Because so many NSS management and organizational structure changes have been implemented, undone, or modified in such a short span of time, the ways in which previous changes had affected the enterprise were not determined before the next ones were initiated; it is not clear whether any of these changes have improved or hindered efforts to deliver enhanced space capabilities and foster better unity of effort.

In the 1990s, major NSS actors included the DoD, USAF, US Space Command (USSPACECOM), and the NRO. With the exception of USSPACECOM, all of these organizations remain key space actors today, but there is greater disorder and uncertainty in the interrelationships between key space policy decision-making structures, both internally and among these organizations, than this single major organizational change would suggest. Contentious issues include whether there is an identifiable and usefully delineated NSS enterprise, what elements should and should not be included within it, and how best to foster better unity of effort and more clear lines of responsibility and authority within this enterprise. Options include: attempting to deepen specific expertise, serve individual users and emphasize separate systems by construing NSS narrowly as separate defence and intelligence space sectors; working to integrate more broadly across all government space activities and between the DoD and intelligence community; and emphasizing growing interdependencies across all space sectors and considering in addition how best to co-ordinate with and protect those civil, commercial and international space capabilities that support national security. Despite the many recommendations and changes designed to improve and assure delivery of space capabilities, foster unity of effort and clarify lines of authority, the problem of 'who is in charge' persists, and today it is even less clear than it was in the 1990s as to which major actors and structures have greatest responsibility and accountability for key NSS decisions.

Due to its sweeping charter and powerful members, the Space Commission has been, to date, the most important and influential examination of NSS issues.[8] Many of the most significant recent NSS changes were made in direct response to the major recommendations in the Space Commission report released in 2001. The USAF and DoD moved quite quickly and effectively to implement at least portions of 10 of the Space Commission's 13 major recommendations, such as making the Commander of Air Force Space Command (AFSPC) a four-star billet that need not be flight rated and moving AFSPC out from underneath USSPACECOM; designating the Under-Secretary of the Air Force as the Director of the NRO, Air Force Acquisition Executive for Space and DoD Executive Agent for Space; aligning the Space and Missile Systems Center (SMC) underneath AFSPC instead of Air Force Materiel Command; and establishing MFP for the NSS budget.[9]

Other major recommendations were beyond the power of the DoD to implement and included: the need for Presidential leadership in recognizing space as a top national security priority; appointment of a Presidential Space Advisory group and establishment of a Senior Interagency Group for Space within the National Security Council structure; and the need for the Secretary of Defense and Director of Central Intelligence[10] to work closely and effectively together on space issues. One of the most important recommendations left undone was primarily within the power of the DoD to implement and called for the creation of an Under-Secretary of Defense for Space, Information and Intelligence. Instead, leadership for space was placed in the Under-Secretary of Defense for Intelligence position created in 2003. Failure to institutionalize centralized authorities and responsibilities for NSS within the Office of the Secretary of Defense (OSD) undermined the Space Commission's vision for organization and management of NSS, helps explain why several important NSS programmes lack unity of effort, and contributes to continuing unhealthy competition between OSD branches as well as overlaps, gaps and unclear lines of authority and responsibility between OSD and the DoD Executive Agent for Space.

Another very significant and probably more important change to the Space Commission's vision for NSS management and organization came in 2002, when USSPACECOM was merged into United States Strategic Command (USSTRATCOM). Although this was originally described as a joining of equals, in practice this major organizational shift quickly amounted to the absorption of USSPACECOM into USSTRATCOM and left very few

National Security Space

vestiges of the original USSPACECOM. Instead of space being the sole focus of one of just nine unified commands, under the new structure space now competes for attention among a very wide array of disparate mission areas, which include global strike, homeland defence, information operations and missile defence. Because unified commands are the warfighters who operate systems and set capability requirements, this change has resulted in less focus on current space operations and future space capability needs. It is difficult to reconcile this organizational change with the Space Commission's overarching recommendation to make space a top national security priority.

Two more recent internal changes in management structures have also significantly slowed steps towards improved NSS integration and unity of effort: movement of milestone decision authority for major NSS acquisitions away from the DoD Executive Agent for Space to the Under-Secretary of Defense for Acquisition, Technology and Logistics in 2005; and separation of the position of Director of the NRO from the DoD Executive Agent for Space. Removal of milestone decision authority, or the authority to make decisions to continue, restructure, or end major acquisition programmes at key decision points, was originally explained as a temporary expedient due to a lack of confirmed USAF leadership able to exercise such authority during that time, but milestone decision authority has never been returned to the DoD Executive Agent for Space. Lack of milestone decision authority undermines the power of the DoD Executive Agent for Space by making it a position with responsibility, but without authority, and contributes to a lack of clarity over who is in charge of NSS acquisition, especially during a period when many major NSS acquisition efforts are troubled and there is not always agreement on the most appropriate corrective measures.

Shortly thereafter, it was announced that the Under-Secretary of the Air Force and DoD Executive Agent for Space would not, as in the past, also be director of the NRO. Although very little public rationale was provided, this 'divorce' was a very significant organizational change, which, like closing USSPACECOM, called into question DoD commitment to some of the recommendations from the Space Commission, as the need for better integration between DoD (military) and intelligence was a major finding of the Commission and it is difficult to understand how two positions and two people can achieve better integration than one. Moreover, the divorce revealed stark inconsistencies in the US approach to NSS management and organization because the argument was made that a separate NRO Director was needed to provide more focused attention on that organization, shortly after USSTRATCOM had absorbed USSPACECOM and created an organizational structure where space could not receive the same focused attention it previously had.

Unlike the major changes just discussed, other developments in management and organizational structures have advanced better integration and unity of effort across the NSS enterprise. These include: publication of an biannual NSS Plan and Program Assessment with resource-constrained priorities specified by the DoD Executive Agent for Space; ongoing NSS Architecture efforts for intelligence, surveillance and reconnaissance (ISR), communications, Position-Navigation-Timing (PNT) and space control; and creation or reinvigoration of a number of co-ordination mechanisms across the NSS enterprise, which include the Space Industrial Base Council, Suppliers Group Council, Space Professional Oversight Board, NSS Science and Technology Council, Space Partnership Council (with meetings several times a year between the DoD Executive Agent for Space and heads of the National Aeronautics and Space Administration (NASA), NRO, USSTRATCOM and AFSPC); and Congressional Space Power Caucus. In addition, the DoD Executive Agent for Space hosts annual meetings with the Chief Executive Officers of commercial satellite communications and remote sensing providers to discuss neighbourhood watch (space situational awareness) and protection issues as well as other best practices. In 2008 the Congressionally-mandated Independent Assessment Panel delivered an important re-examination of NSS organization and management issues, assessed the Space Commission's recommendations and their implementation, and made four major recommendations.[11]

The four major recommendations encompassed: (1) the President should establish and lead the execution of a National Space Strategy and reconstitute the National Space Council at the White House, chaired by the National Security Advisor; (2) establish a National Security Space Authority with the rank of Under-Secretary of Defense and Deputy Director for National Intelligence for Space and authority as the DoD EA for Space (see Illustrations and Documentation: Figure 1.4 Proposed National Security Space Authority); (3) create a National Security Space Organization consisting of NRO, SMC and operational functions of AFSPC and Army and Navy organizations providing space capabilities; and (4) change USAF and intelligence community human resource management policies for space acquisition professionals to emphasize technical competence, experience and continuity.

NATIONAL SECURITY SPACE ACQUISITION PROGRAMMES AND BUDGETS

National security space programmes involve the four military space mission areas mentioned earlier: space support, force enhancement, space control and force application. The vast majority of US national security space efforts today are in the force enhancement mission area.

Space Support

The space support mission includes two main areas: satellite control programmes that provide global communications systems for telemetry, tracking and control (TT&C) of satellites (the USAF Satellite Control Network is one example); and launch and range programmes that maintain and improve the infrastructure at the DoD's two launch sites, the Eastern Range at Cape

National Security Space

Canaveral Air Force Station and the Western Range at Vandenberg Air Force Base (AFB). Under the 2006 Future Years Defense Program (FYDP) and the US Congressional Budget Office's long-term projection, annual investment funding for those activities average about $180m. to 2024.[12] Part of those resources would go toward modernizing the Launch and Test Range System, which provides tracking, telemetry, flight safety and other support for space launches and ballistic missile tests.

NSS space launch capabilities include legacy systems, such as the Delta II, of which production has been phased out,[13] the Evolved Expendable Launch Vehicle (EELV) that the DoD now uses to put many of its satellites into orbit, and the new Operationally Responsive Space (ORS) programme, which is designed to develop concepts of operations, payloads and launchers capable of proactive space operations. The EELV programme has two types of launchers: the Boeing Delta IV that uses RS-68 main engines, and the Lockheed Martin Atlas V that uses Russian RD-180 main engines. Both EELVs are operated by United Launch Alliance, a joint venture between Boeing and Lockheed Martin that began operations in 2006. Each EELV can carry medium-sized payloads from 10 to 15 metric tons, and the Delta IV family now includes an operational heavy-lift variant for putting larger payloads, up to about 25 metric tons, into low Earth orbit (LEO). From 1999 to the end of 2008, there were 57 consecutive successful NSS space launches, including 22 successful EELV launches.

Due to several factors, including a decline in commercial launch demand that was used as justification for the Delta IV and Atlas V, the EELV programme experienced a cost increase of more than 25% in 2004, triggering a Nunn-McCurdy certification breach.[14] The 2006 FYDP implies total funding needs of about $28bn for approximately six to seven EELV launches each year to 2024.[15] A 2002 ORS mission needs statement established the requirement for responsive, on-demand access to, through and from space and led to the establishment of a Defense Advanced Research Projects Agency (DARPA) and USAF joint programme office in 2002. ORS initiatives struggled for a number of years for a variety of reasons, including Congressional concerns about use of ORS for force application missions, changing organizational structures, a lack of significant funding and ill-defined goals. A number of recent improvements, including the creation of the ORS Office at Kirtland AFB in 2007, more funding and clearer goals are likely to lead to better results. Opportunities to develop additional responsive space capabilities are discussed in the final section herein.

Force Enhancement: Military Satellite Communications Systems

Military satellite communications systems provide a range of critical capabilities, including assured connectivity for nuclear command and control, links between commanders and forces globally, and the foundation for network-enabled operations. Satellite communications systems are the single most important military space capability. This NSS mission area has the largest number of programmes and receives large investments, but current projections of the growth of the DoD's demand for satellite communications indicate that even with deployment of much more capable dedicated military systems over the coming years, the DoD will remain reliant on commercial satellite communications systems, especially during wartime. During Operations Enduring Freedom and Iraqi Freedom over 60% and over 80% of military communications, respectively, were carried on commercial systems.[16] Benefits of using commercial systems entail lower acquisition and operations costs as well as greater flexibility, but must be balanced against drawbacks, such as high cost of buying commercial services on the spot market, questionable availability of services, and less secure and protected systems. Nevertheless, the DoD has benefited recently from the widespread availability and relatively low costs of the overbuilt commercial satellite communications sector created during the late 1990s.[17]

The three main types of military satellite communications primarily use different parts of the radio spectrum: (1) wideband on super high frequency (SHF); (2) protected on extremely high frequency (EHF); and (3) narrowband on ultra high frequency (UHF). The Congressional Budget Office estimates for military satellite communications system spending to 2024 a total of $27bn in USAF funding for wideband and protected capabilities and $5bn in Navy funding for narrowband capabilities.[18]

Operational wideband systems encompass the Defense Satellite Communications System (DSCS), a constellation of five primary satellites plus a number of older residual-capability satellites, and the Global Broadcast Service (GBS), which consists of payloads on three Navy UHF Follow-On (UFO) satellites augmented by leased Ku-band transponders. Originally, it was expected that DSCS satellites would have a service life of 10 years; however, they are lasting longer than anticipated, as is the case with many space systems. The DSCS constellation is supposed to remain operational to 2015 and the GBS to 2010. The USAF is in the process of replacing both systems with a constellation of six much more capable geostationary Earth orbit (GEO) satellites known as the Wideband Global System (WGS), a joint Air Force-Army programme with the participation of Australia, which will augment current DSCS X-band and GBS Ka-band capabilities and establish a new two-way Ka-band service. In 2002 the OSD directed the addition of two more WGS satellites as part of the transformational communications architecture; the second block of WGS will support increased bandwidth requirements for airborne ISR and other missions. The first WGS launch had been delayed for over three years and was accomplished in October 2007, while WGS-2 launched successfully in April 2009; the six-satellite constellation is due to be completed in 2013 and remain in service until 2024.[19] In 2007, Australia announced it will pay $823.6m. to purchase WGS 6 as a way to enhance the network-enabled capabilities of Australian forces, leverage the entire WGS constellation, and further the close ties that already exist between the USA and Australia.[20]

Protected satellite communications are now provided by five Milstar satellites, which are expected to be operational until 2014. Like DSCS satellites, Milstar satellites are exceeding their design lifetime and may be available beyond that time. Under current plans, beginning in 2010, a constellation of four to five Advanced EHF (AEHF) satellites will be launched to begin replacing Milstar.[21] AEHF includes cryptography necessary to provide worldwide, durable and anti-jam protected communications for strategic and tactical warfare, as well as much higher data rates than Milstar. Lockheed Martin is the prime contractor on the system and Northrop Grumman is developing the satellite payload. AEHF is a co-operative programme with Canada, the United Kingdom (UK), and the Netherlands, but was not selected as the primary system to provide next generation North Atlantic Treaty Organization (NATO) protected satellite communications capability.

Development of new, complex information assurance products based on the AEHF by the US National Security Agency (NSA) has been a difficult challenge and contributed to cost overruns of approximately $1bn and, thereby, to a Nunn-McCurdy breach at the 15% threshold in December 2004, as well as a two-year launch delay for each satellite, from April of 2007–9 until 2009–11.[22] In addition, the Director of Space and Intelligence Capabilities within the Under-Secretary of Defense for Acquisition, Technology and Logistics is currently leading another Nunn-McCurdy review on AEHF, which was completed in 2008.[23] To supplement the satellites designed to provide coverage of the globe up to about 65 degrees of latitude, the USAF is also pursuing two programmes to improve protected communications over the northern Polar Regions: the first is operational via three low-data-rate Milstar packages on classified host satellites as an interim solution; and an enhanced system will take advantage of AEHF technology and should be available for launch in FY 13 and FY 15.

The most important planned future system for both protected and wideband communications capabilities was the Transformational Satellite Communications System (TSAT) constellation, which was originally scheduled for first launch in 2013, but was extensively restructured and delayed until 2019 then completely cancelled in 2009. TSAT was envisioned as a means to provide DoD with both high-data-rate wideband and protected communications as the space segment of the global information grid (GIG) and key component of the transformational communications architecture. As a result of the US Quadrennial Defense Review (QDR) deliberations during 2005, TSAT was restructured to be acquired incrementally with two blocks of satellites (two in the first block, three in the second, and a spare). The first block was based on reduced requirements for laser communications links and internet-like processing routers; five satellites were planned to establish a laser crosslink ring in GEO.[24] Using these advanced technologies, TSAT would be able to enable communications and networking on-the-move by connecting thousands of users simultaneously through networks, rather than using limited point-to-point circuits.

Planned TSAT laser communications throughput of two gigabytes per second is particularly important to improve network-enabled warfare using future high-data-rate systems, such as ISR sensors on space and airborne platforms.[25] If technologies fail to mature, less-capable technology off-ramps can be used to preserve schedule; these off-ramps would still enhance warfare capabilities significantly and allow advanced technology to be spiral developed into subsequent blocks of TSAT satellites. Following restructuring, delay, or cancellation of other programmes, such as the Army's Future Combat Systems (FCS) and Space Radar, which needed the full communications capability of the original TSAT laser crosslink ring design, in December 2008 the Under-Secretary of Defense for Acquisition, Technology and Logistics issued an acquisition decision memorandum directing the USAF to restructure TSAT again to remove the laser crosslinks, maintain radio frequency crosslinks and send industry new or revised request for proposals (RFP) as soon as possible.[26] According to USAF Deputy Under-Secretary for Space Programs, the schedule to send out the RFP and award a contract to acquire operational satellites has not yet been established.[27]

The US Government Accountability Office (GAO) estimated that TSAT would have a total life cycle cost of $16bn, but Congress also expressed concerns repeatedly about the cost as well as the direction and technical maturity of the programme.[28] Congressional appropriations slashed the President's budget request for TSAT by nearly 50% in FY 06, cut funding by 15% in FY 07, reduced funding by another 15% from $964m. to $814m. in FY 08, and cut funding by almost 9% from $843m. to $768m. for FY 09. These funding cuts along with direction to purchase a fourth AEHF satellite and repeated programme restructurings were clear indications that neither Congress nor the DoD were fully sold on all aspects of the planned TSAT programme, despite its foundational role in future network-enabled force structure. The TSAT programme was cancelled in 2009.

Nine Navy UFO satellites now in orbit (out of 11 launched) provide the current DoD narrowband-communications capability. Beginning in 2011, the UFO constellation is due to be replaced by five Mobile User Objective System (MUOS) satellites managed by the Navy and developed by Lockheed Martin. The capabilities of at least three of the satellites in the UFO constellation are degrading and may fail, leaving little margin for error in the initial launch date for the MUOS constellation.

Force Enhancement: Position, Navigation and Timing

US policy goals for PNT include: maintaining uninterrupted PNT services for all user needs; remaining pre-eminent in military PNT; providing civil services that exceed or are competitive with foreign PNT services; continuing as an essential component of internationally accepted PNT services; and promoting US leadership in PNT.[29] One of the most difficult challenges is posed by the mandate for Navigation Warfare capabilities to: operate the US Global

National Security Space

Positioning System (GPS) effectively despite adversary jamming; deny use to adversaries; not disrupt civil, commercial, or scientific uses unduly outside an area of military operations; and identify, locate and mitigate interference on a global basis.[30] The USAF acquires and operates the GPS constellation which currently (as of 2009) contains 32 satellites developed through a series of block upgrades.

In 2005 the USAF began launching Lockheed Martin block IIR-M satellites, which incorporate two new military signals and a second civilian signal. It plans to start launching Boeing block IIF satellites, which will broadcast a third signal for civilian use, in 2010. The first block of GPS III satellites (GPS IIIA), are scheduled for launch in FY 13 and are being designed to include improvements, like a Galileo-compatible civil signal and increased Earth coverage military code anti-jam capabilities. The GPS IIIA contract, valued at $1.46bn, was let to Lockheed-Martin in May 2008.[31] Improvements planned for Block IIIB include a crosslink command and control architecture and for Block IIIC high power spot beams to increase resistance to hostile jamming.[32] In May 2000, the USA stopped the intentional degradation of GPS signals, and in September 2007 the country announced that this capability, known as Selective Availability (SA), would no longer be procured as a part of future modernizations (GPS III).[33] Based on the FY 06 President's Budget, the Congressional Budget Office projected that the total investment spending on the GPS would be $12.5bn to 2024.[34]

Force Enhancement: Intelligence, Surveillance and Reconnaissance

Many components of the US ISR network are classified, but at least portions of the Future Imagery Architecture (FIA), Space Radar (SR) and commercial remote sensing systems are public knowledge.

In September 1999 the NRO selected Boeing to build its next generation imaging satellites, bypassing Lockheed Martin, the decades-long incumbent for the programme. The original design for the FIA electro-optical (EO) and radar-imaging satellites called for a constellation that split collection functions among smaller, simpler and more numerous satellites to collect more imagery with more frequent revisit rates. FIA, however, soon stumbled badly due to a host of problems plaguing most government satellite programmes during this period, which included overly optimistic initial bids, unrealistic cost caps and lack of management reserves, and granting total systems performance responsibility (TSPR) to contractors while government oversight capabilities and responsibilities languished. In May 2002 a panel of experts reported to the NRO Director that the programme was far behind schedule and would likely cost at least $2bn–$3bn more than the $5bn originally projected.[35] The EO satellites, the most serious FIA problem area, soon fell five years behind schedule and the price tag for FIA had grown from $6bn to as much as $18bn when the Director of National Intelligence (DNI) began his first review of technical programmes in the summer of 2005.[36] As a result of that review, there is a new way ahead for imagery satellites: Boeing will continue developing radar-imaging satellites, but, at the DNI's direction, the EO portion of the architecture was downscaled and reassigned to Lockheed Martin. The Obama Administration reportedly approved a new EO satellite plan in April 2009.

The name of the Space-Based Radar programme was changed to SR in 2005 to highlight a fundamental restructuring of this single, joint programme, which was being designed to meet both national intelligence and joint warfare requirements for a global capability to detect, image and track mobile targets in denied areas during all weather conditions. The plan approved by the Joint Requirements Oversight Council (JROC) in February 2006 called for a constellation of approximately nine satellites with somewhat reduced capabilities, which would be launched beginning in about 2016; earlier plans had envisioned using at least 24 satellites in order to provide near-continuous tracking capability. The system was being designed to enhance horizontal integration through agile, responsive collections using near-real time tasking and data dissemination from an active electronically scanned array capable of providing synthetic aperture radar (SAR) imaging, surface moving target indications (SMTI), and high-resolution terrain information (HRTI). This unclassified SR programme was restructured, then terminated in 2008.

Despite laws and policies designed to create conditions and incentives for the development of a dominant US commercial high-resolution remote sensing industry, this market has not grown as large or as quickly as had been hoped and is not dominated by the USA. Partially in response to continuing concerns about the commercial viability of the industry, President George W. Bush signed a new commercial remote sensing policy in 2003 (National Security Presidential Directive 27) intended to sustain and enhance the US remote sensing industry. This policy and Congressional direction strongly encourage the National Geospatial-Intelligence Agency (NGA) to purchase commercial remote sensing products to augment classified data, and for all parts of the US government to rely to the maximum practical extent on US commercial remote sensing products for all imagery and geospatial needs. US industry in this sector consolidated in December 2005 when Space Imaging and Orbimage merged to form GeoEye. Though not as precise as military reconnaissance satellites, the operating US private sector satellites, Ikonos 2 (GeoEye), QuickBird (DigitalGlobe), Orbview 3 (GeoEye), WorldView 1 and 2 (DigitalGlobe), and GeoEye 1 (GeoEye) can produce imagery with spatial resolutions down to 0.41 metres. Competitors include Canadian, French, German, Indian, Israeli and Russian companies that offer optical and radar imagery with resolution as precise as one metre or less.

Use of commercial remote sensing products has several advantages for the DoD and the intelligence community, including lower costs for acquiring commercial imagery compared to building and operating dedicated systems, availability of wider views of more areas, and easier sharing of data with allies and coalition partners. Drawbacks include widespread commercial availability

National Security Space

of information about ongoing US and coalition military operations as well as concerns about the availability, veracity, reliability and protection of these systems and their data. The NGA has ClearView and NextView contracts with DigitalGlobe and GeoEye, each worth up to $500m., for development of remote sensing systems and acquisition of high-resolution remote sensing data into the next decade.[37]

Force Enhancement: Integrated Tactical Warning and Attack Assessment

The USAF maintains a constellation of GEO satellites, called the Defense Support Program (DSP), to provide warning of ballistic missile launches and some data on the type of attack and the missile's intended target. The last DSP satellite (DSP-23) was successfully launched in November 2007, but failed after less than one year of operations in GEO.[38]

DSP's successor is the Space-Based Infrared System (SBIRS), a programme designed to satisfy operational military and technical intelligence requirements for overhead non-imaging infrared data, provide improved detection and supply foundational assessment capabilities for ballistic missile defence. Lockheed Martin is the prime contractor for SBIRS. The operational SBIRS constellation was originally envisioned to include four GEO satellites, two highly elliptical Earth orbit (HEO) payloads on classified host satellites, and one spare GEO satellite. In addition, the Missile Defense Agency (MDA) is evaluating a Northrop Grumman system formerly known as SBIRS-Low and now named the Space Tracking and Surveillance System (STSS). The MDA launched two research and development satellites in 2009. If these demonstrators work well in tracking missile launches and warheads, an operational system could follow, with a first launch in about 2016–17, although the Obama Administration had not by May 2009 taken a position on these missile defence efforts. The first SBIRS HEO payload was delivered in August 2004 and the first GEO satellite is currently scheduled to launch in 2010.

SBIRS is one of the most troubled NSS acquisition efforts. A 2003 Defense Science Board report called it 'a case study for how not to execute a space programme'. Total cost estimates have jumped to nearly four times the original estimates and the programme has triggered four required reports to Congress for Nunn-McCurdy Act breaches.[39] In December 2005 the Under-Secretary of Defense for Acquisition, Technology and Logistics and the DoD Executive Agent for Space restructured the programme. The restructured programme called for no more than three GEO SBIRS spacecraft and purchase of the third satellite to be contingent on performance of the first. In addition, the restructuring called for the Under-Secretary of Defense for Acquisition, Technology and Logistics to retain milestone decision authority over the SBIRS programme and for the DoD Executive Agent for Space to develop an alternative infrared satellite system (AIRSS). Original goals for the AIRSS programme were to generate viable competition in meeting the requirements for SBIRS GEO 3, exploit alternative technologies, and be ready for launch by 2015. A GAO report in September 2007 emphasized incompatibilities between the AIRSS goals of meeting the requirements and exploiting alternative technologies and also found that delivering an operational system by 2015 was unrealistically optimistic.[40]

This incompatibility in goals for the AIRSS programme, along with successful operation of the SBIRS HEO payload, prompted the USAF to shift the focus of AIRSS to developing alternative capabilities. Earlier in 2007, DoD officials elected to procure the third SBIRS GEO 3, with an option on SBIRS GEO 4.[41] Another SBIRS problem was discussed in a September 2007 memo from the USAF Secretary, indicating that the programme may face a 6–12 month delay and require up to $1bn in additional funding: the 'problem is a safe hold that did not work on a current satellite, causing mission termination, and the design similarity to' SBIRS satellites.[42] This problem and other issues including the failure of DSP-23 led the Under-Secretary of Defense for Acquisition, Technology and Logistics in December 2008 to ask Congress for permission to reprogramme $117m. in FY 09 to hedge against a potential missile warning detection gap that could develop around 2014.[43] Under the 2006 FYDP and Congressional Budget Office projections, investment spending for DSP and SBIRS would total about $11bn to 2024.[44]

Force Enhancement: Environmental Monitoring

The DoD currently uses data from five environmental monitoring (weather) satellites, which are part of the Defense Meteorological Satellite Program (DMSP), plus the data from two National Oceanic and Atmospheric Administration (NOAA) Polar-Orbiting Operational Environmental Satellites (POES). Those systems are to be replaced by three satellites of the National Polar-Orbiting Operational Environmental Satellite System (NPOESS) and one European Meteorological Operational Satellite. NPOESS was created by a May 1994 Presidential Directive that called for a joint project by the DoD, NASA and the Department of Commerce (through NOAA), designed to save money by developing a single, converged system. The USAF (DoD) and NOAA provide equal funding for the programme with Northrop Grumman as the prime contractor. The original programme intent was to buy six satellites, operate three satellites at a time, and be the nation's primary source of global weather and environmental data for operational military and civil use (polar satellites currently provide 90% of the data used in DoD and NOAA weather prediction models) to 2020.

As with SBIRS, development of NPOESS and its complex sensor suite has run into several significant development problems in the past few years. Cost overruns of 15% in September 2005 and 25% in January 2006 triggered Nunn-McCurdy Act reports indicating the programme was more than $3bn over its total original budget of $6.5bn and at least three years behind schedule.[45] Under a restructuring announced in June 2006, the government established new programme goals that included: spending $12.5bn for four less-capable satellites; launching the demonstration satellite known as the NPOESS Preparatory Project (NPP) by 2010; delaying the first NPOESS launch by about four years until 2013; and operating the system until 2026.[46] Cancellation, additional restructuring, or further delay in the NPOESS programme is not likely to present many attractive options, as it is very doubtful other systems can be developed in time to avoid gaps in coverage, especially if there is a failure on launch or on orbit with any of the remaining DMSP or POES satellites. Under current plans, four DMSP satellites remain to be launched, with the last launch rescheduled for April 2012, and the last POES launch rescheduled for February 2009.[47] In February 2010 NPOESS was extensively restructured. Under the new plan DoD will be responsible for the morning orbit and NOAA for the afternoon orbit; they will continue to share ground stations but the satellites will be affected and the NOAA satellite will be based on the NPP design. The Congressional Budget Office projects that the 2006 FYDP would require a total of $3.4bn in investment funding to 2024 for the DMSP and NPOESS programmes.[48]

Space Control

Space control programmes focus on developing ground and space-based sensors to enhance Space Situational Awareness (SSA), which is knowledge of activity of functional space assets and events that are both natural, like space weather, and artificial, like orbital debris, or that could affect circumterrestrial space (space that is 100km to about 80,000km from Earth's surface; GEO is at about 36,000km), improving means of protecting friendly space capabilities from enemy attack, and developing ways of negating enemy space capabilities. The overarching goals of the USA for space control are to enhance deterrence and dissuade development of alternative systems by developing and disseminating high fidelity SSA data, and creating an ability to attribute all activity in circumterrestrial space to natural or human causes. SSA programmes include Spacetrack, which is continuing to develop a world-wide network of radar and optical sensors; the Space-Based Surveillance System (SBSS), an optical tracking system scheduled for launch in July 2009, which will provide much more capability than the MDA Space-Based Visible sensor on the Midcourse Space Experiment (SBV/MSX), which was launched in 1996 and shut down in June 2008;[49] the Commercial and Foreign Entities (CFE) Program, which is designed to disseminate SSA data to authorized entities world-wide;[50] and other ground systems designed to track objects of interest in space. Other space control programmes, such as the Rapid Attack Identification, Detection and Reporting System (RAIDRS) and the Counter Communications System (CCS), focus on developing concepts of operations or technology to protect friendly systems or to disrupt, deny, degrade, or destroy enemy space capabilities.

The DoD Joint Publication 3–14, *Joint Doctrine for Space Operations*, discusses ways to gain or maintain space control by providing freedom of action through protection and surveillance, or to deny freedom of action through prevention and negation.[51] USAF doctrine, by contrast, aligns space control doctrine like air doctrine as offensive counterspace (OCS) and defensive counterspace (DCS). OCS missions would disrupt, deny, degrade, or destroy space systems, or the information they provide, if used for purposes hostile to US national security interests. DCS missions include both active and passive measures to protect US and friendly space-related capabilities from enemy attack, interference, or use for purposes hostile to US national security interests.[52]

Funding for the Orbital Deep Space Imager, a space-based system designed to track objects in GEO, was eliminated from the President's FY 07 budget request. Under the 2006 FYDP, research, development, testing and evaluation funding for space control programmes would increase from $195m. in 2006 to $768m. in 2011.[53] SSA, space control and protection have been areas of particular concern in Congress, as indicated by their asking the Secretary of Defense and Director of National Intelligence in the FY 06 National Defense Authorization Act to report to Congress about these topics in April and July 2006, April 2007 and June 2008.[54] In addition, following the successful Chinese anti-satellite (ASAT) weapons test on 11 January 2007, these issues have aroused even greater concern and the FY 08 Defense Appropriations bill added $100m. to the requested funding for SSA to accelerate efforts for the Space Fence and RAIDRS.[55]

Force Application

Development of systems with the potential to apply force to, in, or especially from, space is the object of intense scrutiny by Congress and attentive audiences world-wide. These concerns are exacerbated by significant difficulties in distinguishing between concepts and technologies being developed for ballistic missile defence, protection, space control and force application, as well as the development of some systems for these missions in classified and special access programmes that have less transparency and oversight. Groups in the USA opposed to space weaponization, such as the Center for Defense Information and the Stimson Center, argue that momentum created by experiments testing space control and force application concepts in space will create 'facts in orbit', driving US policy toward space weapons without debate by either Congress or the public.[56]

It is difficult, however, to see how the USA could continue improving its space protection and ballistic missile defence capabilities without the data provided by conducting these relatively small-scale experiments, how the experiments could appreciably change any facts in orbit, or how it might lead to full-scale space weaponization without triggering significant public debate, especially given the world-wide focus on space weaponization issues and all the space acquisition woes detailed above. Indeed, the cumulative effect of recent NSS acquisition problems has contributed to a small, but perceptible, shift in priorities away from space control and force application. Comparison of recent Space Posture statements to Congress by the Under-Secretary of the Air Force from 2005 to 2007 shows an initial emphasis on assured access to and freedom of action in space followed by an emphasis focused on dealing with the NSS acquisition issues. Following the Chinese ASAT test in January 2007, there were a few calls for reinvigorated development of space force application systems by the USA,[57] but these have been overshadowed by preference in Congress for increased attention to protection, SSA and space control. Moreover, the Obama Administration pledged not to deploy space weapons.

Under the joint Air Force-DARPA programme office created in December 2002, the Common Aero Vehicle (CAV) programme was envisioned originally as means to deliver a variety of conventional payloads that would be launched from intercontinental ballistic missiles (ICBMs) as part of the Force Application and Launch from Continental USA (or FALCON) programme. In response to FY 05 Congressional language, the FALCON portion of the CAV programme was restructured. Redesignated as Falcon (lower-case), the programme focused on the development and transition of more mature technologies into a future weapon system capable of promptly delivering and deploying conventional payloads world-wide. Within the Falcon programme, the CAV was redesignated the Hypersonic Technology Vehicle and all weaponization activities were excluded. The 2006 FYDP called for total funding of less than $100m. per year for those programmes to 2011 and the Congressional Budget Office projection assumes the limited deployment of 40 CAV-equipped ICBMs in about 2015, at which point the demand for investment resources would peak at $600m.[58] The FY 08 Defense Appropriations Bill terminated the original CAV effort by shifting funding from CAV and the Navy's conventional Trident Modification, to provide $100m. for research into promising prompt global strike technologies.[59]

CONCLUSIONS: NATIONAL SECURITY SPACE CAPABILITIES AND MANAGEMENT

The USA will continue to modernize and balance its armed forces for the changing and uncertain security challenges of the global security environment (see Illustrations and Documentation: Figure 2.2 Challenges of the Global Security Environment).[60] Space capabilities will remain essential in building a more advanced future global reconnaissance, precision strike complex which will integrate orbiting platforms with manned and unmanned aircraft, surface forces and other systems, cueing each other automatically and employing multi-level security to allow authorized users to create user-defined operational pictures by pulling whatever information they require, whenever they need it, wherever they are located on the network.

The challenges highlighted in this chapter make it clear that improvements in organizational and management structures are needed to ensure that space capabilities will provide an enduring asymmetric advantage for the USA.

Several areas deserve particular attention: fostering better integration among military operators and with the intelligence community; developing the professional space workforce; improving the acquisition process; creating responsive space capabilities; rethinking export controls (see Chapter 1 for a detailed case study analysis of export controls); and assuring protection of essential space capabilities.

Few space systems have been built from the ground up with technologies or concepts of operations designed to foster improved integration among military operators or between the military and the intelligence community. Emphasizing space-enabled, integrative capabilities, however, has become increasingly important to the new way of war, and essential to the implementation of any comprehensive transformation of future force structure. Space capabilities provide the single most important foundation for future transformed forces, but are not enough alone unless their strengths and vulnerabilities are balanced objectively against complementary terrestrial capabilities, and are then integrated seamlessly and transparently into a single network and concept of operations. Space capabilities are particularly important for point-to-multipoint communications and are essential for the communications and networking on-the-move that must enable lighter and more mobile forces to see and strike first with great precision, such as the Army's recently cancelled Future Combat System.

Beyond enabling better integration between military forces, the NSS community must also work to improve the ties between the 'black' (classified, intelligence community) and 'white' (US military space sector) worlds. This effort requires long-term focus and commitment, adoption of best practices in both directions, and improved co-operation and information sharing at all levels. The need for improved co-operation between the numerous agencies within the intelligence community has been a consistent theme of the DNI and this topic was the primary focus of a 100-Day Plan released by the DNI in May 2007. The Space Radar programme with its goal of creating a single system for all users was a missed opportunity to implement integration from the ground up; its cancellation is a clear indication of the strength of the actors opposed to greater integration.[61]

Because it is likely that US dependence on commercial space services will continue to grow, the US Government needs to become a more reliable and business-savvy partner for the space industrial base. As part of this process, the USA needs to rethink the proper balance between the commercial and national security sectors and adopt clear, long-term, consistent criteria for determining the NSS functions that should be performed by the commercial sector and those that should remain with government.

Developing a space professional workforce is an essential step towards improving the health of NSS capabilities and performance, but is a generational process best implemented in a patient and consistent manner over the long term. Many of the most experienced space professionals in the NSS, civil and commercial sectors are nearing retirement age and more effort is needed to develop, attract and retain top talent to replenish these ranks. The dearth of talent among mid-career professionals who might have chosen a space career, but were instead lured into the burgeoning computer networking sector has been particularly detrimental and will require the space sector to work harder and develop creative ways to attract some of this cohort.

The National Defense Education Program, a programme that targets undergraduate and graduate students studying science, technology, engineering and mathematics (STEM), is one excellent way to address this challenge; its requested budget for FY 08 of $44.4m. is more than double the $19.4m. appropriated in FY 07. Another key area – the development of a more qualified and competent military space cadre – has received growing attention, particularly in the USAF. Congress has commended initiatives such as the National Security Space Institute and Space Education Consortium at Air Force Space Command, as well as the National Space Studies Center at Air University, and recommended the DoD to be more aggressive in developing programmes and partnerships across the US Government, industry and academia.[62] There is also more to be done to improve career development paths and to develop leadership among space officers, problems exacerbated by the USAF's traditional and understandable emphasis on air power and the pilots at its institutional core, which often comes at the expense of its stewardship over space power and space officers.

Improving NSS acquisition processes is probably the most important and difficult of the many challenges currently facing NSS leadership. Because the problems the USA has in this area stem from a number of sources and did not arise overnight, they cannot be resolved quickly or easily; however, they are not intractable. Due to the scale and pervasiveness of the problems during the past few years and the widespread recognition and attention they have received, the DoD has initiated a number of changes and improvements. With recent restructurings and greater emphasis on NSS acquisition, the situation may be turning a corner. Congress remains vigilant as well, as reflected in its close oversight and specific tasking on NSS programmes.

> While the Department has taken positive steps to improve the current space acquisition system, it is not yet apparent what impact these initiatives might have on the performance of space acquisition. As a result, the conferees will maintain this issue at the forefront of congressional interests [...] Additionally, the conferees recommend that the Department develop an alternative and complementary business model for space acquisition and system deployment that will increase the production rate of space systems and lower costs.[63]

The 'back to basics' approach to NSS acquisitions emphasized in a March 2006 Posture Statement by the Under-Secretary of the Air Force indicated intent to focus on addressing these acquisition concerns. This approach is designed to apportion more risk to the earlier stages of the acquisition process, undertake more but smaller and more manageable projects using block buys and spiral development (i.e. bring in technologies in the system development process as they mature), and establish a more constant and predictable rhythm of designing, building, launching and operating space systems.[64] In addition to pushing more risk into the science and technology and technology development stages of the acquisition cycle, the back to basics approach has helped to double the amount of DoD investment in space-related science and technology over the past four years and to bring more discipline and requirements stability into the systems development and systems production stages of technical life-cycle development. Furthermore, the back to basics approach builds on recommendations of previous studies by moving the budget confidence levels for NSS programmes from 50% to 80%, strengthening collaboration between the players in the acquisition process (especially on setting and maintaining requirements), implementing more rigorous system engineering, and improving the recruitment and training of the acquisition workforce.[65]

Developing responsive space capabilities is another key area for improving NSS acquisition and management that holds the potential for creating large, paradigm-changing benefits. There are several viewpoints as well as a number of concepts associated with responsive space capabilities, ranging from relatively minor changes in how quickly satellites can be launched and who controls them, to sweeping transformations in the way satellites and launchers are built; how large, complex, expensive and reliable they should be; how they are operated and serviced on orbit; and who controls them and their potential missions.

Some use the term 'operationally responsive space' to describe relatively minor changes in how quickly satellites can be launched and who controls them. Concepts that are more ambitious would produce standardized, simple satellites on assembly lines, quickly launch them under the control of theatre commanders for specific purposes, and service them with 'plug and play' modules or refuel them on orbit. Paradigm-changing visions for responsive space emphasize flexible distributed architectures and sparse arrays consisting of many networked microsatellites able to perform a range of missions as well or better than missions performed by constellations of single-function satellites and, even more importantly, radically reduce the vulnerabilities inherent in space systems with just a few nodes. Proliferation of the wide range of current and projected threats to all orbital regimes, combined with the fragility of space systems and the predictability of their operations leads inexorably to the conclusion that distributed architectures must at least supplement, if not eventually replace, current architectures if space systems are to remain operationally relevant in the increasingly contested domain of space. Responsive space concepts might help to reduce the costs of developing and launching space capabilities; break the near-monopoly of large aerospace corporations, allow small, space-focused companies to emerge, and get more states and non-state actors involved with a range of space activities including developing space-enabled confidence and transparency capabilities.

For a number of years, Congress expressed disappointment with the pace of progress and concern with the DoD's lack of vision and initiative on responsive space concepts.[66] The FY 06 and 07 National Defense Authorization Act both indicated significant concerns with ORS and tasked the DoD in several areas, including establishing an ORS programme office to 'contribute to the development of low-cost, rapid reaction payloads, busses, spacelift, and launch control capabilities to fulfill joint military operational requirements for on-demand space support and reconstitution; and to co-ordinate and execute operationally responsive space efforts across DoD with respect to planning, acquisition, and operations'; and also assume responsibility for the Tactical Satellite (TacSat) programme from the now defunct Office of Force Transformation.[67] Some Congressional concerns have been allayed by the DoD's delivery of a comprehensive ORS Report to Congress and establishment of an ORS Office in 2007. Current DoD ORS plans call for a three-tiered system to deliver more responsive space capabilities by employing on-demand tasking of existing assets, launching on-call assets and developing new or modified capabilities. Developing truly responsive space capabilities will not be easy and will require major changes in concepts of operations and ways of thinking.

Rethinking export controls might not at first seem an important issue for improving NSS acquisition and management, yet US export control policies are particularly important for the NSS enterprise because they affect the space industrial base, the competitiveness of US aerospace corporations, and run counter to other space policy goals that call for US companies to dominate certain space activities. Present US space export-control policy stems from developments during the past 20 years and has been shaped primarily by Congressional direction and the Executive Branch Department controlling these exports. Between FY 1992 and FY 1998 the Department of Commerce had export licensing responsibilities for most communications satellites and the Department supported these exports strongly. After Hughes and Loral worked with insurance companies to analyze Chinese launch failures in January 1995 and February 1996, a 1998 Congressional review (the *Cox Report*) determined that these analyses communicated technical information to the Chinese in violation of the International Traffic in Arms Regulations (ITAR). As a consequence, the 1999 National Defense Authorization Act transferred all satellites and related items to the Munitions List administered by the Department of State.[68]

Since the return of export controls to State, the US aerospace industry has advocated incessantly for a loosening of these restrictions and has blamed ITAR for business downturns and a decline in market share.[69] European and other satellite manufacturers, including Alcatel Alenia Space and the European Aeronautic Defence and Space Company (EADS), have replaced all US-built components from their communications satellites to make them 'ITAR-free' and avoid these restrictions.[70] Thus, US export controls have clearly created incentives for development of an indigenous foreign high-technology space sector – a counterproductive outcome. There is also considerable merit in the US industry's claims that the current restrictions cost them market share in this strategic sector and do not make common sense distinctions between exports to allies and others, or necessarily keep dual-use technologies thought to be dangerous out of the wrong hands. At the same time, slowing the diffusion of technologies with considerable military potential is a legitimate national security concern and a range of factors beyond ITAR have contributed to the decline in the competitiveness of the US aerospace industry.

US export-control policy must find a better way of balancing these conflicting objectives. Problems that are even more difficult arise when US export controls stifle other space policies designed to create incentives for US industry to dominate certain market sectors. US commercial remote sensing policy is probably the best example of this, but there are a number of other areas, such as communications satellites, where there are obviously conflicting policy objectives. The USA should reevaluate, on a case-by-case basis, which of these conflicting objectives should predominate and then readjust its policies and regulations accordingly. Another excellent starting point for rebalancing export-control priorities would be to implement key recommendations from the recently completed Center for Strategic and International Studies (CSIS) study on this topic, such as removing from the Munitions Controls List commercial communications satellite systems, dedicated subsystems and components specifically designed for commercial use.[71]

The USA must move more expeditiously to institutionalize a range of protection measures to ensure space will continue to provide an enduring asymmetric advantage. The USA must not design its future transformed forces to be reliant on space-enabled capabilities unless it can ensure those capabilities will be available when needed most – during combat operations. The basic problem is that current US space architectures were optimized for performance, rather than built to provide mission assurance despite the types of interference and attacks that are becoming increasingly common and within the capability of more actors. In the past, for a variety of reasons including the widespread perception that space was a sanctuary, each incremental investment almost always went to providing more capabilities rather than better protection of existing capabilities.

It is unclear if space was ever a sanctuary, but, as highlighted by the January 2007 Chinese ASAT test, it is becoming a contested military domain like land, sea or air, where operations face a variety of threats. As the most important first step in implementing specific protection measures, the USA should ensure critical infrastructure protection and continuity of operations by eliminating critical single points of failure on the ground and hardening satellites against total radiation dose failures following high altitude nuclear detonations. A second essential step is to implement and institutionalize the protection standards for all future NSS systems called for in the NSS Protection Strategy Framework signed by the DoD Executive Agent for Space in 2005. As discussed in the responsive space section herein, microsatellite

National Security Space

distributed architectures and sparse arrays should be the foundation for building future architectures that are better protected and more durable.

Increased effort towards this goal is plausible if the Air Force Space and Missile Systems Center and the NRO adopt this approach, but moving the organizations in this direction will be a difficult challenge, as they are at the centre of current NSS acquisition efforts which have evolved, with good reason, towards larger and more capable, but very small numbers of satellites in most current architectures.[72] Other important steps towards better protection that have been initiated or should be undertaken include: funding protection efforts commensurate with their importance; development of a comprehensive space protection strategy and creation of the joint AFSPC-NRO Space Protection Program in July 2008; use of war games and exercises to explore space deterrence concepts and make known specific 'red lines' for deterring potential attacks against satellites that support US national security; and multifaceted approaches to raise awareness about space dependency and vulnerabilities as well as adoption of a 'whole of government' approach to address these interdependent issues.[73] Given all the other NSS acquisition and management problems, it will not be easy to find and sustain the resources required to institutionalize protection, but the NSS community must step up to its responsibilities to ensure space capabilities are at hand when needed most.

Finally, because the USA is becoming increasingly reliant on commercial space services, such as communications and remote sensing, it must also work harder and in more creative ways to assure protection of these services. Better dialogue between the NSS and commercial sectors, and long-term policy consistency are key to improving protection of commercial services. Wherever possible, the US Government also should attempt to shape this sector through favourable licensing decisions or by giving commercial benefits, such as long-term leases or priority in purchasing, to those companies doing the most to ensure protection of their services, but keeping decisions about risk and market forces within the commercial sector. As 'space capabilities are essential at all levels of military planning and operations',[74] addressing NSS political, organizational, management and technical issues is essential for US national security.

NOTES

* The opinions, conclusions and recommendations expressed or implied in this chapter are those of the author and do not reflect the official policy or position of the US Air Force, Department of Defense, or US Government.

1 Programmes within the vMFP have to date been readjusted each February, have not remained constant from year to year, and have not always covered all major space systems. These inconsistencies reduce the utility of the vMFP as a reliable measure of NSS expenditures over time and undercut the primary rationale of the Space Commission in recommending creation of this measure. Section 8111 of US Public Law 110–16 (Fiscal Year (FY) 2008 DoD Appropriations) and Section 8104 of the FY 2009 DoD Appropriations directs the Secretary of Defense to create a 'hard' rather than a virtual MFP for space within the Future Years Defense

Program (FYDP) and to designate an Office of the Secretary of Defense official to provide overall supervision of the preparation and justification of programme recommendations and budget proposals to be included in the space MFP. DoD did not accomplish this action as of the end of 2008.
2 Marcia S. Smith, 'US Space Programs: Civilian, Military, and Commercial', (Washington, DC: Congressional Research Service, 9 August 2005); and Patricia Maloney Figliola, 'US Military Space Programs: An Overview of Appropriations and Current Issues', (Washington, DC: Congressional Research Service, 7 August 2006).
3 Congressional Budget Office, 'The Long-Term Implications of Current Plans for Investment in Major Unclassified Military Space Programs', (Washington, DC: Congressional Budget Office, 12 September 2005): 1; and Marshall Institute Policy Outlook, 'National Security Space FY 2008 Budget: Overview and Assessment', (Washington, DC: George C. Marshall Institute, November 2007).
4 Congressional Budget Office, 'Investment in Major Military Space Programs', 5.
5 Most US Government documents list three rather than four space sectors. See, for example, the 2006 National Space Policy's discussion of civil, national security (defence and intelligence) and commercial sectors. Office of Science and Technology Policy, 'Fact Sheet: National Space Policy' (Washington, DC: The White House, 6 October 2006). For discussion emphasizing four sectors see *Report of the Commission to Assess National Security Space Management and Organization* (Washington, DC: Commission to Assess National Security Space Management and Organization, 11 January 2001): 10–14. Most other major spacefaring states have even less clearly delineated space sectors and place more emphasis on developing dual-use capabilities.
6 Joint Publication 3–14, *Joint Doctrine for Space Operations* (Washington, DC: Joint Staff, Department of Defense, 9 August 2002). Joint Publication 3–14 is being revised, as is required for all joint publications every five years.
7 David E. Lupton, *On Space Warfare: A Space Power Doctrine* (Maxwell AFB, AL: Air University Press, June 1988).
8 The most important previous NSS-related committees and their key space policy recommendations include the following: the 1954–55 Technological Capabilities Panel (TCP) (established the legality of overflight and developed spy satellites); the President's Science Advisory Committee (PSAC), led by Science Advisor James Killian in 1958 (created NASA); the group led by Science Advisor George Kistiakowsky in 1960 (created NRO); the review led by Vice-President Lyndon Johnson in April 1961 (raced the Soviets to the Moon for prestige); Vice-President Spiro Agnew's 1969 Space Task Group (established NASA's post-Apollo goals); the Air Force's 1988 Blue Ribbon Panel led by Maj.-Gen. Robert Todd (integrated spacepower into combat operations); NASA's 1991 Augustine Commission (emphasized scientific exploration over Shuttle operations); and the Air Force's 1992 Blue Ribbon Panel, led by Lt-Gen. Thomas Moorman (emphasized space support to the warfarer and established the Space Warfare Center).

The Space Commission was chaired by former and future Secretary of Defense Donald Rumsfeld and included 12 other members with a broad range of very high-level NSS expertise (listed with the top 'space' job formerly held): Duane Andrews (Deputy Under-Secretary of Defense for Command, Control, Communications and Intelligence); Robert Davis (Deputy Under-Secretary of Defense for Space); Howell Estes (Commander, US Space Command); Ronald Fogleman (Air Force Chief of Staff); Jay Garner (Commander, Army Space and Strategic Defense Command); William Graham (President's Science Advisor and acting NASA Administrator); Charles Horner (Commander, US Space Command); David Jeremiah (Vice-Chairman, Joint Chiefs of Staff); Thomas Moorman (Air Force Vice-Chief of Staff); Douglass Necessary (House Armed Services Committee staff); Glenn Otis (Commander, Army Training and Doctrine Command); and Malcolm Wallop

National Security Space

(Senator). See John A. Tirpak, 'The Fight for Space', *Air Force Magazine* 83 (August 2000): 61.

There were two other major congressionally mandated space studies during 2000: a review of the National Imagery and Mapping Agency (NIMA), *The Information Edge: Imagery Intelligence and Geospatial Information in an Evolving National Security Environment* (Washington, DC: December 2000); and a review of the NRO, *The NRO at the Crossroads* (Washington, DC: National Commission for the Review of the National Reconnaissance Office, 1 November 2000).

9 General Accounting Office, Report to Defense Committees, Defense Space Activities: Organizational Changes Initiated, but Further Management Actions Needed', (Washington, DC: GAO, GAO 03–379, April 2003), 25–27. Secretary Rumsfeld signed a memo on 18 October 2001 directing the DoD to undertake 32 actions to implement Space Commission recommendations and make other changes.

10 This position was strengthened and given specific budgetary authority over the intelligence community by the Intelligence Reform and Terrorism Prevention Act of 2004; it is now known as the Director of National Intelligence (DNI).

11 The Independent Assessment Panel consisted of A. Thomas Young (chairman), Edward Anderson, Lyle Bien, Ronald Fogleman, Lester Lyles, Hans Mark and James Woolsey. The Panel began deliberations in October 2007 and delivered their final report in September 2008. Their charter is in Section 913 of the FY 07 National Defense Authorization Act: Independent Review and Assessment of Department of Defense Organization and Management for National Security in Space. See 'Leadership, Management, and Organization for National Security Space: Report to Congress of the Independent Assessment Panel on the Organization and Management of National Security Space', (Alexandria, VA: Institute for Defense Analyses, July 2008).

12 Congressional Budget Office, 'Investment in Major Military Space Programs', 17.

13 The launch of a Cosmo-Skymed satellite from Vandenberg Air Force Base on 24 October 2008 marked the 84th consecutive successful Delta II launch.

14 Government Accountability Office, 'Defense Acquisitions: Assessment of Selected Major Weapon Programs', Report to Congressional Committees, GAO-06–391, March 2006, 54. Under the Nunn-McCurdy Act (10 USC. 2433), Congress must be notified when a major defence acquisition programme experiences a cost increase of 15% or more. If the increase is greater than 25%, the Secretary of Defense must certify to Congress that the programme is essential to national security, adequately managed, no feasible alternatives exist, and the new cost estimates are reasonable. A comprehensive analysis of EELV issues is provided in National Security Space Launch Requirements Panel, *National Security Space Launch Report* (Washington, DC: RAND National Defense Research Institute, 2006).

15 Congressional Budget Office, 'Investment in Major Military Space Programs', 14.

16 Ronald M. Sega, Under-Secretary of the Air Force, 'Space Posture Statement to Strategic Forces', Subcommittee of House Armed Services Committee, 16 March 2006, 9.

17 This market oversupply has corrected itself.

18 Congressional Budget Office, 'Investment in Major Military Space Programs', 5.

19 Each Block I WGS satellite can route up to 3.6 Gbps (gigabytes per second) of data; the first WGS on orbit provides more throughput than the entire current DSCS constellation. GAO, 'Assessments of Major Weapon Programs', 119–20; and 'America's Wideband Gapfiller Satellite System', *Defense Industry Daily*, 23 October 2006.

20 Turner Brinton, 'Australia to Fund Sixth Wideband Global Satcom Satellite: $823.6 Million Commitment will give Australians Access to Entire WGS Constellation', *Space News*, 8 October 2007, 15.

21 Turner Brinton, 'Senate Appropriators Direct Air Force to Buy 4th AEHF Satellite', *Space News*, 24 September 2007, 12. The FY 08 Defense Appropriations Bill provides US $728.2m. for AEHF ($125m. more than requested) and directs the Air Force to buy a fourth AEHF and maintain an option to buy a fifth satellite. In the FY 08 budget request the Air Force had planned to buy only three AEHFs.
22 GAO, 'Assessments of Major Weapon Programs', 27–28; 'First Advanced EHF Launch Slips to Fall of Next Year', *Space News*, 19 November 2007.
23 Colin Clark, '$400M Overrun for AEHF; OSD Rushes Cost Review', *DoD Buzz: Online Defense and Acquisition Journal*, 19 September 2008 (accessed 14 December 2008).
24 Sega, 'Posture Statement', 13.
25 Bob Brewin, 'Boards Urge Immediate Deployment of Advanced Satellites', *NextGov: Technology and the Business of Government*, 18 December 2008. A Defense Science Board report released in December 2008 strongly recommended deploying TSAT with laser crosslinks as soon as possible and warned that there was no alternative plan or substitute for the foundational communications capability TSAT would provide. See *Report of the Joint Defense Science Board Intelligence Science Board Task Force on Integrating Sensor-Collected Intelligence* (Washington, DC: Defense Science Board, November 2008).
26 Amy Butler, 'Young Gives Nod to TSAT without Laser Links', *Aerospace Daily and Defense Report*, 9 December 2008.
27 Amy Butler, 'TSAT programme said to need more time to skirt pitfalls that doomed other efforts', *Aviation Weeks & Space Technology*, 27 October 2008, 31.
28 See, for example, Government Accountability Office, 'DoD is Making Progress in Adopting Best Practices for the Transformational Satellite Communications System and Space Radar but Still Faces Challenges', GAO-07-1029R, 2 August 2007, 1.
29 Office of Science and Technology Policy, 'Fact Sheet: US Space-Based Positioning, Navigation, and Timing Policy', (Washington, DC: The White House, 15 December 2004), 3.
30 Ibid., 6–7.
31 Breanne Wagner, 'Aerospace World', *Air Force Magazine* 89: 7 (July 2006): 21–22; and Janice Partyka, 'The System – Oct 2007', *GPS World* 1 (October 2007).
32 'SMC Announces Contract Award for Next Generation GPS Space Segment', Press Release 040508 (Los Angeles Air Force Base, CA: Space and Missile Systems Center, 15 May 2008).
33 'Statement by the Press Secretary', (Washington, DC: White House Office of the Press Secretary, 18 September 2007).
34 Congressional Budget Office, 'Investment in Major Military Space Programs', 9.
35 Philip Taubman, 'Failure to Launch: In Death of Spy Satellite Program Lofty Plans and Unrealistic Bids', *New York Times*, 11 November 2007.
36 Ibid.; *Wall Street Journal*, 11 February 2006 asserts that the restructuring of FIA would have added US $8bn to the programme costs and raise the total above $20bn.
37 'Israeli Space-Based Radar Set for Indian Launch as Pentagon's New Commercial Recon Poised for Vandenberg Flight', *Aviation Week & Space Technology*, 17 September 2007, 28; 'DigitalGlobe launches its newest spy satellite', *Denver Post*, 19 September 2007; and 'GeoEye Completes Multi-Million Dollar NextView Contract Modification with NGA', *GISuser*, 10 December 2008.
38 DSP-23 apparently become unresponsive to station-keeping control inputs in September 2008 and may pose a threat to other satellites in GEO. See Andrea Shalal-Esa, 'US satellite failure revives tracking concerns', *Reuters*, 4 December 2008.
39 Report of the Defense Science Board/Air Force Scientific Advisory Board Joint Task Force on Acquisition of National Security Space Programs (Washington, DC:

National Security Space

Defense Science Board, May 2003): 6; GAO, 'Assessments of Major Weapon Programs', 101-2. To date the SBIRS programme has triggered four reports to Congress under the Nunn-McCurdy Act; the programme breached its cost estimates by 25% in 2001, by 15% in 2004 and by 25% twice during 2005.
40 Government Accountability Office, 'Space Based Infrared System and its Alternative', (Washington, DC: GAO-07-1088R, 12 September 2007): 2-4.
41 Jeremy Singer, 'DoD Mulls Missile Warning Options Amid Latest Travails with SBIS', *Space News* 8 October 2007, 1 and 18.
42 Ibid.
43 See Andrea Shalal-Esa, 'US Missile-Warning Satellite Fails', *Reuters*, 24 November 2008; Turner Brinton, 'DSP Constellation Health Concerns Prompt Plan for Gap-Filler Satellite', *Space News*, 24 November 2008; and Colin Clark, 'Missile Warring Worries Spur New Program; Congress Skeptical', *DoD Buzz: Online Defense and Acquisition Journal*, 1 December 2008.
44 Congressional Budget Office, 'Investment in Major Military Space Programs', 12.
45 Jeremy Singer, 'NPOESS Restructuring Plan Trims Satellites, Capabilities', *Space News Business Report*, 12 June 2006.
46 Government Accountability Office, 'Polar-Orbiting Operational Environmental Satellites: Restructuring is Under Way, but Challenges and Risks Remain', Statement of David A. Powner, Director Information Technology Management Issues, Testimony before the Subcommittee on Energy and Environment, House Committee on Science and Technology (Washington, DC: Government Accountability Office, GAO-07-10T, 7 June 2007).
47 Powner 7 June 2007 testimony, 11; and Government Accountability Office, 'Polar-Orbiting Operational Environmental Satellites: Cost Increases Trigger Review and Place Program's Direction on Hold', Statement of David A. Powner, Director Information Technology Management Issues, Testimony before Subcommittee on Disaster Prevention and Prediction, Committee on Commerce, Science and Transportation, US Senate, GAO-06-573T, 30 March 2006, 3-4.
48 CBO, 'Investment in Major Military Space Programs', 14.
49 Staff Sergeant Don Branum, 'Retiring Satellite a Cornerstone of Space Situational Awareness', (Schriever Air Force Base, CO: 50th Space Wing Public Affairs), 29 May 2008.
50 Continuing controversy surrounds implementation of CFE provisions originally authorized by Congress in November 2003 and the pilot programme initiated by AFSPC in January 2005. Commercial operators such as Intelsat claim the USG has not always done enough to share SSA data or responded in timely ways to help them avoid conjunctions while the Joint Space Operations Center (JSpOC), the STRATCOM operations centre charged with this responsibility, emphasizes it has not been given sufficient and dedicated resources to provide all requested services. Congress has extended the CFE programme through 2010. The European Union is developing improved SSA capabilities through the European Space Agency (ESA), although funding for ESA's rather ambitious SSA plans was cut in half to €50m. at the November 2008 ESA ministerial meeting.
51 Joint Publication 3-14, pages IV-5–IV-8.
52 Air Force Doctrine Document 2-2.1, *Counterspace Operations* (Maxwell AFB, AL: Air Force Doctrine Center, 2 August 2004), 25-34.
53 Congressional Budget Office, 'Investment in Major Military Space Programs', 16.
54 Section 911 of the FY 06 National Defense Authorization Act, Space Situational Awareness Strategy and Space Control Mission Review, and Section 911 of FY 08 National Defense Authorization Act, Space Protection Strategy.
55 Conference Report to Accompany H.R. 3222, 'Making Appropriations for the Department of Defense for the Fiscal Year Ending September 30, 2008, and for Other Purposes', Report 110-434 (Washington, DC: GPO, 6 November 2007),

320. Analyses of Chinese counterspace capabilities and the rationale for and implications of their January 2007 ASAT test are: Larry M. Wortzel, 'The Chinese People's Liberation Army and Space Warfare', (Washington, DC: American Enterprise Institute, October 2007); Ashley J. Tellis, 'China's Military Space Strategy', *Survival* 49, no. 3 (Autumn 2007): 41–72; and Philip C. Saunders and Charles D. Lutes, 'China's ASAT Test: Motivations and Implications', (Washington, DC: National Defense University, Institute for National Strategic Studies Special Report, June 2007). For an outstanding proposal on how Congress should respond see Terry Everett, 'Arguing for a Comprehensive Space Protection Strategy', *Strategic Studies Quarterly* 1, no. 1 (Fall 2007): 20–35.

56 Theresa Hitchens, Michael Katz-Hyman and Victoria Samson, 'Space Weapons Spending in the FY 2007 Defense Budget', (Washington, DC: Center for Defense Information, 8 March 2006). Programmes of greatest concern to these groups include the MDA's Space Test Bed, Near Field Infrared Experiment (NFIRE) Kinetic Energy Interceptor (KEI), Multiple Kill Vehicle (MKV) and Airborne Laser (ABL), as well as the Air Force's Experimental Satellite System (XSS), Autonomous Nanosatellite Guardian for Evaluating Local Space (ANGELS) and Starfire Optical Range.

57 See, for example, presentation by Senator John Kyl, 'China's Anti-Satellite Weapons and America's National Security' (Washington, DC: Heritage Foundation, 29 January 2007).

58 Congressional Budget Office, 'Investment in Major Military Space Programs', 16.

59 Conference Report to Accompany H.R. 3222, 240–41.

60 Robert M. Gates, 'A Balanced Strategy: Reprogramming the Pentagon for a New Age', *Foreign Affairs* 88: 1 (January/February 2009).

61 General C. Robert Kehler, Commander of AFSPC, has used recent public presentations to discuss cancellation of the Space Radar, asserting that 'one size does not fit all' and arguing against forced integration between DoD and intelligence community space systems. See Colin Clark, 'Intel, AF Sats Must Go Separate Ways: Kehler', *DoD Buzz: Online Defense and Acquisition Journal*, 16 November 2008.

62 House of Representatives, Conference Report on National Defense Authorization Act for Fiscal Year 2006, H.R. 109–360, H13100.

63 Ibid.

64 Sega, 'Posture Statement', 10–12.

65 Ibid., 13–14.

66 House of Representatives, Conference Report on National Defense Authorization Act for Fiscal Year 2006, H.R. 109–360, H13100.

67 Section 913 of the FY 06 and 07 National Defense Authorization Act, Operationally Responsive Space.

68 The January 1995 failure was a Long March 2E rocket carrying Hughes-built Apstar 2 spacecraft and the February 1996 failure was a Long March 3B rocket carrying Space Systems/Loral-built Intelsat 708 spacecraft. Representative Christopher Cox (R.-California) led a six-month long House Select Committee investigation that produced the 'US National Security and Military/Commercial Concerns with the People's Republic of China' report released on 25 May 1999. The report is available from www.house.gov/coxreport (accessed 11 February 2009).

In January 2002, Loral agreed to pay the US government US $20m. to settle the charges of the illegal technology transfer and in March 2003, Boeing agreed to pay $32m. for the role of Hughes (which Boeing had acquired in 2000). Requirements for transferring controls back to state are in Sections 1513 and 1516 of the FY 99 National Defense Authorization Act. Related items are defined as 'satellite fuel, ground support equipment, test equipment, payload adapter or interface hardware, replacement parts and non-embedded solid propellant orbit transfer engines'.

National Security Space

69 Satellite builders claim that their exports dropped 59% in 2000 and that since March 1999 their share of the global market declined sharply (from 75% to 45%). Evelyn Iritani and Peter Pae, 'US Satellite Industry Reeling Under New Export Controls', *Los Angeles Times*, 11 December 2000, 1. The year 2000 marked the first time that US firms were awarded fewer contracts for GEO communications satellites than their European competitors (they show the Europeans ahead 15 to 13). See Peter B. de Selding and Sam Silverstein, 'Europe Bests US in Satellite Contracts in 2000', *Space News*, 15 January 2001, 1 and 20.
70 Peter B. de Selding, 'European Satellite Component Maker Says it is Dropping US Components Because of ITAR', *Space News Business Report*, 13 June 2005; and Douglas Barrie and Michael A. Taverna, 'Specious Relationship', *Aviation Week & Space Technology*, 17 July 2006, 93–96.
71 'Briefing of the Working Group on the Health of the US Space Industrial Base and the Impact of Export Controls', (Washington, DC: Center for Strategic and International Studies, February 2008).
72 Outstanding and comprehensive technical evaluations of the prospects for moving toward distributed architectures are provided in: Gregory A. Orndorff, Bruce F. Zink and John D. Cosby, 'Clustered Architecture for Responsive Space', AIAA-RS5 2007–1002, (Los Angeles, CA: American Institute for Aeronautics and Astronautics, 5th Responsive Space Conference, 23–26 April 2007), and Naresh Shah and Owen C. Brown, 'Fractionated Satellites: Changing the Future of Risk and Opportunity for Space Systems', *High Frontier* 5: 1 (November 2008): 29–36.
73 US Representative Terry Everett, 'Work Worth Doing', *High Frontier* 5: 1 (November 2008): 2–6.
74 Ronald M. Sega, former Under-Secretary of the Air Force.

Developed Space Programmes

Laurence Nardon

There are several reasons why the USA carries out space programmes. Although military applications have been one of the first and most enduring reasons to develop space in the aftermath of the Second World War, international prestige became the most apparent driver in the 1960s, at a time when competition with the USSR (Union of Soviet Socialist Republics) was rife. Exploration of the universe, starting with the Moon, was a major propaganda goal at that time. As several types of satellite applications became available, space quickly came to include considerations of business as well. Motivations in the USA to develop space also include an important and unique psychological factor. America sees space as a new frontier, an echo of the conquest of the west in the 19th century. This is why space and space exploration are of paramount importance in US contemporary self-representation.

Other space powers in the world do not necessarily entertain similar motivations. Space means different things for different states. Identifying the type of appeal they see in space will tell us about these states and about the programmes they are most likely to pursue in space, including the types of international co-operation in which they may be most interested. Soviet space ambitions in the 1960s were probably a mirror image to those of the USA, but Russian Federation drivers for space are different now. Albeit they operate on much smaller space budgets, European states, India, the People's Republic of China, Japan and Israel have also pursued space programmes for decades. They each have their own set of motivations to develop space programmes. What is it that leads them to devote effort, time and money to space? The focus of this chapter is on the space policy drivers in non-US spacefaring states and state organizations, including Russia, the United Kingdom (UK), France, Germany, Italy, Spain, the European Space Agency (ESA), the European Union (EU), India, China, Japan and Israel (Chapter 4 in this volume discusses developing space programmes and moderate space powers, among which Israel could be included).

RUSSIA

The Russian space programme has known several evolutions since the 1950s. The government motivations for space have changed during that time. With the launch of Sputnik in 1957 and Gagarin's trip to orbit in 1961, it can be argued that the USSR was the first space power in history. Even though the

Developed Space Programmes

USA made it first to the Moon in 1969, the USSR remained a first-rate space power until its fall in 1991. The main driver for space at that time was the fierce prestige competition with the USA. After 1969, the race to the Moon gave way to orbital station programmes, including Almaz, Salyut and Mir. Competition in space exploration ended on the Soviet side when the Buran space shuttle project was cancelled in 1993.

The development of military applications was also a strong motivation for the Soviet space programme. The first observation satellites, which were part of the Kosmos family of satellites, flew in the late 1950s and Soviet engineers quickly explored the full set of military applications. Projects of orbital weapons were developed and anti-satellite weapon (ASAT) tests were conducted in the 1960s. Both were later dropped as inefficient and counterproductive from a strategic point of view. Business considerations were foreign to the Soviet economic system and, therefore, did not represent a motivation at the time.

After the collapse of the USSR in 1991, the 1990s were a decade of political and economic upheaval in Russia. Space appeared as one of the few areas that could yield money and it became a contributor to the economic survival of the country. Ex-Soviet launchers and rocket engine technologies were considered first class and aerospace companies in the USA and Europe were keen to acquire them. Boeing, Lockheed Martin and Arianespace established joint-ventures with Russian and Ukrainian companies to market the Zenit, Proton and Soyuz rockets, respectively. The National Aeronautics and Space Administration (NASA) was interested in the extensive Russian experience of orbital station and space transportation technologies, and arranged for Russia to make contributions to the International Space Station (ISS) programme. Moreover, there was a risk that Russian ballistic engineers, if laid off by their employer, would be offered new jobs in proliferating countries such as Iran or the Democratic People's Republic of Korea (North Korea). Western governments wished to control the risk of ballistic proliferation and strongly supported the deals between Russian and Western agencies and companies.

The third era of Russian space started in the mid-2000s when the rise in the oil price allowed a return to economic stability in the country. With enough money to afford it, today's Russia wants to regain its status as a major space power in the world. Space has again been the focus of public policy. A new Federal Space Programme was presented in 2005, with a corresponding budget.[1] Some US $11bn have been allotted to Roscosmos, the Russian Federal Space Agency, over the 2006–15 period.[2] The goals of Russian space policy are international prestige and a quest for business efficiency.

Federal Space Efforts

Roscosmos lists several areas of focus for the federal space effort:

- Based on the success of the Soyuz and Proton rockets, Russian launcher and engine technologies have been praised for years as highly robust and

reliable. In the aftermath of the fall of the Soviet system, these rockets and derived technologies have been widely marketed in the rest of the world. Former ballistic missiles were turned into smaller launchers under the names Rockot, Dnepr and Volna. Currently, a new launcher called Angara is being developed. Angara will have three versions, allowing payloads ranging from small lift to heavy lift. Angara may also be launched from the Plesetsk launch pad in Russia, in order to reduce dependence on the Baïkonour launch pad that is situated in Kazakhstan. Angara should be operational during the 2010s. Angara is also presented as the first 'environmentally friendly' launcher, using fuel made from oxygen and kerosene, which makes jettisoned boosters less toxic.
- Having fallen into disrepair in the 1990s, many Russian satellite systems must now be rebuilt. The total of operational civilian satellites is slated to more than double in number by 2015. The Glonass satellite navigation system is a particular showcase in the renewal of the Russian space programme. Current plans, as of 2009, call for a Glonass constellation of 30 satellites by 2010 with world-wide coverage. The development of a Russian navigation services industry should follow.
- Earth observation has traditionally been left to the military. According to a new programme, two civilian observation satellites called Resours-P are planned for launch in 2010. They will replace two experimental satellites that are presently in orbit.
- Widespread use of cell phones and the internet in a country as immense as Russia requires development of several telecommunications satellites. As telecommunications are one of the few profitable space applications, investment may not be entirely left to public authorities. Roscosmos will co-ordinate future plans with the private sector.
- Russia is set to resume exploration of the solar system after a 20-year interruption. The Federal Space Programme focuses on the four in-orbit telescopes of the Spektr project, which will be launched by 2015. A lunar probe and a sample return mission to Phobos, one of the two moons orbiting around Mars, are also in the works.
- Russia is also a major actor of human space flight ventures. It contributed many flights to and from the ISS after the Columbia accident of 2003. The Soyuz rocket launches either the Soyuz manned capsule or the Progress cargo module. Space tourism aboard the Soyuz capsule should continue with a new and larger version of the capsule. Russia is also working on a next generation shuttle called Clipper. Co-operation with NASA and other Western agencies is bound to continue in the light of international lunar and Mars exploration efforts.
- Even though less information is available on Russian military space efforts, it is highly probable that military space systems are undergoing a similar kind of renewal as the civilian space systems listed in the Federal Space Programme. Military projects mention all applications, with an emphasis on space monitoring and early warning capabilities.[3]

Restructuring of the Russian Space Industry

Providing an important backdrop to the Russian space programme's current renewal, a major reorganization of the Russian space industry is currently underway.[4] It was launched by the federal administration in 2001. The plan is to make space industries more competitive by concentrating them and changing their legal status. There were approximately 112 space companies in Russia in 2001, employing 250,000 people. The goal is to merge these companies into three or four major holding companies by 2015. Besides, most companies remain federal state unitary enterprises, known as FGUPs, a status created in the 1990s for profit-oriented public companies. Over the last decade, FGUPs have proved to operate with very little transparency. The government's goal is to transform them into shareholder companies with a much more open management system. A number of companies have already been restructured and have seen their legal status modified. Prominent examples include the satellite payload maker NPO PM Rechetnev, the electronics company Reoutov, the engine manufacturer Energomash, and Khrunichev State Research and Production Space Centre, which specializes in launch vehicle and spacecraft development.

Yet, questions have arisen regarding the concentration model chosen by the government. Russian space industries have so far been regrouping around a single production type, such as satellite payload, electronics, engines, or launchers. This dependence on one single product may not be financially sound in the overall space market, where production cycles are long, insurance is costly and investments are capital intensive. The Western model, such as in the examples of Boeing, Lockheed Martin and the European Aeronautic Defence and Space Company (EADS), seeks to integrate the whole sequence of space systems integration and production in a single company and back it with other aeronautics and military productions to distribute risks. The mere size of Russian companies may prevent this type of industrial model for now, but it may be the next evolution.

Another limiting factor of the government plan is the lack of industrial co-operation projects with the West. After a decade when many in the Russian space community had the feeling of being weakened by Western launcher companies, the trend is currently to keep the space industry a national asset. There is also a strong resistance to change on the part of some space stakeholders in the country. The overall industrial reorganization process will be slow.

EUROPEAN SPACE

As of 2009, 18 states (Austria, Belgium, the Czech Republic, Denmark, Finland, France, Germany, Greece, Ireland, Italy, Luxembourg, the Netherlands, Norway, Portugal, Spain, Sweden, Switzerland and the UK) are members of ESA, a body that has recently established links with the EU, which has 27 members as of 2009 (Austria, Belgium, Bulgaria, Cyprus, the Czech Republic,

Denmark, Estonia, Finland, France, Germany, Greece, Hungary, Ireland, Italy, Latvia, Lithuania, Luxembourg, Malta, the Netherlands, Poland, Portugal, Romania, Slovakia, Slovenia, Spain, Sweden and the UK). Members of ESA are the European states most notably active in space activities. They have undertaken a series of space programmes over the years, some of them through ESA and others on a national or multilateral basis. Each country in ESA entertains different motivations for space.

United Kingdom Favours Practical Space

Looking back, the UK could very well have become the leader of European space after the Second World War. As early as 1957, the UK launched the Skylark test rocket from the Woomera launch pad in Australia. It later decided to build a ballistic missile called Blue Streak. However, the programme was deemed too costly and received lukewarm support from the USA. By 1962, the UK had dropped all ballistic missile programmes and bought US Polaris missiles instead. This signalled the end of UK involvement in launcher programmes. Later, its commitment to the Ariane European rocket programme, and to most European space projects, was limited.

The UK space programme is overseen by the British National Space Centre (BNSC), a very small structure that co-ordinates the activities of nine government entities, including: the Departments for Trade and Industry, Education and Skills, Transport, Defence, Environment, and the Foreign and Commonwealth Office, the Natural Environment Research Council, the Science and Technology Facilities Council and the Meteorology Office. The UK approach to space has been utilitarian. Investments are only made if the space system's usefulness to the country is guaranteed from the beginning. This means either benefiting the public and British users, or allowing British companies to create more wealth. In keeping with the former goal, the UK has been very active in telecommunications. While it crosses oceans, the fleet of the British Navy must have sufficient means of communication with the Admiralty back home. The British military communication system Skynet has been operational since 1974. Fifth-generation satellites are currently being launched.

As creating wealth from space remains a challenge, one way around this alternative requirement has been to develop low-cost spacecraft. In this regard, the University of Surrey in the UK pioneered small satellite (smallsat) development. In 1985 the University formed Surrey Satellite Technology Ltd (SSTL) to transfer the results of its research into a commercial enterprise. Since then, SSTL has remained at the forefront of smallsat innovation and development for scientific experiments and commercial use the world over.

Another focus for British space recently emerged. The UK participated in the ESA-led Mars Express mission of 2003. The UK built a Mars lander called Beagle 2, after HMS Beagle, the name of the ship on which Darwin embarked in 1830. The probe failed to reach its target, but received attention in the country. As a result, scientific knowledge has become a new and popular driver for British space. UK politicians have consequently given a degree of attention to space matters. The national space budget reached €200m. in 2005–6. BNSC has been reinforced by advisory and guidance entities, namely the 'UK Space Board' and the 'Space Advisory Council'. A Space Strategy for 2007–10 was formulated as well.

France: Space for National Independence

Similar to the UK, France undertook ballistic programmes after the Second World War, getting its share of German engineers from the German research centre in Peenemünde. However, the similarity ends there. In the post-war context, France sought to assert itself as a significant power on the international scene and considered the conduct of space programmes as an essential status and prestige marker. In that respect, it can be said that space was the twin programme of France's civilian and military nuclear programme, another key marker. Motivations derived from this original one, i.e., reinforcing France's image as a powerful state, have appeared over the years, such as the maintenance of a strong industrial base and the development of independent defence assets.

As little could be done on a national budget, France quickly convinced other states of the European continent to co-operate on space launchers and satellites. This ultimately resulted in the creation of ESA in 1975 and the success of the Ariane launcher in the 1980s. A number of European scientific projects and satellite application programmes followed, in most of which France has a stake.

France continues to be a leading force in the conduct of European space programmes today. With an annual budget of €1.7bn, of which €685m. is directed to ESA, France provides 40% of the global European space budget. The French national space agency, the Centre National d'Etudes Spatiales (CNES), was created in 1962 and is regarded as a key national space actor in the world. There are several CNES facilities, all of which are renowned for their technical excellence. Some even compare in size to NASA centres, including a Launcher Directorate near Paris and a Technical Centre in Toulouse. This allows the CNES Scientific Committee to come up with most of the projects that end up being considered at the European level.

The extent of the CNES workforce and technical credibility also allows it to have dialogue and co-operation with space industry. Following the concentration of defence and aerospace industries in the USA, a similar evolution started in Europe at the end of the 1990s. Prime contractors for launchers and satellites are now European entities. They remain also strongly represented in France, around companies such as EADS, Astrium,[5] and Thalès Alenia Space.[6] Arianespace, the marketing company for the Ariane launcher and Eutelsat Communications, one of the major telecommunications operators in the world, are incorporated in France. A strong network of second-tier and third-tier companies contributes to the high quality of the French space industrial base.

Among the many space programmes in which France participates, Earth observation is considered a key application. The intelligence it can bring in times of regional or international crisis makes it an instrument of strategic independence that appeals to France. The first observation satellite, the Satellite Pour l'Observation de la Terre (SPOT), was launched in 1986. Although oriented towards civilian and commercial use, SPOT put an end to the US-Soviet 'duopoly' on satellite intelligence. France undertook the SPOT project with participation by Belgium and Sweden. SPOT 5 is currently operational; it has three optical instruments on board, each serving a different category of customers: 50cm resolution images are mostly bought by the military; 2.5m resolution images are for mapping; and 10–20m resolution with multispectral capacity serve agricultural and environmental uses.

At the same time, civilian capacity was not considered sufficient to ensure independent intelligence means and France later decided to build a military observation system. The Helios 1A and Helios 1B satellites were launched in 1995 and 1999, respectively. They have an optical resolution of 1–2m. Helios 2A was launched in 2004 and Helios 2B is slated for launch in 2009. This second-generation system offers sub-metric resolution and an infrared instrument for night-time observation. The satellites were designed on the same satellite bus as SPOT. European partners have joined the programme, including Belgium, Italy, Spain and Greece. The successor to SPOT 5 and Helios 2B will be a dual-use system called Pleiades. Two Pleiades satellites will be launched by 2011.

France is the most active European country where military space is concerned. Space is fully taken into account by current thinking on military and strategy issues. The significance of Space Situational Awareness (SSA), for instance, has been recognized. According to a high-level advisory group to the French Ministry of Defence, SSA may bring about a level of deterrence in space, allowing countries such as France to forgo deployment of space weapons in the future.[7]

Regarding acquisition procedures, a recent report commissioned by the Ministry of Defence claims that France must try to build most of the necessary military space system with European partners.[8] Only the most sensitive technologies should be developed nationally. Accordingly, a team of six European countries, the Helios partners plus Germany, are designing the successor for Helios and Pleiades, called the Multinational Space-based Imaging System (MUSIS). The design studies for MUSIS are only just beginning. The system should not be operational before 2015 to 2017. Meanwhile, France builds more sensitive demonstrators on its own. An Essaim electronic intelligence demonstrator was launched in 2004 with Helios 2A and a second one is being launched with the Helios 2B in 2009. Their follow-up, the Electronic Intelligence Satellite (ELISA), will be launched in 2010. An early-warning demonstrator was launched in 2008 and another demonstrator for a laser link between satellites and unmanned aerial vehicles (UAVs), called the Liaison Optique Laser Aéroportée (LOLA), is being tested.[9]

Germany: Space for Trade and Industry

Germany boasts some of the earliest space realizations. Along with Konstantin Tsiolkovski of Russia and Robert Goddard of the USA, Hermann Oberth from Germany is considered one of the pioneers of modern astronautics. Under the supervision of his once-assistant Wernher von Braun, a research centre and launch pad were developed under the German Nazi regime in Peenemünde. This is where the first human foray into space was realized, with the launch of a sounding rocket in October 1942. Peenemünde was also the birthplace of the V2, the first ballistic missile. The facility was dismantled in 1945 at the conclusion of the Second World War.

After the Second World War, the German space effort focused on nonmilitary programmes. The first German space programme was adopted in 1967 and included a co-operation project for a telecommunications satellite with France. The USA launched the two French and German-built Symphonie satellites in 1974 and 1975. Several scientific projects were also undertaken with NASA, such as the Helios solar probes and the Viking Mars probes. The German space agency, known as Deutsches Zentrum für Luft and Raumfahrt (DLR), was created in 1969. It administers the current €850m. space budget of Germany, of which €550m. goes to the ESA. The remaining part of the budget is augmented by DLR direct revenue, so that the actual national space budget as of 2009 is €450m.

A primary driver of the German space programme is the establishment of a strong industrial base and of significant technology development capacities. The German Ministry for Economy and Technology is currently in charge of space policy and dictates the German space programme. German prime contractors for space are EADS Astrium, with facilities in Bayern and in Bremen, and OHB Technology. The latter was reinforced by the acquisition of Kayser Threde in June 2007. The downstream service industry is expected to develop and derive the best from the national space effort. A strong lobbying role is played by an industry representative body called Bundesverband der Deutscher Luft-und Raumfahrtindustrie (BDLI), which includes 90% of space industries in Germany.

In order to provide a sustained stream of orders for its industries, Germany is attached to ESA's 'Fair Industrial Return' financing mechanism. The Fair Return method ensures that for a given programme, one country's industries receive contracts for an amount that is roughly equivalent to the budget their government has agreed to provide. This mechanism has prompted many European countries to participate in ESA's space programmes over the years, ensuring that enough funding is available. On the other hand, it has also required that countries with top-of-the-range technologies, such as France, transfer some of their know-how to less advanced European partners, so that the latter's companies can produce the required elements.

Germany has participated in many ESA-led space programmes over the years. It is, for instance, the second contributor to the Ariane 5 programme. Germany is also very keen to take the lead in the management of the Galileo satellite navigation programme, preferably under ESA financing rules. Another area in which Germany is trying to push its industries forward is the ISS programme. Germany provides 41% of the ESA budget for the ISS and is in charge of a major European contribution. The Columbus laboratory module, which was built in Bremen, was launched in 2008 and is now a functioning part of ISS. Additionally, the Automated Transfer Vehicle (ATV) called Jules Verne, an operational cargo transfer vehicle to the ISS, was developed by the EADS facility in Les Mureaux, France and then manufactured in EADS Bremen.

In keeping with Germany's dedication to environmentally friendly policies, another space motivation worth mentioning lies with environmental monitoring. Germany is involved with the TerraSar civilian Earth observation programme and the ESA Earth Observation Envelope Programme (EOEP) 3. It plans to play an important role in the future Global Monitoring for Environment and Security Programme (GMES) led by the EU.

Even though there is no significant military space programme in Germany, the need for independent intelligence means was recognized in recent years. Of note is that observers have claimed that intelligence supplied by the USA to Germany was insufficient during some crucial episodes of the war in Kosovo. This led to the design and launch of the SarLupe military system, constituting five radar satellites, with 50cm spatial resolution. The full SarLupe system was operational as of 2008. Additionally, a data exchange agreement with France's Helios was signed in July 2002.

Italy: Ambitious Space Actor

Italy boasts a tradition of co-operation with the USA on high-technology programmes. Italian physicists, such as Enrico Fermi, Nobel Prize winner in 1938, emigrated to the USA in the 1930s and participated in the Manhattan project. When the Italian space programme started in the late 1950s, Italy drew on that experience of high-technology co-operation to launch bilateral co-operation programmes with NASA. The first such programme linked NASA with an aerospace laboratory at the University of Rome. Together, they built the San Marco launch platform on a former Italian oil platform near the coast of Kenya. It was used to launch sounding rockets and scientific satellites aboard small launchers, such as the American Scout launcher. The San Marco launch pad remained in use until 1988. The programme has been followed by other bilateral co-operation efforts with NASA. In September 2007, for instance, an imaging spectrometer built by Italy was launched aboard the NASA space probe Dawn, which explores the asteroid belt. Italy still calls itself 'America's best European friend in space' today.

Italy's space capacities have also grown beyond the scope of NASA co-operation. Although the Agenzia Spaziale Italiana (ASI), or Italian Space

Agency, established in 1988 remains small, laboratories and universities play a large role in space. Furthermore, owing to the industry concentration effort of the last decade, Italy now has a few prime contractor-level actors, like Alcatel Alenia Space (satellites), Telespazio[10] (services), and Avio (engines). An association of small and medium-sized aerospace enterprises in Italy, Associazione Italiana per l'Aerospazio (AIPAS), was started in 1998 to serve the needs of Italian companies operating in the space sector. The Italian space budget has grown over the 2006–8 period, reaching €989m. in 2008. The national programme represents 54% of that sum and ESA receives the remaining 46%.

Italy has serious ambitions in space. It wants to play a leading role in the Galileo programme, which has led to many disputes with Germany over recent years. It has also claimed a bigger role in European launcher programmes. As France would not let Italy expand its role in the Ariane programme, it has agreed to the launching of the Vega small launcher programme, funded up to 65% by Italy. Italy is also active in Earth observations. The Constellation of Small Satellites for the Mediterranean Basin Observation system (Cosmo-Skymed) is a dual-use radar satellite funded by the Italian Ministries of Research and of Defence. The first satellite was launched in June 2007 and the next three will be launched before the end of 2009. In addition, a data exchange agreement with Helios was signed in June 2005.

Spain: Newcomer

Spain is a more recent actor in space. The Instituto Nacional de Técnica Aeroespacial (INTA), or national institute for aerospace technology, has conducted research in the field of aeronautics since 1942, but only turned to space in the late 1960s. It is linked to the Spanish Ministry of Defence. Additionally, the Centre for the Development of Industrial Technology (CDTI), an organization under the Ministry of Industry, Tourism and Trade, actively promotes the technology know-how of the Spanish industry. The CDTI is in charge of all agreements between the Spanish space industry and ESA.

The first Spanish satellite, the experimental Intasat, was launched in 1974 and Spain was a founding member of ESA in 1975. Significant growth of the space sector was noted from the 1980s. It was brought about by geographical return programmes under the leadership of ESA. Today, the Spanish space industry is characterized by the predominance of small and medium-sized enterprises. The country's place in the overall European production chain is at subcontractor level.

Satellite communications were deemed important with the creation in 1989 of Hispasat, an operating company managing satellite communication systems. Spain is also present in the field of space exploration. One current focus of the Spanish space effort is in Earth observation. As Spain was excluded from exclusive data exchanges set up between France and Germany (Helios/SarLupe), and France and Italy (Helios/Cosmo-Skymed), it devised its own

Earth observation project, hoping to be in a better position to negotiate its place in the MUSIS dual-use observation project as well as in the EU-led GMES project.

Spain is developing two indigenous satellite systems.[11] The optical sensor-based Spanish Earth Observation Satellite (SEOSAT) will be launched in 2011; and the Spanish Earth Observation Synthetic Aperture Radar (SEOSAR) satellite is expected to be launched by 2015. The country is injecting corresponding funds. Funding increased after 2005, with the Spanish space budget reaching €200m.

Other European Spacefaring States

Sweden is a founding member of ESA and has conducted research and experiments since 1964 from the European Space and Sounding Rocket Range (Esrange) facility. Esrange is located 200km north of the Arctic Circle and is ideal for satellite tracking, sounding rocket launches and other space-related experiments. Sweden has shown a long-standing interest in space observation. It participated in SPOT funding at the very beginning of the programme. Several policy shifts have led Sweden to pay more attention to the military aspect of space since 2000, and it may now join the MUSIS project.[12]

Former members of the Warsaw Pact with long experience in space have become members of the EU. ESA lately established links with countries such as the Czech Republic, Hungary, Poland and Romania, which have the most promising space industries and knowledge. This indicates that these new EU members will eventually play a role in Western European space programmes and ESA.

EUROPEAN SPACE POLICY DYNAMICS

This overview of European countries' motivations for space is not complete without consideration of European space policy dynamics. As the scope and budgets of space programmes are growing, there is less possibility that a space programme can be fully conducted on a national basis in Europe. European entities in charge of space policy, ESA and the EU, tend to play a larger role today and the issue of 'European space governance' has been a topic of debate in recent years.

At the same time, France seems to have taken a step back since the 1990s. Its budget for civil and military space has been stable for a decade or so. It was unsuccessful in engaging military space European co-operation projects in the 1990s.[13] Meanwhile, the EU has taken a step forward. For the first time, it took the lead in major space projects, namely the Galileo navigation system and GMES. A rapprochement with ESA was the logical consequence and a framework agreement between the two entities was signed in 2004, allowing better co-ordination.

ESA and the EU have developed space policy motivations of their own, which come into play in the general dynamics of European space. According to a 2005 study,[14] policy goals of the European space effort focus on the preservation of high-technology industry and jobs in Europe and the offer of top-of-the-range services to the European citizen. Equally high on the list is the continuation of a century-old tradition of scientific discovery. Starting with Aristotle and continuing with Copernicus, Galileo and Newton, this tradition of scientific excellence lives on today with the British Martin Rees and Stephen Hawking.

On the other hand, and in contrast to the USA, manned exploration does not enjoy strong historical and cultural references in Europe. Moreover, the recent partnerships with NASA on the ISS have been fraught with difficulties. European partners have often felt that their contribution to the ISS was overlooked. This experience does not encourage new co-operative ventures in the field of human space exploration.

The EU and ESA have more recently shown interest in security-related space projects. The EU is now in charge of the Torrejón Satellite Centre, a space imagery interpretation centre located at Torrejón near Madrid, Spain. The centre was created in 1992 and was transferred to the EU in 2002, where it serves the 'Common Foreign and Security Policy'. Both GMES and Galileo may have military users in the future. More recently, ESA launched an industrial study in 2006 exploring the feasibility of European SSA architectures. This emergent interest in European security is a new motivation for spacepower at the European level.

CHINA

It is difficult to assess Chinese space activities, because secrecy is at their core.[15] Whether announced space achievements and programmes are bona fide is regularly disputed by foreign observers. The role of the different space actors, be they commissions, academies, corporations, or military entities, is also difficult to ascertain.

What can be said with certainty is that Chinese space policy is driven by one defining element – China wants to shed its image as a developing country and gain the status of a great space power. International and domestic prestige are the strongest drivers of the Chinese space effort. It is in that respect very similar to early space programmes in the former USSR and the USA. This leads China to develop high-visibility human-in-space and exploration programmes, in addition to military space programmes. Space programmes conducive to the economic development of the country are also pursued.

China's national human space flight programme was developed on its own, without co-operation, although Chinese *taikonauts* trained at the Russian Star City complex. The Shenzhou human space programme was launched in 1992. Successful Chinese human space flights occurred in October 2003, October 2005 and September 2008, when a three-man crew conducted space walks.

Future Shenzhou missions are planned, including the development of a space station. The policy goal of the Shenzhou programme has been fully reached, as China is now seen as the third space power in the world, i.e., the third nation independently able to launch humans into space.

Plans for lunar exploration are also in the works. The China National Space Administration (CNSA) has established a lunar exploration programme. China's recent success with a lunar robotic probe in 2007, Chang'e 1, was a first step in this direction. China has also stated ambitions for a series of Chang'e probes, human missions to the Moon, and the establishment of a human-tended lunar base in the 2020s.

Participating from the same goal of self-assertion on the international scene, China has undertaken an important military space programme. Military programmes encompass space application systems such as observation, navigation and telecommunication. Chinese satellite systems are not technically on a par with their Western equivalents for now, but this may change in the future. China has also undertaken an ambitious programme of space weaponization. Indeed, China conducted two tests in recent years: the temporary laser blinding of a US satellite in August 2006, followed by the kinetic destruction of a Chinese meteorological satellite in January 2007. The scope and reality of this weaponization programme remains difficult to ascertain, as exaggerating the Chinese threat may serve legitimate space weaponization policies in the USA, and given China's political support for the Prevention of an Arms Race in Outer Space (PAROS) process through the United Nations Conference on Disarmament. More than likely, Chinese military space programmes are intended to assert the country as a first-rate and legitimate space power in the world vis-à-vis the USA and Russia.

Commercial space could become a big asset of the Chinese space effort, at a time when Chinese exports are on the rise everywhere in the world. Nevertheless, the Long March launcher is partially excluded from the international launch market by US sanctions adopted in 1998 in relation to Chinese proliferation activities. The Long March rocket suffered a string of failures at the beginning of the 1990s, culminating with the 1996–97 failed launches of Lockheed Martin and Boeing satellites, but since then the Chinese rocket has boasted a string of successes. The development by European companies of satellites that are free of US-made components and, therefore, free of any US trade restrictions and export controls, is allowing the Long March to renew its presence in the international launch market.

China has used space as a tool of soft power foreign policy interests. It embarked on a co-operation programme with Brazil in the late 1980s, developing the China-Brazil Earth Resources Satellite (CBERS) series. CBERS-1 was launched in 1999, CBERS-2 in 2003 and two others are planned. More recently, China sold telecommunications satellites to Nigeria and Venezuela in exchange for oil. It has also endeavoured to organize a regional space forum: the Asia-Pacific Space Cooperation Organization (APSCO) was created in 2005. Current signatories are Bangladesh, China, Indonesia, Iran, Mongolia,

Pakistan, Peru and Thailand. The goal of APSCO is to promote co-operation and exchange under Chinese supervision. The other two major Asian space powers, Japan and India, are not members of APSCO, which underlines the geopolitical tensions of the region.

JAPAN

The main actor of the Japanese space effort is the Japan Aerospace Exploration Agency (JAXA). It was set up in 2003 by merging the Institute of Space and Astronautical Science, the National Aerospace Laboratory of Japan and the National Space Development Agency of Japan. In addition, a number of other institutions continue to play a role in space. The different ministries responsible for the Economy, Territory Management, Science, Telecommunications and the Environment, as well as the Prime Minister's Office, each have a small budget to engage in space activities. JAXA's budget in 2008 was €1.4bn. There are indications that the budget will grow in the future.

A defining feature of the Japanese space effort is the fact that, to this day, the Japanese Constitution strictly limits military developments in the country. Article 9 of the 1947 Japanese Constitution states that Japan renounces war and that it cannot keep military forces, except for self-defence. There is consequently no official military space actor. Yet, due to geopolitical developments in Eastern Asia, Japanese space policy has undergone a major evolution in the last decade, with emerging responsibilities in that area attributed to the Prime Minister's Office.

In August 1998, North Korea launched a TaepoDong 2 intercontinental ballistic missile (ICBM) over Japan. This came as a shock to Japan, who had always felt vulnerable to North Korea and China, and now wished to ascertain some rights to self-defence. In the wake of the TaepoDong test, there has been a strong debate in the country on the possible modification of Article 9 of the Constitution. A direct consequence of the North Korean launch has been the addition of military goals to the Japanese space programme.

This has translated into the fast-track development of a radar satellite reconnaissance system called Information Gathering System (IGS). The fourth and last satellite of the system was launched in February 2007. The national launcher H2A was used for reasons of independence and secrecy. After satellites 3 and 4 of the IGS system, which would have added optical sensors to the system, were lost during a H2A failed launch in November 2003, the Japanese government chose to launch the subsequent satellites one by one. The system is, therefore, extremely expensive. This does not seem to matter, as IGS serves national security purposes and comes under the Prime Minister's budget and authority. Replacement optical satellites with 60cm resolution are planned for launch in 2009.

Despite the recent military space interest, the primary objective of the Japanese space programme when it started in the 1970s was scientific excellence rather than national security. Space has been used as a channel for Japanese technological ambitions. This trend is somewhat similar to that of the German space programme. Japan has engaged in a number of scientific programmes, such as the Kaguya/Selene lunar mission.[16] The Moon polar orbiter was successfully launched on a H2A rocket from the Tanegashima launch pad in September of 2007. It sent back data on the origin and evolution of the Moon.

The goal of excellence has led Japan to try and develop all space systems indigenously. Autonomous access to space is the reason why Japan developed the national launcher H2A. This launcher was not developed for business reasons because, for one thing, the Japanese launch pad of Tanegashima Island operates under very strict environmental constraints. However, Mitsubishi Heavy Industries, the company that develops the H2A, became part of an innovative agreement with Boeing and Arianespace in 2003. The Launch Service Alliance allows each company to call on the others' launchers for back-up in case of system failure. The goal is to ensure that the companies' launch manifests will know no interruption. The Japanese launcher programme saw a number of set-backs in the 1980s, but is now reliable.[17]

Japan has an active Earth observation programme. Key systems of the programme include the Advanced Earth Observing Satellite (ADEOS) and the Advanced Land Observation Satellite (ALOS). ADEOS 1 was launched in 1996 and ADEOS 2, launched in 2003, was subsequently lost. ALOS was launched in January 2006 and is currently operational. Images taken by ALOS were released after the July 2007 earthquake off the coast of Japan. Other Earth observation systems are under development.

Japan is developing the Quasi-Zenith Satellite System (QZSS), a regional navigation system. It is an 'augmentation' system of the US Global Positioning System (GPS). Three satellites will enhance the capacities of GPS over the country, where mountainous areas and the many high-rise buildings in the cities degrade GPS performance and accuracy.

Japan is also keen to participate in international co-operation. Because of its close political ties with the USA, many co-operation programmes are undertaken with NASA. There is a JAXA office in Florida and Japanese astronauts have flown on US space missions. US-Japanese co-operation focuses on manned space and the ISS programme. Japan has built the Japanese Experimental Module (JEM), Kibo, which was docked to the ISS in 2008. The JEM features two areas, one that is pressurized and another one, 'the terrace', that is fully exposed to the space environment. It is serviced by a robotic arm.

Major scientific co-operation is also underway with ESA. Amongst these, the Bepi-Colombo space probe will be launched in 2013 and explore the planet Mercury. At the same time, political constraints and geopolitical issues prevent co-operation with Asian neighbours, such as China.

INDIA

India has been active in space since the early 1960s. The Indian Space Research Organization (ISRO) was created in 1969. The management of Indian space activities is fairly transparent, with most budgets appearing on the website of the Ministry of Finance. The budget devoted to space was €1bn in 2008. The governing principle for Indian space programmes seems to be mostly inspired by the 'Gandhian' philosophy of national self-reliance. Autonomy and development are the main justifications of the Indian space policy.

The first aspect of the Indian space policy is the quest for autonomy. In the aftermath of a two century-long colonization by the British Empire, India was keen to limit its dependence on other countries. During the Cold War, India became one of the leading 'neutral and non-aligned' countries of the Third World. Initial co-operation with NASA was not actively pursued. However, some links were established with the USSR and the first Indian astronaut flew with a Russian mission in 1985.

The goal of self-reliance implies efforts to acquire independent access to space. The Polar Satellite Launch Vehicle (PSLV) is operational and the Geostationary Satellite Launch Vehicle (GSLV) has been under development since 2001. India bought several cryogenic stages for the GSLV from Russia. The country also tried to acquire from Russia the know-how to build that element on its own, but the deal was not made, due to the opposition from the USA.

Scientific schools in India are high level and renowned for their computer programming and cyber technology education programmes. This translates into top-level space technologies, as well as a very advanced scientific space programme. ISRO presented a report to the 36th Scientific Assembly of Cospar[18] in 2006 detailing its many high-level scientific programmes and capacities. With an estimated US$83m. budget, the Moon programme Chandrayaan is the main scientific programme currently being undertaken by India.[19] Chandrayaan was launched in 2008 with flight instruments provided by NASA and ESA.

The fight against the economic underdevelopment of the country is the other chief objective of the Indian Government. Like all other areas of public policy, space must participate in the development of the country and serve the Indian citizen. Most scientific programmes are, therefore, meant to enhance India's development. Megha-Tropiques, for instance, is a research programme undertaken in co-operation with CNES, which will study the monsoon phenomenon. Launched in 2004, it benefits agriculture and helps to anticipate floods and droughts. India is recognized for its innovative education and medicine programmes by satellite. These satellite-beamed services reach citizens in rural areas of the country playing an important role in development.

By contrast with the Chinese, international preponderance and prestige seem to remain secondary driving forces of the Indian space programme. India entertains little ambition to become a major international power. Although the metre-based spatial resolution observation satellite Cartosat may prove very useful to monitor activities on the Kashmiri border with Pakistan, the development of military space applications is not at the forefront. Similarly, even though manned space is officially pursued, the loss of an

American astronaut of Indian origin in the 2003 Columbia tragedy has dampened the enthusiasm for such prestige-driven missions in India.

Today (2009), the Indian space programme is on the verge of an important evolution. The nuclear test campaign conducted by India in 1998 signalled an all-time low in India-USA relations. At the time, ISRO was listed on the 'entity list' of the US Department of State and subject to export control restrictions. This list enumerates the entities with which US companies are forbidden to do business for security reasons. When it became necessary to enhance strategic balance in the region, due to the events of 11 September 2001, the USA expressed willingness to foster better relations with India, and to start significant and visible civil space co-operation. An official conference was organized in Bangalore in June 2004 by ISRO and the US Department of State. This, in turn, led to space co-operation between India and the USA.

ISRAEL

The Israeli space programme started in earnest in the 1980s. In 1983, two years after the programme started, the Israel Space Agency (ISA) was created. The main contractor for Israeli space projects is Israel Aerospace Industries (IAI), an aerospace and defence company created in 1953. IAI primarily contracts to the Israeli Ministry of Defence, but also supports satellite development for the commercial space sector in Israel.

National security is the primary raison d'être of the Israeli space programme. Procuring intelligence about hostile neighbours in the Middle East is of crucial importance to the country and Israeli space policy focuses on developing intelligence gathering systems. Israel receives routine intelligence information from the USA, but never succeeded in acquiring direct and real-time data access to US reconnaissance satellites. Moreover, after the peace treaty with Egypt was signed in 1979, Israel had to let go of significant intelligence systems in the Sinai and renounce the possibility to conduct air reconnaissance missions in the area.

Built by IAI, the first military observation satellite, 'Ofeq, was launched in 1988 aboard the Israeli-developed Shavit launcher. The last in the 'Ofeq series ('Ofeq 7) launched in 2007, with plans underway for a next generation optical reconnaissance system. These satellites must be launched westward over the Mediterranean Sea to avoid flying over Arab and Islamic states east of the country. These launches have to fight the rotation of the Earth and are at a disadvantage. In order to avoid this problem, the first radar observation satellite, TechSAR, was launched on the Indian PSLV in 2008.

A second motivation for Israel to conduct space programmes is the development of technologies. Israel has practically no natural resources and, thus, technology is a key asset for the country. Space is a promising sector in the Israeli technology-based economy. The approach of the space community in Israel is to undertake research that leads to commercial applications, to look for niches and open up new markets. For example, the main telecommunications

Developed Space Programmes

satellite system, Amos, which is a commercial endeavour owned and operated by the Israeli company Spacecom, has three satellites in geostationary Earth orbit (GEO), with plans for a fourth satellite by 2010. To add to this, two commercial Earth Resources Observation Satellites (EROS), EROS A and EROS B, owned and operated by the Israeli company Imagesat and built by IAI, were launched from the Russian Svobodny launch pad in Siberia in 2000 and 2006; EROS C is planned for launch in 2009.

CONCLUSIONS

Going over the cases of Russia, the European countries, China, Japan, India and Israel, it appears that there are only a few typical reasons to undertake space programmes. They can be listed as follows: reinforcing the prestige and prominence of the country; business and industrial considerations; the economic development of the country and well-being of its citizens; the pursuit of science and knowledge of the cosmos; and security considerations.

These motivations are not equally distributed amongst space powers. Whichever comes first depends on the history and current situation of the country and the mindset of its people. The different motivations also interact with each other. The need to foster development programmes, such as satellite-based education and medicine, appeals to developing countries, such as India, should in time change into the need to set up a strong aerospace industry. Besides, one can claim that scientific programmes are just as prestigious as human exploration of space. Last, but not least, some military programmes may be undertaken for prestige reasons more than out of real security concerns, though Israel is an exception to this.

The reasons that drive space programme development in space powers – science and education, the economy, security, technology and group psychology (prestige) – show how space covers many different aspects of policy. Other major programmes undertaken by powerful countries, such as nuclear programmes, do not encompass that many reasons and issues. This is why space appears as one of the policies that is most revealing about society. The set of reasons for justifying space programmes in developed states, identified herein, can be applied to investigating other emerging space powers, such as Brazil and the Republic of Korea (South Korea).

NOTES

1 Presentation of the Federal Space Agency on the Roscosmos website in the Russian language. See www.roscosmos.ru/SpaceProg.asp and for English see www.roscosmos.ru/SpaceProg.asp?Lang=ENG (accessed 10 February 2008).
2 The budget in rubles is 305bn. See www.roscosmos.ru (accessed 10 February 2008).
3 'Russia: Space Troops Commander Says Satellite System Improving', *Vestnik Vozdushnogo Flota* (Moscow) 5/04 (31 October 2004), FBIS Document CEP20050118000212.
4 Laurence Nardon and Tatiana Kastouéva-Jean, *La restructuration de l'Industrie spatiale russe*, Note de l'Ifri, April 2007. See www.ifri.org (accessed 10 February 2008).

5 The EADS Astrium subsidiary is owned by EADS NV (incorporated in the Netherlands). As of July 2007, the shareholders for EADS NV were the French state and the French company Lagardère for 27.38%, the German company Daimler Chrysler for 22.41%, and the Spanish company SEPI for 5.46%. The rest are free-floating shares.
6 Created in March 2007, Thalès Alenia Space is owned by the French company Thalès for 67% and by the Italian Company Finmeccanica for 33%.
7 Personal Correspondence, Paris, France, February 2007.
8 *Donnons Plus d'Espace à la Défense*, report of the Groupe d'Orientation Stratégique de la Politique Spatiale de Défense (GOSPS), released by the French Ministry of Defence in February 2007.
9 Peter de Selding, 'European Firms Embark on Dual-Use Satellite Laser Communications Project', *Space News*, 1 October 2007.
10 Telespazio is owned by the Italian company Finmeccanica for 67% and by the French company Thalès for 33%.
11 Robert Wall and Michael A. Taverna, 'France is Stepping up European Milspace Push', *Aviation Week and Space Technology*, September 2007.
12 Lars Höstbeck, 'Swedish Military Space Policy', presentation to the 3AF Conference *Military Space: Questions in Europe*, Paris, France 17–19 September 2007.
13 Laurence Nardon, 'Où va le Programme spatial français', *Politique Etrangère* 2 (2007), Ifri, Paris, France.
14 Laurence Nardon et Maïté Jauréguy, *Europeans in Space*, Note de l'Ifri, September 2005. See www.ifri.org (accessed 10 February 2008).
15 Joan Johnson-Freese, *The Chinese Space Program, A Mystery Within a Maze* (Malabar, FL: Orbit Books, 1998).
16 See www.jaxa.jp/projects/sat/selene/index_e.html (accessed 10 February 2008).
17 Interview with Pierre-Henry Pisani, European Space Policy Institute, Vienna, November 2006.
18 COSPAR is the Committee on Space Research. It was created in 1958 under the auspices of the International Council for Science.
19 See www.isro.org/chandrayaan/htmls/home.htm (accessed 10 February 2008).

Moderate Space Powers

WADE L. HUNTLEY

This chapter reviews the space policies and activities of three moderate space powers – the Republic of Korea (South Korea), Brazil and Canada. The term moderate power applies loosely – although these states are minor space actors in contrast to many developed spacefaring states and groups of states (e.g. the USA, Russia, the People's Republic of China, Europe, Japan and India), they are important regional space actors with notable global roles, whose relatively diminutive status in space activities flows from the significant resource barriers to entry and advancement in this arena.

Moderate powers' space policies and programmes share certain attributes stemming directly from the relatively limited resources these states have available for these efforts. Beyond these circumstantial similarities, however, the approaches and achievements of these states are quite heterogeneous. The experiences surveyed in this chapter, and those of many other moderate space powers, suggest that there is no such thing as a 'typical' moderate power with respect to space ambitions (see Illustrations and Documentation: Table 4.1 Moderate State Civil Satellites; Table 4.2 Moderate State Military Satellites; and Table 4.3 Moderate State Space Programmes).

This does not mean that these divergent paths are idiosyncratic or inexplicable in terms of the forces of power, politics and ideas shaping the international behaviour of states more generally. However, moderate powers, with relatively limited resources, are sensitive to the constraints of circumstance, and so the diversity of the circumstances within which moderate powers find themselves articulates these forces into divergent behaviours and outcomes.

In an effort to offer a foundation for understanding broader patterns in the space interests and activities of moderate powers, the following country reviews focus on the key features of each country's story that most reflect its subjective choices in adapting to its objective circumstances. Concluding comments identify certain traits common to moderate powers' space outlooks and activities, and present a set of generalizations contrasting the 'great power' and 'moderate power' ideal types in the space arena.

SOUTH KOREA

South Korea's interests and activities in space have historically been overshadowed by the political context of the hostile division of the Korean peninsula and its alliance relationship with the USA. That relationship has presented both opportunities and constraints with respect to South Korea's pursuit of space-related goals. The wider political context, including South Korea's transition from military rule to democratic governance in the 1980s, has also been instrumental in shaping the country's space goals. Given this context, those goals have generally focused on commercial and industrial components, while also incorporating political and military elements, with an underlying strong motivation to be at the forefront of capabilities in both of these areas.

The Cold War era security conflict on the Korean peninsula drove both Korean states' interests in developing ballistic missile technologies. In South Korea, this interest, tempered by US reactions, became expressed in the two principal development goals that defined South Korea's space programme in the post Cold War era: (1) production and placement of a fleet of highly capable data collection satellites; and (2) creation of an orbital launch capability.

South Korea's explicit interest in developing a space programme emerged in the 1980s. In 1985 the Ministry of Science and Technology (MOST) issued a 10-year space development plan; in 1987 the government authorized creation of the Korea Aerospace Research Institute (KARI) for the purpose of promoting full-spectrum aerospace development. KARI, established in 1989 as a government-funded institute and independent since 1996, leads the country's space development activities. KARI supports commercial, research and governmental space activities with an emphasis on economic benefits and scientific advancement. Development of military aerospace applications is separately entrusted to the Agency for Defence Development (ADD) under the responsibility of the Ministry of Defence.[1]

In 1995 South Korea issued its National Space Development Plan, with a 20-year time frame comprising satellite development, rocket development and scientific applications. In 2001 South Korea set a goal of becoming one of the world's 10 leading space powers by the year 2015, committing new investment totalling US $4.26bn. South Korea has been steadily increasing its ongoing financing and in 2008 KARI had an annual budget of approximately $250m., with overall spending totalling some $337m.[2]

The effort to accelerate the pace of activities reflects South Korea's sense of itself as a latecomer to the development of a top-tier space programme, particularly in contrast to immediate neighbours, Japan and China. The prioritization of satellite and space launch development represents a strategy of concentrating on specific niches that combine strong needs with unique resources, in an effort to bypass stages of development, rather than achieve a minimal but broader proficiency.[3] The strategy depends crucially on international collaboration and technology transfers. The additional importance of achieving high standing in the international spacefaring community drives South Korea to supplement these two core developmental goals with efforts to participate at the leading edge of space science and exploration, for symbolic as well as material benefit.

Satellite Series

South Korea's satellite development efforts began with KARI. The programme emphasized industrial development and commercial applications, reflecting the public and governmental prioritization of the rapid economic advancement taking place in South Korea in this period also evinced in the country's industrial policies for other core economic sectors. Policy-makers envisioned that making South Korea a vital player at the forefront of space technologies and achievements would generate significant broader economic benefits. Focusing on satellite development also offered prospects for direct benefits, leading to initiation of three separate programmes: (1) the experimental Kitsat (Uribyol); (2) the communication Koreasat (Mugunghwa); and (3) the multi-purpose Earth observation Kompsat (Arirang). The National Space Development Plan, as revised in 2000, envisioned the production of seven Kitsats, five Koreasats in geostationary Earth orbit (GEO), and eight Kompsats by the year 2015.[4]

The Kitsat series, a small and lower-cost experimental satellite intended to provide South Korea's programme with experience in technology development, reached fruition first. The 50kg Kitsat-1, carrying two computer-controlled display (CCD) cameras for Earth imaging, was developed with the help of the University of Surrey, UK, and launched by a European Space Agency (ESA) rocket in August 1992. Kitsat-2, a replica of its predecessor, but assembled indigenously in South Korea, was launched in September 1993. The 110kg Kitsat-3, carrying a 15m resolution CCD and a set of space physics sensors, was launched in 1999.[5]

Koreasat, the second Korean satellite series to reach space, was designed primarily for broadcasting and telecommunications services in GEO. The 830kg satellites carry a set of Ku-band transponders. Koreasat-1 and Koreasat-2, manufactured by Lockheed Martin, were launched from Cape Canaveral in August 1995 and January 1996, respectively. Koreasat-3, also built by Lockheed Martin, but with greater assistance from Korean contractors, was launched by Arianespace in 1999.[6]

The Kompsat series, providing electronic-optical Earth observation, is regarded by KARI as the centrepiece of its satellite development programme. The multi-purpose design creates a capability to use the same satellite bus for a range of observation options, including both low Earth orbit (LEO) and GEO applications. Development of the satellite began in 1994 in co-operation with TRW, which was acquired by Northrop Grumman in 2002, with multiple Korean subcontractors providing some 60% of the parts manufacturing. The 500kg Kompsat-1, launched in 1999 into a sun-synchronous orbit at 685km altitude, carried a panchromatic camera of 6.6m resolution (adequate for minimal military mapping applications), a six-band ocean-scanning multispectral camera with a 1km resolution and 800km swath, and a science physics sensor.[7]

Kompsat-2, carrying a panchromatic camera of 1m resolution provided by El-Op, an Israeli firm, and a multispectral camera of 1m panchromatic and

4m four-colour resolutions, was launched aboard a Eurockot launcher from Russia's Plesetsk facility in July 2006. While officially for civilian uses, the resolution quality of the cameras suggested possible use for intelligence, surveillance and reconnaissance military missions. The dual-use functionality generated new questions concerning technology transfers to South Korea and provided the basis for North Korean claims that the vehicle was a spy satellite.[8]

South Korea's first satellite with an overt military mission is the dual-use Koreasat-5. Built by the French firm Alcatel, the satellite was designed to provide SHF and Ka-band military communications and Ku-band civilian telecommunications and broadband applications. The satellite is owned partly by ADD and the Korean Armed Forces operate its military transponders as part of South Korea's newer Space Communications System. After delays in its preparation, the 4.5-ton Koreasat 5 was placed in orbit in August 2006 by the Sea Launch consortium.[9]

Space Launch Capability

The South Korean space launch vehicle (SLV) programme originated in ballistic missile development efforts that began in the early 1970s. These efforts were catalyzed by the combination of the Democratic People's Republic of Korea's (North Korea) contemporaneous advancements and rising doubts about the reliability of the US security commitment – the latter fuelled by US troop withdrawals from the peninsula in the context of US President Nixon's 'Guam Doctrine' calling for greater security self-reliance among US Asia-Pacific allies. Feeling a sense of urgency, the South Korean military in 1974 developed a plan to quickly develop indigenous capabilities by reverse engineering the Nike Hercules ballistic missile developed by the USA in the 1950s and deployed in South Korea in 1961.[10] The first version of this missile, the 'Paekkom', was successfully tested in 1978; a subsequent improved version, the 'Hyonmu', went into full-scale production in 1986 for surface-to-surface missions and remains deployed near Seoul.[11]

South Korea wanted as much foreign assistance as possible. The USA was willing to provide technological, material and component support of the Paekkom and Hyonmu initiatives, but in exchange required South Korea to enter into a then-confidential 1979 bilateral agreement limiting any ballistic missiles to a range of 180km and a payload of 500kg.[12] A consequence of this arrangement was to deepen South Korea's reliance on the USA as its sole technological support provider.

The emergence of the South Korean Government's initial interest in developing a space programme in the early 1980s, by incorporating space launch, provided an alternative outlet to rocket development interests otherwise curtailed by the limitations on military missile applications. KARI launched its first single stage sounding rocket, the KSR-1, twice in 1993. A two-stage solid-fuelled sounding rocket, the KSR-2, was successfully launched in 1997 and 1998. The final version in sounding rocket series, the KSR-3, utilized a

Moderate Space Powers

liquid-fuelled propellant. Launched in 2002, this rocket tested propulsion, control and guidance technologies to be utilized for an orbital launch rocket.[13] Efforts to develop that rocket, the Korea Space Launch Vehicle (KSLV-1), began at this time.

With a space launch programme in development, South Korean policy-makers increasingly chaffed at the restrictions of the 1979 bilateral missile agreement with the USA, which many saw as anachronistic and which restricted wider access to global technology support. Following several rounds of negotiations, in 2001 the USA agreed to supplant the bilateral agreement and endorse South Korea's entry into the Missile Technology Control Regime (MTCR), formed in 1987 to co-ordinate and bolster ballistic missile technology export controls. MTCR admittance acknowledged the shift of South Korea's efforts to its civilian space programme and opened the door to a wider range of international partnerships in support of its space launch programme.[14]

One of the most important newer partnerships has been with Russia, which South Korea has contracted as a primary supporter for development of the KSLV-1. KARI initially planned to have a launcher capable of lifting a 100kg payload into LEO by 2005. Facing significant problems with rocket engine development, KARI increasingly sought Russian technical advice. In 2005 the agency announced a fundamental redesign of the KSLV, making it significantly larger with a first stage consisting of a liquid-fuelled engine based on Russia's new Angara design; the second stage would utilize South Korea's own solid-fuelled KSR design. In 2008 KARI announced completion of work on the second stage, but delays on both the South Korean and Russian sides put off the first launch to the second half of 2009. KARI plans subsequently to begin development of the KSLV-2, intended to be a larger booster based on exclusively indigenous technology.[15] South Korea also contracted with the S.P. Korolev Rocket and Space Corporation Energia of Russia (RSC Energia) for construction of the new Naro Space Centre in Goheung, including a launch pad, control centre, tracking station, vehicle assembly building, test stands and simulators. The 1,000 acre facility, from which the KSLV-1 is launched, was completed in June 2009.[16]

Recent Developments

Moving forward, KARI planners appreciate that achieving the ambitious goals of South Korea's space programme will require expanding international collaboration, not only with the major spacefaring leaders, but also with a range of developing countries.[17] In addition to the ubiquitous relationship with the USA and the crucial collaboration with Russia, South Korea established early space technology co-operation with China, France and Poland, and forged more recent agreements with Japan, Germany, the UK, Australia, Austria and Italy, including initiating participation in the European Union's (EU) Galileo satellite navigation system in 2005.[18] Other new initiatives included expanding co-operation with the USA; in July 2008 KARI officials reached an agreement for South Korean participation in the NASA-led International Lunar Network, an initiative including seven other countries planning to land robotic stations on the lunar surface for exploration of the lunar environment and resources.[19]

In April 2008 the first South Korean travelled in outer space. On the basis of an agreement with Russia concluded four years earlier, the South Korean rode a Soyuz rocket to the International Space Station (ISS) to conduct a series of experiments and promote Korean culture.[20] The trip evoked the expressed desire of South Korea's space programme to participate at the highest levels, epitomized the country's readiness to find international partners toward these ends, and expressed the importance of the symbolic and material function of the space programme for the country as a whole.

Meanwhile, South Korea has also continued its efforts to reach its core satellite and space launch objectives. In May 2008 plans were announced for the development of Koreasat-6. The 2,750kg satellite, scheduled for launch in late 2010, will have 30 active Ku-band transponders and a design lifetime of 15 years. Korea Telecom contracted with Thales Alenia Space to manage the design, manufacture, testing, launch and early operations of the telecommunications satellite, with Orbital Sciences Corporation of the USA providing the satellite platform.[21]

Advancement of the Kompsat series also continues. Kompsat-3, still under development, will improve on the imaging capabilities of its predecessors, providing high-resolution images required for geographical information systems and other environmental, agricultural and oceanographic monitoring applications. In 2008 KARI announced a partnership with Japan's Mitsubishi Heavy Industries to launch the satellite from Japan's Tanegashima Space Centre by 2012. Development also proceeded for Kompsat-5 (Arirang-5), intended to be the country's first satellite with synthetic aperture radar (SAR) capable of imaging objects through cloud cover and at night. The SAR satellite will undertake military missions principally related to North Korea, but can also image underground or undersea features for mineral exploration and other civilian and commercial purposes. In 2007 completion and launch of the satellite was delayed from 2008 to 2010 due to funding shortfalls.[22]

Beyond these satellite capacities, South Korea had not pursued more active military space capabilities and explicitly eschewed the kind of partnership the USA has with Japan for missile defence development.[23] Recently, however, South Korea has taken steps to acquire a modern short-range missile defence system specifically oriented towards the North Korean threat. In 2007 South Korea finalized purchase of 48 US Patriot Advanced Capability-2 (PAC-2) missiles from Germany, the first of which were deployed in September 2008. South Korea has also initiated construction of its own Korea air and missile defence (KAMD) network, expected to be initially operable by 2010 and fully in place by 2012, in which the PAC-2 systems would serve as the core. The KAMD network would also involve missile defence systems based on Aegis-equipped KDX-III destroyers due to go into service in 2012. Notably, in 2006

South Korea requested purchase of SM-2 missiles for deployment on the Aegis destroyers, and was considering procurement of more advanced SM-6 missiles when available – capabilities at the centre of US-Japan missile defence collaboration. South Korea has maintained its intentions are only to counter North Korea's missile threats and that it chose not to obtain more advanced PAC-3 systems partly for this reason.[24] This behaviour is consistent with the trend in South Korean defence policy to increase self-reliance and autonomy, not deeper integration with the US alliance.[25] Hence, South Korea's new interest in obtaining its own missile defence capabilities does not *ipso facto* augur a shift in longer-term thinking concerning joint missile defence development with the USA and military space applications more generally.

South Korea's space launch aspirations have been seriously set back by the failures of its initial two KSLV-1 launch attempts, the first in August 2009 and the second on June 10, 2010.[26] Nevertheless, South Korea continues to embrace ambitious long-term space programme goals. KARI plans to launch a 1.5-ton, multipurpose satellite in 2017, marking a milestone of indigenous capability.[27] KARI still maintains that development for a second generation Korean space launch vehicle (KSLV-2), capable of lifting 300 tons, will be complete by 2018. This rocket would be utilized to launch an indigenously developed lunar probe from the country's Naro launch centre by 2020. A second lunar probe launched in 2025 would land on the Moon's surface and then return to Earth. The projected cost for the lunar exploration programme is $3.9bn.[28] These plans are evidently intended to keep South Korea's space achievements abreast of its neighbours, with Japan and China both in the business of launching lunar probes. Whether these lofty goals are realistic, in light of budgetary constraints and technological setbacks, remains to be seen. Yet the dramatic economic, industrial and political advancement of the country since the 1980s is a track record that should not be discounted.

BRAZIL

Brazil was the earliest Latin American country to initiate space activities and it sustains one of the most advanced space programmes among developing states. The origins of government-backed efforts date to the 1961 formation of the Organizing Group of the National Commission for Space Activities (GOCNAE), the earliest antecedent to the current Instituto Nacional de Pesquisas Espaciais (INPE), or national institute of space research. In 1972 Brazil became the third country in the world, after the USA and Canada, to install a complete Landsat Satellite Ground Station.[29] This focus on capacity for utilization of Landsat imaging products reflected a particular interest in environmental monitoring, such as data collection on Amazon deforestation, which has been a sustained driving motivation of Brazil's space activities.[30]

Planning for space initiatives coalesced in the 1970s, culminating in 1979 with the promulgation of the Missão Espacial Completa Brasileira, or Brazilian complete space mission (MECB). This long-term plan established the objectives to develop a set of four small satellites, two for data collection and two for remote sensing, along with an indigenous satellite launch vehicle (VLS). Institutionally, these objectives were entrusted, respectively, to INPE and the Aerospace Technical Centre (CTA).[31]

In the 1970s Brazil's military government provided strong support to the space programmes. In 1985 Brazil's return to civilian rule included the creation of its first Ministry for Science and Technology. The impact on Brazil's space programme was divided. On the one hand, in the tumultuous political climate of the early years of civilian rule, prioritization of space activities diminished. However, INPE's transfer into the new Ministry of Science and Technology helped to stabilize resource availability and reinforced the civilian nature of the space programme at a time when Brazil's efforts were increasingly impeded by international restrictions on export of dual-use technologies.[32]

Gradually, governmental attention and commitment to the space programme re-emerged. In the 1990s Brazil undertook additional measures to demonstrate the civilian and pacific nature of its space ambitions. The Brazilian Space Agency (AEB) was formed in 1994 as a civilian organization within the Ministry of Science and Technology. This event corresponded with the election of Fernando Henrique Cardoso as President, initiating a period of political and economic stability providing a firmer foundation for space activities.[33]

In the 1990s, with commitment to the core space objectives of the original MECB, Brazil's indigenous satellite development went forward. The first data collection satellite (SCD-1) was launched on a Pegasus rocket by Orbital Sciences Corporation in 1993, four years after the intended date. The satellite was designed to relay valuable meteorological and environmental data on Brazil's remotest regions to hundreds of ground stations operated by several different organizations for an variety of purposes.[34] The successor SCD-2 was also launched on a Pegasus rocket in 1998.

Space Launch Capability

Brazil's interest in developing its own space launch capability stems in part from its unique geographic location near the Earth's equator. Due to the Earth's rotation, eastward launches enjoy an initial velocity greater than that at higher latitudes, providing a fuel saving and payload surplus of up to 30%.

Brazil took advantage of this circumstance early on, establishing a launch centre at Barreira do Inferno (CLBI) in 1965, which has since been utilized for over 2,000 launches of sounding rockets as well as satellite tracking.[35] The larger Alcantara Launch Centre (CLA) was completed in 1986. Located only 2.4 degrees south of the equator to maximize the geographic advantage, the centre was conceived as a future major international launch site. The site offered a further advantage in that CLBI, located further east, could provide tracking throughout the entire first stage of a launch. Into the 1990s, this potential began to be realized as a number of countries utilized the site.[36]

Developing an indigenous space launch capability was intended to enable Brazil to capitalize on its fortunate geography. However, the effort has encountered significant setbacks. The first VLS launch in 1997 failed just over one minute into its flight. The second launch in 1999 also failed within minutes, destroying one of Brazil's domestically produced satellites in the process. The third failure, in August 2003, was catastrophic; a pre-launch rocket engine explosion collapsed the launch pad, killing 21 top scientists and technicians. A government inquiry traced the cause of the explosion to poor funding and lax management, a conclusion reinforced by an independent *New York Times* investigation.[37]

These failures highlighted a key feature of the rocket programme: its institutional separation within the Brazilian Air Force. Lack of transparency in the programme fuelled suspicions of Brazil's intentions to develop ballistic missiles, which, in turn, curtailed Brazil's access to international collaboration and technology assistance. These dynamics were fuelled by Brazil's orientation to global non-proliferation regimes more broadly. In the 1980s Brazil had become a major arms exporter; from 1985 to 1989, for example, 40% of these exports were to Iraq, making Brazil a major supplier in its war with Iran. Plans for development of advanced weapons systems included incorporating ballistic missile capabilities garnered from development of a satellite launch vehicle.

Brazil had also developed an active nuclear programme, including acquisition of two West German reactors in 1975 that were not under International Atomic Energy Agency (IAEA) safeguards and so were utilized to support a secret nuclear weapons programme, code-named 'Solimies', also initiated in that year. By 1987 that programme had been revealed and Brazil was publicly pursuing uranium enrichment technologies. However, in 1988 Brazil's new constitution banned the military uses of nuclear technologies and Brazil subsequently undertook a series of steps further restricting its nuclear activities, culminating in a set of agreements in 1991 with Argentina, including mutual establishment of full IAEA safeguards for both countries' nuclear facilities. In 1994 Brazil joined the Treaty of Tlatelolco (banning nuclear weapons in Latin America), and in 1998 both joined the long-standing Nuclear Non-Proliferation Treaty (NPT) and signed the newly minted Comprehensive Nuclear Test Ban Treaty (CTBT).[38]

In this climate, Brazil decided to abandon its ballistic missile programme and began seeking to join MTCR, in hopes of both acquiring support for its domestic launch programme and opening up technology co-operation opportunities more generally. Brazil unilaterally adopted MTCR rules by 1994 and the following year it established missile export control regulations sufficient for the USA to agree to allow Brazil to join the MTCR despite controversy over Russian sales of technology to Brazil to support the VLS programme. The issue remained contentious, with critics charging that allowing Brazil to join the MTCR without cancelling its VLS programme could allow Brazil to maintain a surreptitious ballistic missile programme, compromising the credibility of the regime. The contentiousness was evinced by the tumult surrounding the 2000 US-Brazil Technology Safeguard Agreement, which authorized US companies to launch satellites from the Alcantara facility. Features of the agreement that the USA saw as advancing nonproliferation and export control objectives sparked protests by Brazilian critics perceiving the provisions as impositions on national sovereignty; the agreement has yet to be implemented. Over time, a consensus emerged that Brazil's space launch efforts did not harbour ballistic missile ambitions, but questions concerning the sufficiency of its export control systems still lingered.[39]

Despite the sequence of launch failures, Brazil's commitment to a space launch capability was sustained. The country quickly repaired the Alcantara facility and successfully launched a joint Brazilian-German sounding rocket in October 2004. Plans moved forward for reconstruction of the assembly tower, required for the VLS programme, and development of a second-generation VLS, based on conversion of the VLS solid motor technology to liquid propellant rocket engines, reportedly with further Russian assistance. The VLS-2 is intended to be capable of launching 400kg–1,000kg satellites into orbits up to 2,000km. At the same time, the failures in the VLS programme combined with a more receptive international climate encouraged Brazil's search for international collaboration. From 2001 to 2004 it concluded a series of agreements with Ukraine to upgrade Alcantara to accommodate launches of its Cyclon-4 rocket, and in 2004 Brazil entered into a memorandum of understanding with Russia to inaugurate the Southern Cross Programme to jointly develop by 2022 a launch vehicle family of five rockets capable of carrying larger satellites into orbit, as well as securing Russian technology to produce a liquid-propellant powered version of the VLS. Brazil's more recent National Space Activities Programme (PNAE), covering 2005–14, reportedly envisions eventual development of a launcher able to reach GEO carrying 800kg.[40]

Brazil has also become increasingly interested in joint projects and partnerships in areas beyond developing a commercially viable space launch capability. Brazil has pursued participation in the EU's Galileo navigation system, joint atmospheric monitoring with India, and other new collaborations on related space activities. However, Brazil's most enduring and important global collaboration has been with China.

China-Brazil Earth Resources Satellite Programme

Brazil's collaboration with China on development of a set of jointly utilized Earth imaging satellites has evolved into a centrepiece of Brazil's space programme. It represents the evolution of the orientation of Brazil's approach to space activities away from one nationalistically focused on indigenous self-sufficiency, and towards one more internationally interactive and technologically open, embracing collaboration not just for tactical technical assistance, but for the development of strategic partnerships.[41]

Brazil's transition to civilian rule in 1985 accelerated efforts to forge collaboration with China across a range of fields that were already underway. At this time, difficulties in realizing the objectives of the MECB, either independently or with the support of principal developed countries, motivated Brazil to seek greater support for its space programme outside those circles. China, at an early stage in the development of its own programmes, saw similar opportunities in collaboration with Brazil, particularly in speeding up development of satellite application capabilities.[42]

Discussions on space activities collaboration began in 1986, culminating two years later in an initial agreement between the INPE and the Chinese Academy of Space Technology (CAST), establishing the China-Brazil Earth Resources Satellite (CBERS) programme. The agreement allocated project costs 70% to China and 30% to Brazil – a considerable savings to Brazil in comparison to the independent satellite production envisioned in its MECB.[43]

The initial objective of the CBERS project was to develop two remote sensing satellites to map the Earth resources of the two countries, and eventually other countries. Each satellite would include three different cameras: a CCD camera with 20m resolution and stereoscopic capabilities; an infrared multispectral scanner of 80m resolution; and a wide-field imager camera of 260m resolution. The satellites also include a transponder to retransmit environmental information gathered and broadcast from ground stations. Both countries saw these satellites as serving specific needs for gathering information across large and remote land masses, and as providing alternatives to the then-dominant US Landsat and European Satellite Pour l'Observation de la Terre (SPOT) systems, as well as satisfying data needs for which those systems were not optimal, such as the CBERS wide-field imager's 910km swath width enabling single-shot images of large land areas.[44]

The initial years of the collaboration were plagued by Brazilian shortfalls due to lack of governmental support in the midst of a sequence of presidential transitions. Crucially, China responded patiently, for which it was subsequently rewarded. Following the impeachment and resignation of President Fernando Collor de Mello in December 1992, successor Itamar Franco supported the CBERS initiative strongly, and from 1993 Brazilian resource commitments were generally sufficient and timely.[45]

CBERS-1, launched in 1999 from the Chinese base in Taiyuan, was both countries' first indigenously developed Earth imaging satellite. The satellite operated until August 2003, outlasting its two-year life expectancy and delivering over 230,000 images applicable to environmental, social and economic functions. That year marked the launch of its successor, CBERS-2, sustaining the flow of images; this satellite also outlasted its expected life, operating until October 2008. The success of the programme to this point led to agreement to a new protocol in 2002 envisioning the development of CBERS-3 and CBERS-4 with more advanced imaging capabilities, planned at that point to include a panchromatic camera of 5m spatial resolution, a multispectral camera of 20m resolution, an infrared system with four spectral bands capable of

40m resolution, and a Brazilian-developed wide-field imager with four spectral bands of 73m resolution and a swath width of 866km.[46] Notably, the protocol called for China and Brazil this time to equally share project costs, and for CBERS-4 to be launched from Brazil's Alcantara facility, potentially utilizing the joint Brazilian-Ukrainian rocket.[47]

While these new satellites were under development, the collaboration continued with the construction of CBERS-2B, launched from Taiyuan in September 2007. This satellite was a near replica of its then active predecessor, though substituting for the infrared camera a panchromatic camera with 2.5m resolution for black-and-white images, potentially applicable to some military intelligence missions.[48]

Some observers initially suspected that CBERS-2 included a 3m–5m resolution sensor providing rudimentary 'strategic targeting' capabilities, but verification of these reports was lacking. However, the CBERS programme has not been without its military implications. In January 2003 China launched its second Zi Yuan (ZY-2) photo-reconnaissance satellite, with 10cm–20cm resolution capabilities, adapted from the CBERS programme.[49]

Recent Developments

Brazil, in an agreement with NASA, signed on to participate in the ISS by contributing equipment components to support experiments relating to long-term space exposure and advancement of remote sensing.[50] Meanwhile, a Brazilian pilot, Marcos Pontes, was also selected for training by NASA, beginning in 1998, for both Space Shuttle and ISS missions. Under a 2005 agreement with Russia, Pontos became Brazil's first man in space, joining a Soyuz mission to the ISS in 2006, where he oversaw several nanotechnology experiments before returning to Earth.[51]

In December 2008 INPE announced progress on the testing of the next generation of CBERS satellites. The satellites, CBERS-3 and CBERS-4, are now scheduled to be launched in 2011 and 2013, respectively. Meanwhile, Brazil and China agreed to begin providing CBERS-2B satellite images and requisite data interpreting software to Latin American and African countries at no charge.[52] In 2008 Brazil signed an agreement with Germany to develop a satellite with SAR imaging capabilities for observing the Amazon region at night and through cloud or smoke cover. The 500kg satellite will be built by Brazil's INPE by 2010 and launched in 2013, at a projected cost of $157m.[53]

Brazil's collaboration with Ukraine for development and launching of the Cyclone-4 rocket has faced significant delays, but remains active. The first launch from Alcantara had been scheduled for 2006, but Ukraine's production of the rocket was interrupted by financing shortfalls rooted in political uncertainty. Plans in 2008 envisioned quantity production of the rocket to begin in 2009, with its launch facilities at Alcantara ready by 2011. Brazil is also moving forward on its Southern Cross collaboration with Russia, forging an updated agreement in 2008 on the development of a family of launchers

Moderate Space Powers

(incorporating Russia's liquid-fuelled Angara rocket design for the first stage), the largest of which would enable Brazil to offer commercial launches from Alcantara of satellites up to 4 tons into GEO. The Brazilian Government plans to allocate $1bn for the project to 2015 in anticipation of generating $60m.–$100m. income per year.[54] In October 2008 the AEB announced that it had successfully tested the second stage of the VLS-1, now designed with four stages, with the hope of launching a basic version of the rocket on a suborbital trajectory by 2011.[55]

CANADA

Canada's interests and activities in space are inevitably overshadowed by its proximity to the USA, both geographically and politically. Canada's crucial collaboration with the USA on continental defence through the North American Aerospace Defense Command (NORAD) opened opportunities, but also introduced tensions as Canada's commercial space interests developed. Over time, the evolution of the Canada-USA space relationship progressed, but also became more complicated as Canada emerged as an important global actor in both satellite imaging technologies and diplomatic efforts to promote stronger global restraints and global governance on military uses of space.[56]

In the earliest years of the space age, Canada possessed neither the industrial scale nor the indigenous launch capability required for an independent space programme, but it did have the technological expertise for close co-operation with the USA. Through such co-operation, Canada hoped to realize important autonomous public policy and commercial opportunities, while also sustaining high-priority defence links. These interests converged in the Arctic. Satellites offered new opportunities to better understand Arctic atmospheric phenomena and to develop reliable communications vital to both providing early warning of Soviet attack and to linking the small and isolated communities of the north to Canada's main population centres.[57]

Based on a proposal submitted to NASA in 1958, Canada became the third state in the world to design, build and orbit its own satellite when in 1962 the USA launched the 145kg Alouette-1 into a 1,000km orbit. The satellite mission to measure the electron density of the ionosphere was dual-purpose in nature, aiming to improve ground-based radio communications for both military and civilian needs. As its one-year design life stretched into a 10-year mission, Alouette-1 contributed tangibly to research on the ionosphere and other areas of space activity. This collaboration success led the partners to extend it into a fuller programme, International Satellites for Ionospheric Studies (ISIS), which developed and launched three further satellites: Alouette-2, the refurbished back-up, was launched in 1965, and ISIS-1 and ISIS-2, more technically advanced, including better data-collection capabilities and added optical sensors, were launched in 1969 and 1971, respectively. This programme continued into the 1980s.[58]

Canada pursued independent efforts in other areas. Given its extensive territory and widely dispersed population, the potential utility of satellite communications for Canada was apparent, and developing this capacity indigenously became a core focus of Canadian space activities. This led to the Anik satellite series, creating a purely national communications satellite system, Telesat Canada, independent of the then US-dominated Intelsat. The first fruit of this effort, the 560kg Anik-A1, launched in 1972 carrying 12 C-band transponders, was the world's first commercial, domestic, GEO communication satellite and remained in service for 10 years. Anik-F2, the 14th in this series, was launched into GEO in 2004 by Arianespace. Carrying 38 Ka-band, 32 Ku-band and 24 C-band transponders, the 5,950kg satellite over its 15-year lifetime will provide wireless broadband internet connections, telemedicine, teleteaching, teleworking and e-commerce to the most remote regions of Canada, advancing the goal of bringing high-capacity internet access to all Canadian communities.[59]

An additional opportunity for civil space collaboration with the USA was Canada's role as the developer of the remote manipulator arm for the Space Shuttle. The 'Canadarm' gave Canada a role in the core US space plans of the period and a visibility previously lacking for its space programme. The manipulator arm effort facilitated Canadian astronauts entering the Space Shuttle programme, the first of whom flew in 1984 on the Space Shuttle Challenger. Following initiation of the ISS programme that same year, US-Canadian space co-operation deepened further. Canada's eventual development of the Mobile Servicing System manipulator units for the ISS was the first time the USA had allowed any country a role in the 'critical pathway' – project elements that absolutely must be accomplished for the programme to succeed.[60]

This period saw the growth of Canadian space science and Earth imaging efforts. Canada developed or contributed components to more than 20 scientific launches, including several environmental monitoring missions, a series of co-operative initiatives with the ESA, and a role in a Japanese Mars exploration effort. Canada began contributing imaging components to several other nations' satellites in the 1980s, including instruments for measuring Arctic-related phenomena, such as auroral energy patterns and ozone depletion.[61]

Interestingly, only in 1989 did Canada constitute the Canadian Space Agency (CSA), despite several decades of diversified space activities and long-standing calls for creation of an institutionalized co-ordinating body. The CSA is located within the Ministry of Industry, underscoring the civilian and commercial basis of Canada's space intentions, and operates as a governmental department, with the CSA President the equivalent of a Deputy Minister. The CSA's principal mission is to lead development and application of space technology and knowledge across all pertinent Canadian sectors. Its strategy, formally articulated in 2003, entails focusing on Earth observation, satellite communications, space science, public education objectives, and facilitating intra-governmental integration and expansion of external partnerships,

Moderate Space Powers

especially with other countries.[62] As a result of its later formation, the CSA fills a different organizational role within Canadian space activities than its NASA counterpart, serving less to direct the activities of established governmental units, commercial enterprises, and research entities, than act as a 'lead partner' to shepherd them. This looser, less hierarchical architecture has, over time, presented both co-ordination challenges and adaptive flexibility.[63]

The 1990s saw development of Canada's premier indigenous earth imaging capabilities, the Radarsat series. Radarsat-1, launched from the Space Shuttle in 1995, featured a C-band SAR capable of producing military-quality, all-weather images. Multiple beam widths provided resolutions ranging from 8m to 100m, capture swaths of 45km–500km and incidence angles of 10 degrees to 60 degrees. Radarsat-1 was placed into a sun-synchronous orbit at 793km–821km altitude and 98.6 degrees inclination, covering the Arctic daily and most of Canada every 72 hours, depending on instrument targeting.[64]

The successor Radarsat-2 continued to push the frontier of global imaging capabilities. Launched in 2007 after several years' delays into an identical orbit to its predecessor, the newer satellite contained numerous technical enhancements, including an 'ultra-fine' 3m resolution imaging mode and multiple polarization mode options, providing one of the best commercially available satellite SAR imaging systems. Radarsat-2 also has an experimental Ground Moving Target Indicator mode supporting use of the SAR for detecting and monitoring the movement of large vehicles.[65]

While principally intended for commercial use, the high quality of Radarsat imagery offers military applications. The US Department of Defense (DoD) became a customer of Radarsat-2 data as soon as it was launched.[66] In 2005 Canada's Department of National Defence initiated Project Polar Epsilon, an approximately $60m. project to develop dedicated ground stations to receive Radarsat-2 information and integrate it with other sources to create space-based wide-area surveillance of Arctic and maritime regions and enhance situational awareness of Canadian Forces operations globally. Construction of two ground stations was contracted in 2007; the project is due for completion in 2011.[67]

Shifting Partnerships

Although Canada was a relatively early entrant to space, it has never seriously pursued the launch capability requisite to a fully independent programme. Canada's space efforts have, therefore, embraced international collaboration and project partnerships as a permanent mode of operations.

As long as the USA was the only provider of access to orbital space for Canada, Canada focused on developing initiatives that appealed to the USA. Canada's lack of interest in developing an independent space launch capability fit US preferences, but some Canadian commercial interests created friction. Further, the USA tended to limit co-operation on civil space efforts to a project-by-project basis, resisting establishment of a deeper continuing relationship. By the 1970s the establishment of the ESA and development of the European Ariane booster gave Canada an alternative space launch option, providing an escape from the constraints of potential US resistance to specific Canadian payload developments. Consequently, Canada reached out to the ESA as a broader partner in space activities, in part as a response to difficulties in doing so with the USA.[68]

As partnership opportunities expanded, Canada was able increasingly to realize its own prioritization of social and economic benefits in space activities. Accordingly, Canada's incipient space strategy – maximizing opportunities for US collaboration that realized national goals – blossomed into a concentration on staying at the leading edge of space technologies and leveraging these capabilities for long-term national economic competitiveness.[69] Development of both Canadarm and Radarsat marked the maturity of Canada's strategy to become a 'niche' technology provider with unique value both to larger states' efforts and in global commercial markets. The strategy has paid off, with the value of Canada's exports in space technologies surpassing governmental space activity budgets.[70]

As this strategy matured, military space activities declined in relative importance; a trend bolstered by growing US insularity in its own military space activities. Canada became increasingly concerned about securing peaceful uses of space and a world leader in efforts to construct a legal regime controlling military space activities. For example, Canada joined in drafting the first international space treaties in the earliest proceedings of the United Nations Committee on the Peaceful Uses of Outer Space (UNCOPUOUS), and has actively led efforts at the United Nations Conference on Disarmament to develop a treaty for the Prevention of an Arms Race in Outer Space (PAROS), sometimes collaborating with China and Russia.[71] These policy directions introduced friction into the still-central relationship with the USA, but Canada's demonstrated independence also served to reinforce growing co-operation with others actors.

Exemplifying this emergent tension, the USA in 1999 withdrew its exemption for Canada (and other US allies), to the restrictions of the US International Traffic in Arms Regulations (ITAR) export control regime. The withdrawn exemption imposed significant new limitations on US-Canada space and defence technology exchanges. Canada's implementation of parallel export controls under its Controlled Goods Program (CGP) restored some access, but, in turn, impinged on technology co-operation with non-US partners. Canadian space industries' success at working around these restrictions has come at a considerable cost in money and time.[72]

These constraints obstructed Canada's development of Radarsat-2. ITAR restrictions compelled Radarsat-2's Canadian builder, MacDonald, Dettwiler and Associates (MDA), to sever its US provider contract for the primary satellite platform. Meanwhile, pressures inside the USA to resist aiding foreign competition in Earth imaging capabilities derailed plans for a US launch of the satellite. Radarsat-2 was instead developed with European collaboration

Moderate Space Powers

and launched by Russia in Kazakhstan.[73] To assuage concerns over access to Radarsat-2's precision products, Canada subsequently adopted the Remote Sensing Space Systems Act, establishing governmental 'shutter control' over Radarsat imagery, enabling limitations on collection and dissemination for national security or foreign policy purposes and priority access in major crisis circumstances. The legislation was intended to implement the 2000 Canada-US Intergovernmental Agreement concerning the operation of commercial remote sensing satellite systems, and sustain Canada's adherence to the 1967 Outer Space Treaty's obligations to regulate citizens' space activities.[74]

Given the ongoing intimacy of US-Canada defence relations, broader co-operation on space activities remains vital to Canada. However, large asymmetries of capacity make the relationship necessarily skewed, and US restrictions on dissemination of certain space technologies meaningfully impinge on Canada's own efforts. At the same time, the emergence of alternative partnership options for Canada, most developed with the ESA, but also with Japan, Russia and potentially China, has enabled Canada's space programme to mature and become an independent force on the global stage.[75]

Recent Developments

The CSA's programme planning continues to emphasize the civilian sectors and the necessity of co-operation with other countries, while sustaining national capabilities.[76] This prioritization of commercial viability and national autonomy in Canada's space policies posture was highlighted by two political controversies. First, was the contentious debate over whether to join emerging US continental missile defence development efforts, under NORAD auspices. This debate culminated dramatically with the decision of the Canadian Government in 2005 not to participate in development efforts, largely on the basis of concerns over missile defence connections to space weapons planning. However, Canada had previously agreed to the use of NORAD early-warning data in any US missile defence system, which some observers considered more important for the USA than Canadian participation in the programme itself.[77]

The second controversy involved Radarsat. In 2008 MDA, by then Canada's largest space technology firm, announced that it had agreed to sell its space division, including rights to the data from the just-launched Radarsat-2, to US-based Alliant Techsystems. Some critics worried that the turnover of Canada's showcase independent space achievement would be an irreparable national loss, especially given the degree of taxpayer investment in the satellite's development. Alliant's role as a US defence contractor and the high quality of Radarsat-2's imaging data suggested to others a US military space planning dimension. Some long-time observers of Canadian space activities lamented less the immediate loss of control over Radarsat-2 than the longer-term impact that the loss of the space division as a whole represented for Canada's capacity to lead future space technology development. With these concerns coalescing, the Canadian Government in 2008 invoked the Investment Canada Act to block the sale on national interest grounds – the first such veto after some 1,600 reviews of over 10,000 foreign takeovers.[78] The episode once again highlighted the tensions between co-operation with the USA and maintaining independent capabilities and policies that has defined Canadian space activities for decades.

In addition to being a global leader in the development of independent space surveillance capabilities, Canada is pioneering microsatellite technologies. The CSA and Defence Research Development Canada (DRDC) are jointly developing the Near Earth Object Surveillance Satellite (NEOSSat), planned for launch in 2010. The 65kg satellite will carry a powerful optical telescope to track and identify asteroids from Earth's orbit – the world's first designed to track asteroids as well as satellites. NEOSSat and the preceding Microvariability and Oscillation of Stars (MOST), a 60kg satellite for measuring the age of galactic stars, are among a series of Canadian-built microsatellites designed to demonstrate their cost efficiency. NEOSSat utilizes Canada's Multi-Mission Microsatellite Bus, developed by the CSA's Space Technology branch to capitalize on technology developed for the MOST project.[79] The CSA and DRDC are also continuing development of the Maritime Monitoring and Messaging Microsatellite (M3MSat) utilizing proprietary automated identification signals, broadcast from large ships for navigation and identification purposes, to create a comprehensive view of global shipping traffic.[80]

Related to these efforts, the CSA's plans for the next stage of the Radarsat programme is the Radarsat Constellation Mission (RCM), a deployment of smaller satellites utilizing Radarsat technology, while extending geographic coverage. In 2008 the CSA contracted with MDA to begin designing the initial three-satellite configuration, scheduled for launch in 2012, with plans to add three more.[81]

Canada is developing the Sapphire system, which will be its first dedicated space-based military system. The satellite, operating in a circular, sun-synchronous orbit at an altitude of approximately 750km, will utilize an electro-optical space surveillance sensor to detect objects in high Earth orbits, from 6,000km to 40,000km. The spacecraft, anticipated to cost $96.4m., is planned for launch in 2011 and to operate for a minimum of five years. Information from the sensor will contribute to NORAD's Space Surveillance Network (SSN), with Canada's Department of National Defence developing a Sensor System Operations Centre (SSOC) to interface with the US Air Force Joint Space Operations Center (JSpOC).[82] Canada's intention is to secure, through its contribution, access to the entirety of the US-controlled SSN database.[83]

Canada's increasing interest in military space applications, relative to its long-burgeoning non-military interests, evokes the ongoing complexity of Canada's relationship with the USA in two ways. First, efforts to sustain the intimacy of the defence relationship – Canada's very public decision not to participate in continental missile defence more exceptional than indicative –

Moderate Space Powers

extend increasingly into space activities, in contrast to the more competitive nature of the two countries' commercial space activities. Second, these tensions also exist within the commercial and military space sectors: despite commercial competitiveness Canadian enterprises depend on US partnerships, while Canadian initiatives such as Polar Epsilon and Sapphire evince Canada's preparation to establish more autonomous military space capabilities.

CONCLUSIONS

The past and planned space activities of the national space efforts reviewed in this chapter display a fair share of diversity. Differences in the motivations behind space-related pursuits, priorities among programme options, chosen partners and anticipated benefits are all in evidence. Differences in geography, international environment and domestic circumstances all influence choices. Similar reviews of other, smaller countries with space activities would expand this mosaic (see Illustrations and Documentation: Table 4.1 Moderate State Civil Satellites; Table 4.2 Moderate State Military Satellites; and Table 4.3 Moderate State Space Programmes).

A more comprehensive survey of moderate powers' space ambitions and achievements would enable us to systematically map the key variables shaping their diversity. The preceding reviews provide a basis for identifying certain shared characteristics among countries with smaller space programmes, which also distinguish them as a group from the leading spacefaring states. Such generalizations and distinctions between the ideal types of 'moderate powers' and 'leading powers' do not fully capture the attributes of specific countries in either category – most states' orientations toward space exhibit elements of both typologies. However, the relationships between the ideal types posited below do capture important dynamics of international relations relative to the role of smaller space programmes within the totality of humanity's increasing space presence.

One trait common to moderate powers is the emphasis on social and economic benefit. In each case above, and among smaller space-active states generally, social and economic goals are paramount and the payoffs expected to be tangible. Brazil exemplifies that for developing countries this priority is decisive, and South Korea and Canada demonstrate that it is central to more securely developed moderate powers as well. Security concerns can play important roles and symbolic motivations can influence decisions at decisive moments. For moderate powers, though, these factors are not the principal drivers.[84]

A second key shared trait is the search for distinctive specialization. This is clear in the Canadian emphasis on excelling in specific robotic and SAR technologies, and Brazil's commitment to developing launch capabilities to exploit the opportunity of its geography. The quest to be very good at a few things rather than adequate across the board is an example of the familiar economic principle of comparative advantage. The concept underlying this principle applies more broadly, such as in Canada's interest in sustaining a meaningful contribution to North American defence or South Korea's desire for stand-out achievement relative to its East-Asian neighbours.

This 'niche strategy' involves a number of risks. Rapid or unanticipated changes in technology may render one's specialty obsolete. Rivals may take over the niche. Larger and leading states may deliberately undermine moderate powers' niche strategies to eliminate competition, or do so inadvertently out of ignorance and inattention.

Yet, developing niche specializations is also vital to realizing the third shared trait of moderate powers: the quest for collaboration and partnership. For moderate powers, joint projects with peers or with more advanced and leading spacefaring states open technology and information flows that are a primary means to accelerate programme development. Participation in multilateral programmes, such as the ISS, presents prime opportunities to develop specific roles that entail an even wider range of tangible and intangible benefits. Developing niche specializations, if possible, enables moderate powers to become attractive participants in such endeavours, positively reinforcing the strategy.

These traits converge to evince a distinctive viewpoint of moderate powers with respect to their participation in the growing human extension into space: for moderate powers, the key to interactions with other states is to develop relationships. These relationships may be specific and direct, such as a joint project or bilateral partnership, or general and multilateral, such as a multination project or global regime. In all cases, the development of structured relationships enables moderate powers to achieve goals unrealizable independently and to make certain that great and leading powers continue to recognize their interests over time.

This viewpoint is explicable in terms of the basic asymmetries of capabilities between large and moderate powers. Generally speaking, leading powers reckon their positions in relation to other leading powers, and tend to see those other powers as rivals, if not adversaries. Hence, they consider interactions with other states to be competitive, if not conflictual, and to measure smaller powers only to the extent that they meaningfully add to one or the other side of the adversarial ledger. Moderate powers do not similarly focus on their own peer smaller powers, but rather on the leading and great powers. Further, they regard the great powers not just as allies or adversaries, but also in terms of sheer attention, so as not to find their interests impinged out of ignorance or apathy. The greater interest in building relationships with both larger powers and peers, rather than competing autonomously, flows from this perspective.

This assessment is essentially 'realist' in conception, i.e. the tendencies of moderately sized states like Canada to seek solutions in international cooperation more readily than the world's dominant states stem from practical considerations of relative power. This observation is sometimes obscured by the more colloquial notion that moderate powers are intrinsically more idealistic and 'globalist' in orientation. In fact, the basic asymmetries of capabilities moderate powers typically face in the international system drive them towards developing enduring relationships, regardless of any idiosyncratic dispositions. Importantly, alliances as well as regimes can serve this relationship-seeking imperative.[85]

This 'balance of power' dynamic is exaggerated in space-related issues, for several reasons. One is the unique ubiquity of the relevance of space to all nations. Within all the other three domains of human interaction and organized conflict – land, sea and air – states are territorially positioned. Interests of any given state in a particular issue are conditioned by geographic proximity (see Illustrations and Documentation: Figure 4.1 Geopolitics of Space). All states, however, have equivalent 'proximity' to space, driving the world-wide conception of space as a shared global commons. As all states increasingly benefit from space-based capabilities – even the smallest states with no programs at all utilize satellite communications and imaging – space is taking on the character of a public utility of equivalent interest to all states independent of relative capabilities to influence its disposition (see Illustrations and Documentation: Table 4.1 Moderate State Civil Satellites).

A second factor exaggerating relative power distortions is the enormous relative power consequences of varying levels of technological prowess. In a terrestrial battle, moderate powers can conceivably contribute forces proportionate to populations; but several technology levels, most notably space launch capability, are absolute barriers to entry to space-based activities, anointing the few states with those capabilities very powerful gate-keeping roles.

These factors influence the relative importance to leading and moderate powers of security and non-security aspects of space activities. The few states with significant spacefaring capabilities can pursue their interests independently and, typically of great powers, tend to regard space as a realm of competition with security concerns at the forefront. Conversely, the tendency of moderate powers to pursue their interests through relationships is exaggerated by the clear barriers to fielding independent security-related space capabilities, inducing them to prioritize civil and commercial space activities, where potential collaborations and niche roles offer opportunities for tangible participation and influence.

Moderate powers are not immune to space-related security concerns. On the contrary, they are highly sensitive to both their exposure to the consequences of space-based conflict and their abject inability to protect themselves independently. Moreover, because the factor of territorial proximity is absent in the space domain, states tend to view the consequences of conflict in space in absolute rather than relative terms, which is more akin to nuclear conflict than conventional war. This condition has consequences for moderate powers' choices among alliances or regimes as relationship vehicles: whereas establishing patrons or entering alliances often well serves terrestrial security needs, the absolute nature of the consequences of space conflict induces moderate powers to seek multilateral constraints. Hence, near-universal support for space regime enhancement, such as the PAROS treaty effort, flows directly from moderate powers' perception of their material interests and relative capabilities.

The appeal of space spans humanity. The communication and environmental monitoring benefits satellites provide are indispensable. However, dreams of space exploration are also global, well evinced by the satisfaction in each of the countries reviewed herein to have had their citizens in orbit. People outside the leading spacefaring countries embrace space visions as eagerly as those within them. Ironically, while other states may have only minor roles in realizing humanity's greatest hopes for future space development, they are in many ways more sensitive to the peril that space-based conflict poses to those hopes. Well attuned to the value of well-formed relationships in an anarchic world, smaller space-active states also have much to offer in guiding international efforts away from potential conflict and toward realization of those hopes. Leading spacefaring states would be wise to welcome the contribution of this niche specialization as well.

NOTES

1 Paik Hong-yul (KARI), 'Space Development in Korea', in Dana J. Johnson and Ariel E. Levite, eds, *Toward Fusion of Air and Space: Surveying Developments and Assessing Choices for Small and Middle Powers* (Santa Monica, CA: RAND, 2003), 107–8; Daniel A. Pinkston, 'North and South Korean Space Development: Prospects for Cooperation and Conflict', *Astropolitics: The International Journal of Space Politics & Policy* 4.2 (2006): 211, see dx.doi.org/10.1080/14777620600919168 (accessed February 2009); Peter Marquez, 'South Korea: A Space Power by Proxy', in Rebecca Jimerson and Ray Williamson, eds, *Space and Military Power in East Asia: The Challenge and Opportunity of Dual-Purpose Space Technologies* (Washington, DC: Space Policy Institute, December 2001), see www.gwu.edu/~spi/spacemilch6.html (accessed March 2009).
2 Korea Aerospace Research Institute, www.kari.re.kr/english; 'Countries with advanced-launch capabilities', Center for Nonproliferation Studies (CNS), see cns.miis.edu/research/space/index.htm (accessed March 2009); Pinkston, 'North and South Korean Space Development', 221; 'South Korea', Secure World Foundation (SWF) Country Profiles, see 75.125.200.178/~admin23/index.php?id=88&page=South_Korea (accessed March 2009).
3 Kiran Krishan Nair, *Space: The Frontiers of Modern Defence* (New Delhi: Centre for Air Power Studies and Knowledge World, 2006): 182–88.
4 Pinkston, 'North and South Korean Space Development', 214, 216; and Paik, 'Space Development in Korea', 111.
5 Pinkston, 'North and South Korean Space Development', 215; Paik, 'Space Development in Korea', 111–12; and Marquez, 'South Korea' (SWF).
6 Pinkston, 'North and South Korean Space Development', 215–16; Paik, 'Space Development in Korea', 112; and Marquez, 'South Korea' (SWF). See www.globalsecurity.org/space/world/rok (accessed March 2009).
7 John Pike, 'The Military Uses of Outer Space', in *SIPRI Yearbook 2002* (Oxford: Oxford University Press: 2002): 641; Paik, 'Space Development in Korea', 113–14; *Space Security 2008* (Spacesecurity.org, 2008), 127.
8 'Countries with Advanced Launch Capabilities' (CNS); Marquez, 'South Korea' (SWF); and Theresa Hitchens, 'Military Satellites 2006: International Satellite

Moderate Space Powers

     Innovation and Cooperation', Center for Defense Information, 18 April 2006, see www.cdi.org/PDFS/Microsat%202006.pdf (accessed March 2009).
9 *Space Security 2008*, 127; 'Countries with Advanced Launch Capabilities' (CNS); 'South Korea and Satellite Communication Systems', see www.globalsecurity.org/space/world/rok/comm.htm (accessed March 2009).
10 In similar fashion and at roughly the same time, the Democratic People's Republic of Korea (North Korea) began its indigenous ballistic missile development programme by acquiring and reverse-engineering 1950s-era Soviet Scud missiles. 'North Korea Profile', Nuclear Threat Initiative, see www.nti.org/e_research/profiles/NK/index.html (accessed March 2009).
11 Pinkston, 'North and South Korean Space Development', 208–10; 'Countries with Advanced Launch Capabilities' (CNS).
12 Pinkston, 'North and South Korean Space Development', 212; 'South Korea Profile', Nuclear Threat Initiative, see www.nti.org/e_research/profiles/SKorea/index.html (accessed March 2009).
13 Paik, 'Space Development in Korea', 111; 'South Korea', SWF.
14 Pinkston, 'North and South Korean Space Development', 212; *Space Security 2008*, 105; cf. Cha, Victor D., 'Strategic Culture and the Military Modernization of South Korea', *Armed Forces & Society* 28: 1 (2001): 111–13.
15 'Development of Korea's first KSLV-I Upper Stage Completed', Korea Aerospace Research Institute, 14 April 2008, www.kari.re.kr/english (accessed March 2009); 'Launch of Korea's First Space Launch Vehicle scheduled for the Second Half of 2009', Korea Aerospace Research Institute, 9 August 2008, www.kari.re.kr/english (accessed March 2009); Kim Tong-hyung, 'Rocket Launch Delayed to Next Year', *The Korea Times*, 8 August 2008, www.koreatimes.co.kr/www/news/nation/2008/08/133_29009.html (accessed March 2009); Mark Wade, 'KSLV-I', Encyclopedia Astronautica, www.astronautix.com/lvs/kslvi.htm (accessed March 2009); Pinkston, 'North and South Korean Space Development', 213; and 'South Korea', SWF.
16 'Russia to Build Space Launch Pad in South Korea', *Russia in Global Affairs*, 29 October 2004, eng.globalaffairs.ru/printver/685.html (accessed March 2009); and 'South Korea', SWF.
17 Paik, 'Space Development in Korea', 116.
18 'International Cooperation', Korea Aerospace Research Institute, 31 December 2006, www.kari.re.kr/english/02_cms/cms_view.asp?iMenu_seq=107 (accessed March 2009); 'South Korea Negotiates with EC on Galileo', *SatNews Daily*, 23 May 2005, www.satnews.com/stories2005/758.htm (accessed March 2009); *Space Security 2008*, 84; and 'South Korea', SWF.
19 'KARI signs a Statement of Intent for participation in the ILN', International Space Fellowship News, 12 August 2008, spacefellowship.com/News/?p=6327 (accessed March 2009).
20 'Korea's first astronaut returned to Earth in April 2009, with mission accomplished', Korea Aerospace Research Institute, 25 April 2008, www.kari.re.kr/english (accessed April 2009); and 'First S Korean astronaut launches', BBC News, 8 April 2008, news.bbc.co.uk/2/hi/science/nature/7335874.stm (accessed March 2009).
21 'South Korea and Satellite Communication Systems', www.globalsecurity.org/space/world/rok/comm.htm (accessed March 2009).
22 'Preferred bidder selected for the launch of KOMPSAT-3', Korea Aerospace Research Institute, 31 October 2008, www.kari.re.kr/english (accessed March 2009); 'South Korea to Launch First Radar Spy Satellite by 2010', *SatNews Daily*, 22 March 2007, www.satnews.com/stories2007/4167 (accessed March 2009); 'Korea Multi-purpose Satellite (KOMPSAT)', www.globalsecurity.org/space/world/rok/kompsat.htm (accessed March 2009).
23 Pinkston, 'North and South Korean Space Development', 216.

24 Jung Sung-ki, 'Seoul Begins Deploying Patriot Missile Interceptors', The Korea Times, 16 September 2008, www.koreatimes.co.kr/www/news/nation/2008/09/205_31122.html (accessed March 2009); and 'South Korea', SWF.
25 See Victor D. Cha, 'Strategic Culture and the Military Modernization of South Korea', *Armed Forces & Society* 28.1 (2001), 118–19.
26 Stephen Clark, "South Korea says rocket likely exploded after liftoff," *Spaceflight Now*, June 10, 2010 (http://www.spaceflightnow.com/news/n1006/10kslv/, accessed June 2010); "South Korea rocket 'explodes' moments after take-off," *BBC News*, June 10, 2010 (http://news.bbc.co.uk/2/hi/10281073.stm).
27 Kim Tong-hyung, 'Rocket Launch Marks First Test for Korean Space Ambitions', *Korea Times*, 20 October 20 2008, www.koreatimes.co.kr/www/news/nation/2008/10/133_33007.html (accessed March 2009).
28 'Lunar probe ready by 2020: KARI', *JoongAng Daily*, 21 August 2008, joongangdaily.joins.com/article/view.asp?aid=2893926 (accessed March 2009); Cho Jin-seo, 'Korea Plans to Send Moon Orbiter in 2020', *The Korea Times*, 20 November 2007, www.koreatimes.co.kr/www/news/tech/2007/11/133_14085.html (accessed March 2009); and 'South Korea eyes moon orbiter in 2020, landing 2025', Reuters, 20 November 2007, www.reuters.com/article/scienceNews/idUSSEO24596320071120 (accessed March 2009).
29 Demetrio Bastos-Netto (INPE-CES), 'Dilemmas in Space Strategy for Regional Powers: A Brazilian Perspective', in Dana J. Johnson and Ariel E. Levite, eds, *Toward Fusion of Air and Space: Surveying Developments and Assessing Choices for Small and Middle Powers* (Santa Monica, CA: RAND, 2003), 119; José M. Filho 'Brazilian-Chinese Space Cooperation: An Analysis', *Space Policy* (1997): 161.
30 Thelma Krug, 'Space Technology and Environmental Monitoring in Brazil', *Journal of International Affair* Spring (1998): 663
31 A. T. Furtado and E. J. C. Filho, 'Assessing the Economic Impacts of the China-Brazil Resources Satellite Program', *Science and Public Policy* 30: 1 (2002): 26; and Krug, 'Space Technology and Environmental Monitoring in Brazil', 657.
32 Furtado, 'Assessing the Economic Impacts', 26; Krug, 'Space Technology and Environmental Monitoring', 657; Filho, 'Brazilian-Chinese Space Cooperation', 158.
33 Victor Zaborsky, 'The Brazilian Export Control System', *The Nonproliferation Review* (2003): 124.
34 Krug, 'Space Technology and Environmental Monitoring in Brazil' 658; Helio K. Kuga and Rama R. Kondapalli, 'Satellite orbit determination – A first-hand experience with the first Brazilian satellite SCD-1', *International Astronautical Congress*, Graz, Austria, 16–22 October 1993.
35 Bastos-Netto, 'Dilemmas in Space Strategy for Regional Powers', 120.
36 Ibid; 'Brazil is Going into Space', *ISTOE* (Brazilian Weekly), 9 August 1995, www.terra.com.br/istoe (accessed March 2009), as reproduced at www.fas.org/news/brazil/lat95188.htm (accessed March 2009).
For example, in 1994, NASA launched 33 satellites from Alcantara.
37 'Countries with Advanced Launch Capabilities' (CNS); Rohter, Larry, 'Brazil's Soaring Space-Age Ambitions are Shy of Cash and Sapped by Calamity', *The New York Times*, 23 January 2004, www.nytimes.com (accessed March 2009).
38 Zaborsky, 'The Brazilian Export Control System'124–26.
39 Ibid., 127–28 and 132–34; and Filho, 'Brazilian-Chinese Space Cooperation', 168.
40 'Countries with Advanced Launch Capabilities' (CNS); Zaborsky, 'The Brazilian Export Control System', 127; and Rohter, 'Brazil's Soaring Space-Age Ambitions'.
41 Filho, 'Brazilian-Chinese Space Cooperation', 153; Furtado, 'Assessing the Economic Impacts of the China-Brazil Resources Satellite Program', 27.
42 Knowledge of organizational techniques and project management may have been even more important to China than technology transfer. Brazil shared its methods,

Moderate Space Powers

documentation standards and technical procedures, bolstering CAST's then-precarious administrative capacity. See Furtado, 'Assessing the Economic Impacts of the China-Brazil Resources Satellite Program', 28.
43 Filho, 'Brazilian-Chinese Space Cooperation', 154 and 161–62; Zhao Yun. 'The 2002 Space Cooperation Protocol between China and Brazil: An Excellent Example of South-South Cooperation', *Space Policy* 21 (2005): 213; and Furtado, 'Assessing the Economic Impacts of the China-Brazil Resources Satellite Program', 37.
44 Theresa Hitchens, 'Military Satellites 2006: International Satellite Innovation and Cooperation', Center for Defense Information, 18 April 2006, www.cdi.org/PDFS/Microsat%202006.pdf (accessed March 2009); Normile, Dennis and Ding Yimin, 'Science Emerges from Shadows of China's Space Program', *Science* 296: 5,574 (7 June 2002); Furtado, 'Assessing the Economic Impacts of the China-Brazil Resources Satellite Program', 27; and Krug, 'Space Technology and Environmental Monitoring', 660–62.
45 Filho, 'Brazilian Chinese Space Cooperation', 162–68. Brazil ultimately provided 27% support for the CBERS satellites, contracting its other 3% obligation back to China. See Furtado, 'Assessing the Economic Impacts of the China-Brazil Resources Satellite Program', 28.
46 Hitchens, 'Military Satellites 2006'.
47 Zhao, 'The 2002 Space Cooperation Protocol between China and Brazil', 214–15; Krug, 'Space Technology and Environmental Monitoring', 660; and Rohter, 'Brazil's Soaring Space-Age Ambitions'.
48 Stephen Clark, 'China and Brazil team up to launch remote sensing craft', *Spaceflight Now*, 19 September 2007, www.spaceflightnow.com/news/n0709/19cbers/ (accessed March 2009); and *Space Security 2008*, 132.
49 Joan Johnson-Freese, 'China's Manned Space Program: Sun Tzu or Apollo Redux?' *Naval War College Review* 56: 3 (Summer 2003); and globalsecurity.org (accessed March 2009).
50 *Space Security 2007* (spacesecurity.org, 2007), 64.
51 Center for Defense Information Space Security Update #3, 12 February 2006; and Center for Defense Information Space Security Updates #10, 26 October 2005.
52 *Space Security 2008*; 'Brazil begins Mechanical Tests on Satellites', *Xinhua*, 22 December 2008, news.xinhuanet.com/english/2008-12/22/content_10542136.htm (accessed March 2009). "Brazil, China To Postpone Joint Satellite Launching To 2011," Space Travel, February 16, 2010 http://www.space-ravel.com/reports/Brazil_China_To_Postpone_Joint_Satellite_Launching_To_2011_999.html (accessed June 2010).
53 'Brazil, Germany to Develop Night-Vision Radar Satellite', Xinhua News Agency, 18 March 2008, www.spacemart.com/reports/Brazil_Germany_To_Develop_Night_Vision_Radar_Satellite_999.html (accessed March 2009).
54 Yury Zaitsev, 'Russia Begins Elbowing Ukraine out from Brazil's Space Program', *RIA Novosti*, 16 September 2008, via Space Daily, www.spacedaily.com/reports/Russia_Begins_Elbowing_Ukraine_Out_From_Brazil_Space_Program_999.html (accessed March 2009).
55 'Brazil hopes to launch satellite rocket in 2011: report', AFP, 21 October 2008, www.space-travel.com/reports/Brazil_hopes_to_launch_satellite_rocket_in_2011_report_999.html (accessed March 2009).
56 For one review, see Roger Handberg, 'Outer Space as a Shared Frontier: Canada and the United States, Cooperation Between Unequal Partners', *American Behavioral Scientist* 47: 10 (June 2004).
57 Christopher Gainor, 'Canada's Space Program, 1958–89: A Program without an Agency', *Acta Astronautica* 60: 2 (2007).
58 'Satellites', Canadian Space Agency, www.asc-csa.gc.ca/eng/satellites (accessed March 2009); Jocelyn Mallett, 'Canada's Space Programme', *Space Policy* 6

(1990); and Andrew Godefroy, 'Canada's early space policy development 1958–74', *Space Policy* 19 (2003).
59 'Satellites', Canadian Space Agency, www.asc-csa.gc.ca/eng/satellites (accessed March 2009); and Fergusson, James and Stephen James, 'Report on Canada, National Security and Outer Space', Canadian Defense and Foreign Affairs Institute, June 2007.
60 Handberg, 'Outer Space as a Shared Frontier', 1,257.
61 'Satellites', Canadian Space Agency, www.asc-csa.gc.ca/eng/satellites/ (accessed March 2009).
62 Government of Canada, *The Canadian Space Agency, 2007–2008 Estimates: Report on Plans and Priorities* (2007), www.space.gc.ca/asc/pdf/FINAL_RPP_07-08_e.pdf (accessed March 2009); 'About the Canadian Space Agency', Canadian Space Agency, www.asc-csa.gc.ca/eng/about/default.asp (accessed March 2009); and Canadian Space Agency (CSA), *The Canadian Space Strategy* (Saint-Hubert: CSA, 2003).
63 Roger Handberg, 'Dancing with the Elephants: Canadian Space Policy in Constant Transition', *Technology in Society*, 25: 1 (January 2003); Fergusson and James, 'Report on Canada, National Security and Outer Space'.
64 Canadian Space Agency (CSA), *Radarsat, Annual Review 1998/99* (Saint-Hubert: CSA, 1999); and 'Radarsat-1', Canadian Space Agency, www.asc-csa.gc.ca/eng/satellites/radarsat1 (accessed March 2009).
65 'Radarsat-2', Canadian Space Agency, www.asc-csa.gc.ca/eng/satellites/radarsat2 (accessed March 2009); David Pugliese, 'Canada to Launch Design Work on Radarsat', *Defence News*, 29 September 2008; and *Space Security 2008*, 135.
66 'Radar Love', *Aviation Week and Space Technology*, 29 September 2008; and Barbara Opall-Rome, 'DoD Looks Abroad for Space Radar', *Defense News*, 1 September 2008, www.defensenews.com/story.php?i=3700661 (accessed March 2009).
67 'Polar Epsilon Project', *National Defence and the Canadian Forces*, 10 January 2008, www.forces.gc.ca/site/news-nouvelles/view-news-afficher-nouvelles-eng.asp?id=2546 (accessed March 2009); Stephen Thorne, 'Feds Look to Satellite to Assert Arctic Sovereignty', Canadian Press, 28 August 2005; David Pugliese, 'Canada Focuses Military Space on Continental, Homeland Defense', *Space News* 4 April 2005, www.space.com/spacenews/archive05/Milcan_040405.html (accessed March 2009).
68 Lydia Dotto, *Canada and the European Space Agency: Three Decades of Cooperation* (Norodwijk, Netherlands: European Space Agency Publications Division, 2002); Canadian Space Agency, *National Paper: The Canadian Space Program* (Saint-Hubert: CSA, 1999); John Kirton, 'Canadian Space Policy', *Space Policy* 6: 1 (1990).
69 Roger Handberg, 'Outer Space as a Shared Frontier: Canada and the United States, Cooperation Between Unequal Partners', *American Behavioral Scientist* 47: 10 (June 2004).
70 Canadian Space Agency, *State of the Canadian Space Sector 2000* (Saint-Hubert: CSA, External Relations Directorate, 2002).
71 'Space Security', Canadian Department of Foreign Affairs and International Trade, 15 July 2008, www.international.gc.ca/arms-armes/non_nuclear-non_nucleaire/space_security-securite_spatiale.aspx?lang=en&menu_id=120&menu=R (accessed March 2009).
72 The move was precipitated by the return of US export control authority on dual-use space technologies from the Department of Commerce to the Department of State, in part due to then-palpable worries over increasing Chinese efforts to surreptitiously obtain US high technology capabilities. See Eric Choi and Sorin Nicelescu, 'The Impact of US Export Controls on the Canadian Space Industry', *Space Policy* 22 (2006): 29–34.
73 Canadian Space Agency, *Radarsat, Annual Review 1998/99* (Saint-Hubert: CSA, 1999); cf. Dave Caddey, 'Radarsat-2: A cautionary tale', *Aerospace America*, January

2001, www.aiaa.org/Aerospace/Article.cfm?issuetocid=45&ArchiveIssueID=9 (accessed March 2009); Handberg, 'Dancing with the Elephants;' and J. Bates, 'Canadian Military Mulls Tandem Radarsat Mission', *Space News*, 13 May 2002.
74 'Bill C-25: An Act Governing the Operation of Remote Sensing Space Systems', prepared by Lalita Acharya, Science and Technology Division, Library of Parliament, Government of Canada, 20 December 2004, www.parl.gc.ca/common/Bills_ls.asp?Parl=38&Ses=1&ls=C25 (accessed March 2009); and *Space Security 2008*.
75 Dotto, *Canada and the European Space Agency*; Daniel Sorid, 'Japan to Sign Space Accord with Canada', space.com News, 10 September 1999, www.space.com/news/japan_canada.html (accessed March 2009).
76 *The Canadian Space Agency, 2007–2008 Estimates: Report on Plans and Priorities*, Government of Canada (2007), www.space.gc.ca/asc/pdf/FINAL_RPP_07-08_e.pdf (accessed March 2009): 11 and 20.
77 See James Fergusson, 'Shall We Dance? The Missile Defence Decision, NORAD Renewal, and the Future of Canada-US Defence Relations', *Canadian Military Journal*, Summer 2005, www.journal.forces.gc.ca/vo6/no2/inter-01-eng.asp (accessed March 2009).
78 Andrew Mayeda and David Akin, 'Ottawa blocks sale of space agency to US firm', Canwest News Service, 10 April 2008, www2.canada.com/montrealgazette/news/story.html?id=6426dd8d-ab10–426b-9f60–358c1fea7b9b&k=35463 (accessed April 2009); 'Government confirms decision to block sale of MDA space division', CBC News, 9 May 2008, www.cbc.ca/money/story/2008/05/09/alliant-sale.html; Jessica West, 'Radarsat-2: Launched and lost?', The Ploughshares Monitor 29: 1, Spring 2008, www.ploughshares.ca/libraries/monitor/monm08d.pdf (accessed March 2009); and Michael Byers, 'For Sale: Arctic Sovereignty?' *The Walrus* June 2008, www.walrusmagazine.com/articles/2008.06-technology-for-sale-arctic-sovereignty-radarsat-mda-michael-byers (accessed March 2009).
79 'Canadian Asteroid-Hunting Satellite A World First', *Space Daily*, 7 July 2008, www.spacedaily.com/reports/Canadian_Asteroid_Hunting_Satellite_A_World_First_999.html (accessed March 2009); 'MOST, Canada's First Space Telescope', University of British Columbia, www.astro.ubc.ca/MOST/ (accessed March 2009); and *Space Security 2008*, 36, 44 and 136.
80 'Com Dev wins micro-satellite contract', *CBC News*, 23 June 2008, www.cbc.ca/technology/story/2008/06/23/comdev-m3msat.html (accessed March 2009); and *Space Security 2008*, 136.
81 'CSA Announces Design Contract with MDA for RADARSAT Constellation', *SpaceMart*, 18 November 2008, www.spacemart.com/reports/CSA_Announces_Design_Contract_With_MDA_For_RADARSAT_Constellation_999.html (accessed March 2009).
82 Paul Maskell and Lorne Oram, 'Sapphire: Canada's Answer to Space-Based Surveillance of Orbital Objects', Canadian Forces Surveillance of Space Project, n.d. (2008), appspacesol.com/pdf/sapphire.pdf (accessed March 2009); Chris Wattie, 'Canada will launch own spy satellite: Project Sapphire', *National Post*, 14 November 2006, A6; David Pugliese, 'Canada Focuses Military Space on Continental, Homeland Defense', *Space News*, 4 April 2005, www.space.com/spacenews/archive05/Milcan_040405.html (accessed March 2009); and *Space Security 2008*, 36 and 143.
83 'CDI Space Security Update #11', Center for Defense Information, 3 June 2004, www.cdi.org/friendlyversion/printversion.cfm?documentID=2247#4 (accessed March 2009).
84 *Space Security 2008*, 12–13.
85 For an elaboration of these observations, see Wade L. Huntley, 'Smaller State Perspectives on the Future of Space Governance', *Astropolitics* 5:3 (Fall 2007).

Commercial Space Actors

DAVID D. CHEN AND MOLLY K. MACAULEY

This chapter focuses on contemporary commercial space actors in the USA whose visions have been influential in the politics of space. The effectiveness of these actors has been wielded most notably in pushing the frontiers of space legislative and regulatory policy to accommodate technological innovation and to nurture, rather than retard, business opportunities. The individuals described here fought two battles: one for innovation and one for commercial success in an industry traditionally dominated by government.

GEOSTATIONARY COMMUNICATIONS SATELLITES

Of the many benefits gained from space enterprises, telecommunications was the first commercially viable industry to develop. Many of the goods and services of space technologies are difficult to market, either by reason of production cost, or by the very nature of the service providing a 'public good' that cannot be easily bought and sold, such as national security. Satellite-based communications, however, could be commercialized. In the 1950s global communications were growing rapidly.[1] In connecting callers, telecommunications companies found a marketable good in space. They could charge callers a fee for service, collect revenue and, ultimately, convince policy-makers and investors to build a global satellite communications system.

The creation of a satellite communications framework was not a foregone conclusion due to technical problems associated with reaching geostationary Earth orbit (GEO) and the time delay of signals to and from that orbit. Simultaneously, the ownership and management of what was envisioned to be a global communications network was similarly contested over the question of government ownership, regulated monopoly, or private competition. The system that resulted would be the product of the competitive innovation of engineers and politicians, and of scientists and corporate executives.

Underwater and Overhead

While GEO was first described by the father of space flight, Konstantin Tsiolkovsky, in 1895, and Guglielmo Marconi was the first to establish a commercial intercontinental radio network in the early 1900s, the synthesis of these two ideas was not published until 1945, by Arthur C. Clarke.[2] Clarke proposed placing three 'satellite stations' with an orbital radius of 42,000km around the Earth. At this distance, the period of the orbit is 24 hours – GEO. By using this special orbit, any observer on Earth would be able to maintain continual sight of the satellite as it would appear to hover in the same position in the sky throughout the day and night. If such a system could be built, global telecommunications would be within reach.

Clarke's proposal was not the only, or necessarily the best, technical solution to providing a global communications network. The problems of reaching GEO were not trivial. New and more powerful launch vehicles were needed before the venture could even be entertained. Such rockets certainly did not exist in 1945 and would not be available until the 1960s. In the meantime, other options were being proposed and implemented. The first transatlantic telephone cable went into use in 1956, though with the limited capacity of only 36 calls at any one time.[3] Undersea cable networks had already been well established by the United Kingdom (UK) and other states, but growth in demand motivated the search for lower-cost alternatives. A RAND Corporation study in 1960 estimated that a low Earth orbit (LEO) communications satellite (comsat) system would cost about US $8,500 per channel per year, or less than a third of the costs of oceanic cables.[4] The question of laying more undersea cables or using space to satisfy the world's growing demand for telecommunications would favour satellites in the era of copper wires, but the question would re-emerge to overturn that assessment in the era of fibre optics.

Sky High in Degrees

Dr John R. Pierce, director of American Telegraph and Telephone (AT&T) Bell Telephone Laboratories, was at the forefront of experiments involving the use of LEO and medium Earth orbit (MEO) satellites to provide a radio relay network spanning the globe. In 1960 his team launched the first passive radio satellite experiment, ECHO-1. Essentially a 100-foot diameter Mylar balloon, coated in reflective aluminium, the satellite was designed to act as a passive reflector for radio signals from the ground. The problem with this design, however, was that signals to the reflector had to be very powerful and receivers of the reflected signal had to be very sensitive. The signal reflected back to receiving stations on the ground was only 'a millionth of a millionth of a millionth of the energy sent up to it'.[5] Utilizing an active repeater design instead reduced the power requirements by many orders of magnitude. A repeater works by taking the received signal and re-broadcasting it on the same or another frequency, amplifying the signal, if necessary.

Several repeater satellite designs were in the works by 1961. Telstar, a design by John Pierce's team, used an LEO microwave relay carrying telephone, television and data communications. Relay, constructed by the Radio Corporation of America (RCA), was also an LEO telephone, television and data relay. As both designs utilized LEO, any individual satellite would travel eastward across the sky at 4 miles per second, quickly vanishing over the horizon from the vantage point of ground stations. A constantly-in-view network would require dozens of satellites arrayed like a string of pearls, following one another, so that as one passed over the horizon in the east, another was rising in the west. This approach presented huge costs in terms of manufacturing the multitude of satellites, launching them, taking into consideration replacement costs for launch failures, and including the added number and complexity of ground stations. An additional consideration for politicians and executives was speed of delivery, as the first to establish the network would enjoy first-mover advantage in terms of national prestige and market position.

Even as Telstar and Relay were lauded as examples of American ingenuity by transmitting television broadcasts across the Atlantic, a Hughes Aircraft design team led by Harold Rosen and Don Williams was building a repeater satellite of its own. Their approach was different: it utilized GEO as envisioned in Clarke's 1945 paper. Rosen's design solved the weight problem by using gyroscopic spin-stabilization to keep the satellite on station and properly oriented, eliminating the need for many of the control thrusters otherwise required.[6] In 1963 Syncom 2 successfully demonstrated that this design could work. Echo control techniques mitigated the effect of signal delay on telephony to an acceptable level for public users, and live and taped television was successfully relayed across the Atlantic and Pacific Oceans.[7] With these demonstrations, the policy options for constructing a global satellite-based communications network were now between a multitude of active LEO repeaters or a few GEO repeaters. The political economy of satellite telecommunications presented as complex a set of problems, if not more so, as the engineering and rocket science.

Policy Dilemma

With the technology proven, and the milestone of Yuri Gagarin's human space flight, comsats became a potential tool in the technological and political competition of the Cold War between the USA and the USSR (Union of Soviet Socialist Republics). In 1961 President John F. Kennedy challenged the US satellite community to establish 'at the earliest practicable time operational communications satellites'.[8] Establishing an international satellite communications network, with US technology, by US companies, became a national priority. Kennedy instructed that the system should operate 'on a global basis, giving particular attention to those of this [Western] hemisphere and newly developing nations throughout the world'.

Comsats would be a useful beacon on the world stage, giving the USA increased national prestige and an important advantage in an emerging industry. With time pressure being a decisive factor in implementing an operational system, the US Federal Communications Commission (FCC) and US telecommunication companies favoured a policy that treated the satellite segment as part of the ground network, thus privileging the existing telecommunications companies. The US Congress, unwilling to bequeath federal research to private companies, proposed an alternative to either market control or government control under a 'Comsat Authority'. What resulted was the Communications Satellite Act of 1962, which created a chartered corporation called the Communications Satellite Corporation, or Comsat Corporation, which was neither completely corporate nor completely governmental, but a mix of private and public control.

Comsat Corporation was designed essentially as a government mandated monopoly, composed of Class A shareholders, like telecommunications companies AT&T, International Telephone & Telegraph (ITT), RCA and Western Union, and Class B shareholders, with shares sold to the public. It was thus that the Comsat Corporation was conceived from US technology, delivered by bureaucratic and corporate interests, and intended to fulfil Kennedy's mandate of establishing an international satellite communications system as rapidly as possible. The advantage of such a de facto monopoly was speed. Comsat was able to move swiftly in establishing an international satellite communications market, reassured in having US telecommunication companies as customers, having National Aeronautics and Space Administration (NASA) support for at-cost satellite launches, and the full backing of the US State Department in international treaties and regulatory forums. The executive branch of the US Federal Government would maintain oversight of the Comsat Corporation through the FCC, the State Department and the White House Office of Telecommunications Policy, as well as the power of the President to appoint three of Comsat's 15 board members. Some 50% of the corporation's stock could be owned by the telecommunications companies, with the remaining 50% available to the public.[9]

Comsat faced immediate challenges, the first technical and the second commercial. In the years leading up to that time, AT&T had advanced the Relay system of satellites in LEO and MEO as the profile of choice for an international satellite network. The competing plan to place fewer satellites higher up in GEO was also under development, under Hughes' Syncom design. Comsat engineers recommended funding an experimental deployment of two Syncom satellites, estimating that GEO would result in cheaper and simpler operations on the ground and faster deployment in orbit. The 'experimental-operational' HS-303 satellite was launched in April 1965, becoming the first commercial telecommunications satellite in GEO;[10] GEO proved successful beyond expectations, and became the orbit of choice for telecommunications. Pierce, conceptual designer of the Telstar and Relay systems, even commented that in hindsight that it was remarkable that any other orbit had been considered.[11]

Comsat, representing the interests of US telecommunications companies, and the US Government, immediately set out to create an international consortium of countries that would participate in a non-governmental, non-profit satellite telecommunications international body, the International Telecommunications

Satellite Organization (Intelsat). The USSR, along with Eastern bloc countries, rebuffed the idea at the outset, while the Europeans warily joined, knowing that the restrictions on creating competing satellite telecommunications systems would keep them from developing their own regional networks for some time. Intelsat was created in 1964 and the first order of '60 voice-grade circuits' across the Atlantic was made by AT&T, beginning the first international commercial telecommunications via satellite.[12] Intelsat priced service independent of route, essentially using heavy traffic routes to subsidize lighter traffic ones, usually to developing states.[13] In this manner, Comsat and Intelsat balanced commercial and geopolitical interests to bring into existence a new model of telecommunications that has proven vital to the modern international community.

By the late 1980s and 1990s a world-wide trend had emerged toward deregulation of telecommunications and other industries. Industries with cost structures that had previously been seen as appropriate for regulated monopoly – electricity, natural gas, in addition to telecommunications – were now being reorganized to significantly increase the role of private markets. Riding this trend was Rene Anselmo, co-founder of PanAmSat, which in 1988 launched its first communications satellite. Anselmo was outraged that his company could not supply services to the public telephone companies of states who were signatories to the Intelsat consortium. As editors of one of the space industry's newspapers put it, 'with extraordinary vision, an in-your-face-style and the determination to do whatever it took to gain a foothold in the satellite business, Anselmo almost single handedly shattered Intelsat's global monopoly on satellite communication services'.[14]

The deregulatory trend and the persuasiveness of Anselmo led the US Congress to pass in 2000 the Open-market Reorganization for the Betterment of International Telecommunications Act (ORBIT Act). The ORBIT Act called for Intelsat to transfer its satellites and financial assets to private investors and become largely independent of former signatories, who were typically government-owned companies; Intelsat became a private company in 2001.[15] With this move, the architecture of international communications had fully evolved, from the success of commercial visionaries who pioneered early technological developments, to those who worked to establish and then later restructure the role of government's relationship to the industry.

COMMERCIAL SPACE TRANSPORTATION

Commercial space transportation is the second largest economic activity in space, by annual revenue, after communications satellites. Traditionally, launch services have been provided by large aerospace companies such as Lockheed, Martin Marietta (in 1994 these two companies merged to become Lockheed Martin), and Boeing. Orbital Sciences joined the market in 1990 with its air-launched rocket, Pegasus, and more recently, a host of new companies vying to break into the LEO launch market and potentially beyond, have pushed the frontier of space transportation services and government regulatory policy.

Big Players Face Uncertainty

The birth of commercial launch was an offshoot of the original purpose of heavy-lift rockets, developed for intercontinental ballistic missiles (ICBMs). Manufacturers sold heavy-lift rockets to NASA, the US Department of Defense (DoD) and eventually to commercial customers for satellite payloads. In 1972 the Nixon Administration proposed a new Space Transportation System, the Space Shuttle programme, as the nation's primary space transportation system for all civilian payloads. Threatened with cancellation by the Carter Administration in 1978, Congress passed a law designating the Space Shuttle as the 'sole provider of American civilian launch services'.[16] The new policy had the Shuttle launching government payloads that otherwise could have been launched commercially. In addition, the fee charged by NASA to commercial payload operators was much less than fees charged by commercial space transportation suppliers, enabling the Shuttle to capture this market as well.[17] Private companies that had been producing expendable launch vehicles (ELVs) had no choice but to begin phasing out their operations.

The US Air Force (USAF) was particularly uneasy with relying on the Space Shuttle alone for access to space and so under special dispensation was allowed to maintain a 'complementary ELV' programme of Titan launch vehicles for guaranteeing 'assured access to space'.[18] The need for a backup launch system to the Space Shuttle was brought home with the tragic loss of the Space Shuttle Challenger in January 1986. Combined with failures in two of the remaining ELV models later that year and the grounding of the Shuttle fleet, the USA effectively lost all access to space. Following the Challenger accident, the Space Shuttle fleet remained grounded for over two years, resulting in mission delays and cancellations, particularly for those 'shuttle-unique' payloads that had been engineered to fit the Shuttle's payload bay in compliance with the 1978 policy.

Faced with the stark consequences of relying on one space launch system, President Reagan issued a national directive to abandon the previous policy of the Shuttle's launch monopoly. The new directive sought a 'balanced mix of launchers' using ELVs and the Shuttle, and additionally prohibited the launching of 'commercial and foreign payloads' on the Shuttle, unless there were no other alternatives.[19] This opened the commercial launch market, removing the policy barrier to entry as well as a major economic barrier to entry, government-subsidized launch services on the Shuttle. As a result of the shift in policy, orders for Titan IV ELVs grew from '10 vehicles over five years to 23 over the same period'.[20] To support the new industry's financial success, Congress legislated a licensing process for commercial ELVs in the 1984 Commercial Space Launch Act and since then has evolved the legal regime through amendments; for example, the first set of amendments were legislated the 1988 Commercial

Space Launch Act Amendments providing that the US Government would indemnify space transportation suppliers for a portion of third-party claims associated with harm during launch and launch operations, and also specified an upper ceiling on the third-party insurance companies needed to purchase.[21]

New Space Pathbreakers

New regulatory challenges have emerged since the introduction of privately developed space transportation enterprises. In 2004 aerospace designer Burt Rutan and his Scaled Composites team successfully developed a crewed launch and re-entry vehicle, and reached suborbital altitude twice within two weeks, winning the $10m. Ansari X-Prize. Subsequently, in Congressional testimony, Rutan urged the Federal Aviation Administration (FAA) to create 'a proper research environment to allow innovation', particularly by avoiding over-regulating the industry.[22] To balance the safety concerns with the need to let the industry innovate, Congress gave the FAA 'authority to begin regulating for passenger safety in eight years', or if a serious accident occurs before then.[23] As new players test new systems, the regulatory framework for conducting such innovation will become increasingly important. Along with Rutan, a spate of other small and pioneering companies has emerged to attempt a revolution in lowering costs and increasing reliability of accessing space.

A more ambitious goal is to reach orbital altitudes, at which point government customers would pay for space transportation services from satellite delivery, to replenishing, or even servicing the International Space Station (ISS). These ventures benefit from the capital of 'angel investors' such as Microsoft's Paul Allen and the Virgin Group's Richard Branson in the case of Scaled Composites. Branson's Virgin Group established Virgin Galactic, which plans to launch space tourists to suborbital altitudes, and is making plans to launch satellites of 50kg–100kg into LEO using an unmanned rocket for less than $2.5m.[24] Elon Musk, inventor of PayPal, founded his own space enterprise, Space Explorations Technology, or SpaceX, with the explicit goal of reaching LEO and lofting satellites, re-supplying ISS through the NASA Commercial Orbital Transportation Services (COTS) programme, and carrying commercial space flight passengers, for less than half the cost of current space transportation.[25] As of 2010, SpaceX had achieved success with its Falcon 9 heavy-lift launch vehicle, and secured a $1.6 bn. contract commitment from NASA to deliver supplies to the ISS. Commercial customers, such as Iridium, have followed with orders of their own. The retirement of the STS programme and shift in US Administration policy toward relying on commercial space flight providers has created more market space for private contractor companies, like SpaceX.

As the field of potential space transportation suppliers grows, so does the potential for innovative breakthroughs in engineering and in business models. The challenge to policymakers, engineers and investors alike is to have the vision to navigate high costs, risky systems and regulatory roadblocks. Success in this navigation will provide the 'new space' industry the chance to succeed in making access to space routine and affordable.

COMMERCIAL REMOTE SENSING SATELLITES

Space-based remote sensing is the business of capturing information – images in photographic or other form – from the unique vantage point of space, for use in meaningful ways back on Earth. In the late 1950s remote sensing had its origin in one of the most carefully guarded of strategic businesses, military reconnaissance. The first remote sensing satellites were the Corona series of 'keyhole' spy satellites. These beginnings in military secrecy and classified information foreshadowed a long and difficult road before remote sensing would become a commercial enterprise. In the half century since the first remote sensing satellite was launched, the work of industry, government and technical visionaries, helped along by an information revolution, would be required to open up new markets and revolutionize the way people understood their geospatial planet.

A Rocky Start

The first civilian remote sensing satellite programme was a collaborative project led by NASA, in consultation with the DoD. Agencies such as the US Department of Agriculture were also involved in the development of what would become the Land Remote Sensing Satellite (Landsat) programme. However, the DoD was wary of a civilian programme that might draw unwanted attention to the ongoing reconnaissance satellite programme, and discussions with NASA on the appropriate limits on resolution for the civilian versions were an issue of contention. Delays in the interagency process frustrated many of the analysts who wanted satellite imagery to support their work, especially as new imaging technologies were becoming available.

In 1966 Dr William Pecora, Director of the US Geological Survey (USGS), was growing increasingly frustrated with the delays. He proposed to jumpstart the process and suggested what amounted to a bureaucratic coup. He announced the US Department of the Interior's own satellite-based Earth observation programme, despite the fact that Interior had never run a satellite programme before or had the budget to develop one on its own.[27] However, the Secretary of the Interior agreed, and the news conference that launched the Earth Resources Observation Satellites (EROS) programme was enough to provoke NASA into action. After several more years of bureaucratic wrangling, the first Landsat satellite was launched in 1972 and immediately began providing data with detailed imagery.

In Search of a Market

While many Landsat users in government and industry understood the value of these new images to land-use management and other applications, the wider commercialization of Landsat data was not immediately forthcoming. Delays were due in part to the inadequacies of the data processing and distribution system. The vast amounts of data that were being downloaded to ground stations in Alaska and California had to be mailed to NASA's Goddard Space Flight Center to be processed, often taking 'twenty to forty days' before the images became available to users. For many applications, such as real-time monitoring of snow runoff, range-land conditions, or ice in shipping lanes, the time factor was crucial.[28] The limitations of the data processing system and the information infrastructure caused 'perishable' information goods to quickly lose their value. Landsat was an information resource ahead of its time. A convergence of policy reform, technological revolution and entrepreneurial vision in the 1990s would spur a new generation of commercial imagery satellites.

Geopolitical forces would prove key in developing the commercial imagery market. In the late 1980s and early 1990s the Russians, strapped for foreign currency, began selling imagery from their remote sensing assets, alarming the DoD.[29] The thaw in the Cold War opened up a race for market share of the commercial imaging business.

Operations Desert Shield and Desert Storm showcased the US military's use of defence and civilian space assets to achieve a 'force multiplier' effect. In the run-up to war, civilian remote sensing satellites provided detailed imagery for producing and updating maps of the region, critical for operational planning and search-and-rescue missions. The satellite operators also co-ordinated with US force commanders to co-ordinate their assets to complement military reconnaissance assets, imaging specific targets more frequently.[30] The thaw in the Cold War opened up a race for market share of the commercial imaging business.

These international developments underscored how important the commercial space segment had become. In the final months of his term, President George H. W. Bush, as part of the space legacy of his Presidency, signed the Land Remote Sensing Act of 1992. This policy opened up satellite remote sensing to commercial licensees and several companies launched new satellite programmes to take advantage of the opportunity. These included Space Imaging Ikonos series (1994) and Orbital Science Orbview series (1995), which merged to form GeoEye in 2007, and DigitalGlobe Quick Bird (2001).[31] The infrastructure was now in place in the USA to produce high-quality commercial images, though pricing and distribution models, as well as value-added products and applications, were still evolving.[32]

Revolution

In the 1990s an information revolution was occurring, a combination of advances in network, processing and storage capacity, the rapid development of the internet and, in many sectors, the growing use of geographic information systems (GIS). Two such innovators of this revolution founded DigitalGlobe.

Commercial Space Actors

In 1992 Walter Scott, formerly of Lawrence Livermore National Labs, and Doug Gerull, former executive vice-president for GIS at Intergraph, recognized that the market for GIS was expanding rapidly and believed the demand for the underlying remote sensing information would also expand.[33] The company formed by Scott and Gerull would be a supplier of such satellite imagery. At that time, and continuing to the present, tension between national security concerns and commercial applications surrounded the resolution, i.e. level of detail in spatial, spectral or temporal qualities, to be permitted in the issuing of licences for commercial remote sensing. Initially, resolution restrictions in the USA were much tighter than those allowed to be practised by the commercial systems of other states. This has changed recently with the rapid advance of the high-resolution commercial remote sensing sector. As of 2008 the US Government has authorized spatial resolutions as low as 0.25m.

A related issue, not unlike the challenge presented to the commercial space transportation market when Space Shuttle pricing undercut commercial services, has been the pricing of remote sensing imagery. Like most information products, the reproduction of multiple copies of remote sensing information is quite inexpensive. The supply of Landsat data, for example, has historically been close to 'the cost of reproduction' without any attempt to recoup amortized costs of the spacecraft and the other equipment used to acquire the imagery. This pricing policy inhibited the early days of attempts by the commercial market to establish a profitable business that properly accounted for amortizing the investment in the space assets.

Obtaining licences, planning for a competitive price and, then, acquiring the imagery itself would be the first steps in the production chain; the next steps would be processing and presenting the data. In 2001 John Hanke co-founded a company to do exactly that. Keyhole's flagship product was a programme called Earth Viewer, which streamed satellite images from Keyhole servers to a desktop interface that was a virtual earth, giving users the illusion of flying anywhere on the Earth. The revolutionary idea in Earth Viewer was to allow users to mark places of interest to them using a simple descriptive language called keyhole markup language (KML).[34] The potential of Earth Viewer and KML would attract the attention of other innovative entrepreneurs, specifically the founders of Google.

Google bought Keyhole in 2004 shortly after Google co-founder Sergey Brin discovered Earth Viewer.[35] The programme was repackaged as Google Earth and distributed through free download with subscriptions for upgraded versions. Mapping products had been around for at least a decade when Google Earth was published, and other satellite imagery systems were already on offer for free, including Microsoft Corporation's TerraServer, unveiled in 1998. However, the ground-breaking vision behind Earth Viewer and Google Earth bears striking similarity to the revolutionary potential of the first web browsers, a decade before. KML, like the hypertext markup language (HTML) on which the web is built, is flexible and easy to use, giving users the ability to add information, overlay content and engage the world in new ways.

Apart from the novelty of finding one's own house from space, amateur photo interpreters have raised the ire of foreign governments, as in the case of the People's Republic of China, by identifying a secret tank training base previously hidden out of sight in inland deserts and the secret latest-model submarines docked in harbour. Other communities of enthusiasts and professionals alike have taken to annotating the virtual world in as many ways as there are interests.

In conjunction with navigation and positional services like the Global Positioning System (GPS), the potential for applications of remote sensing data seems quite large. Planetary browsers, like Google Earth, TerraServer and Microsoft's Virtual Earth are on an evolutionary path towards a more complete virtual representation of the Earth, tying the world of the internet to the world of streets and mountain trails, all through the view of satellites. These commercial developments have proceeded apace and on a tack increasingly differentiated from Landsat. Meanwhile, the US Government is planning its next generation of Landsat spacecraft, called the Landsat Data Continuity Mission (Landsat 5 and 7 were operational as of 2009, although both are functioning in degraded capacity), in response to a community of researchers and other experts who argue the need for long-term continuity of the oldest series of publicly provided remote sensing data.

GLOBAL SATELLITE POSITIONING

Since ancient times, humans have relied on the regular and predictable patterns of the sun and stars to navigate the globe. Following those same principles, the invention of radio and satellites allowed for new navigation technologies, and eventually GPS. The US policy of opening the GPS signal as a public utility to commercial and civilian users translated into a booming commercial industry for positioning, navigation and timing services.

Military Origins

Many scientific and technological visionaries contributed to the system that would eventually be known as GPS. Beginning in the 1920s, radio beacons were used as navigation guides for ships and airplanes. These radionavigation techniques proved useful for military purposes during the Second World War, to guiding bombers to their targets, to land in poor weather or at night and for ship navigation. After the war, terrestrial systems, like the Long Range Navigation (LORAN) system in the North Atlantic, implemented more sophisticated time differential signals for navigation in coastal waters, and eventually across the continent.

While tracking Sputnik in 1957, William Guier and George Weiffenbach at The Johns Hopkins University Applied Physics Laboratory (APL) observed that the Doppler shift of the satellite passing overhead could be used to calculate the satellite's orbital path accurately.[36] This scientific insight, reasoned APL's Director, Frank McClure, could be applied in reverse. By observing the signals of a satellite in a known orbit, the observer's ground location could be calculated.[37] This concept had immediate applicability to the US Navy for locating ballistic missile submarines and APL was commissioned to begin development of such a space-based navigation system.

Richard Kershner led the design of the first operational satellite navigation system, Transit. The system consisted of a series of polar orbiting satellites that gave positional fixes to within 200m accuracy. The first successful Transit launch occurred in 1960 and civilians were allowed access to the system in 1967.[38] The widespread use of Transit receivers by pleasure craft and commercial vessels alike was a foreshadowing of the popular use of satellite navigation for civilian purposes.

As the armed services began to realize the utility of satellite-based positioning and navigation services, each service sponsored its own programme. Timation, a second Navy satellite radionavigation programme, began in 1964 under the direction of Roger Easton. Timation perfected the use of high precision clocks, which greatly improved the prediction of satellite orbits, and was to include 21–27 satellites in MEO, eight-hour orbits.[39] In the meantime, the USAF was developing its own radionavigation satellite system, System 621B. The Aerospace Corporation, under a project led by Ivan Getting, brought new innovations to satellite navigation. System 621B would include four or five clusters of satellites in GEO, configured to provide overlapping hemispherical coverage. While this feature would not be included in GPS, the use of a more robust jam-resistant 'pseudorandom noise' signal and the ability to track altitude were adopted from System 621B.[40] The US Army was simultaneously developing its own system, called Sequential Correlation of Range (SECOR). Eventually, a joint task force reconciled the competing systems and using the best features of each, designed an operational navigation system for the military.

In 1968 the Navigation Satellite Executive Group (NAVSEG) was established as the steering committee responsible for co-ordinating the military effort.[41] After years of testing and development, the best design features of the Navy and Air Force systems were selected and incorporated into the final design of the GPS satellites. In 1974 two refurbished Timation satellites were launched under the GPS programme, carrying the first atomic clocks into space.[42] In addition to the navigation payload, GPS also includes DoD nuclear detonation detectors as part of the Comprehensive Test Ban Treaty verification programme.[43] The final system plan included 24 satellites and three on-orbit spares, in MEO.

Global Utility

Throughout the design and deployment of the GPS system, the military was the foremost customer in mind. Military contractors produced GPS receivers for airplanes, ships and ground vehicles. Commercial deployment of GPS was a result of events involving the military uses of GPS and the vision to offer GPS as a global utility.[44]

In 1983 a Korean Airlines passenger jet, Flight 007, inadvertently ventured into Soviet airspace and was shot down by Soviet interceptors, killing all on board. The airliner had failed to make navigation checks that would have informed the flight crew that they had strayed from the planned course and into Soviet territory. President Reagan, in response to the incident, announced that the GPS signal would be made available to civilian users as soon as the system was fully operational.[45] President Reagan's announcement opened a new horizon of potential applications for GPS and even before all the satellites were in place, land surveyors began using the GPS Standard Positioning Service signal to good effect.

The 1990–91 military operations of Desert Shield and Desert Storm in the Persian Gulf showcased the utility of GPS and precision navigation. In preparation for the invasion, demand for GPS receivers outstripped the capacity for military production, compelling the DoD to purchase civilian receivers for the troops who needed them. As a result, the Selective Availability 'degradation feature' of GPS was deactivated for the duration of the war, providing a magnitude improvement in accuracy from 100m to tens of metres to all users for the first time.

The experience of full access to the most accurate signal spurred the debate on eliminating Selective Availability altogether. The US Coast Guard had already been experimenting with Differential GPS systems to counter the artificial errors introduced into the signal by Selective Availability.[46] The FAA also argued forcefully for access to the non-degraded signal. In 2000 President Clinton terminated the use of Selective Availability.[47]

The explosion in GPS commercial technology that followed could hardly have been predicted. The incorporation of GPS receivers in cell phones was originally mandated to comply with emergency services policy, but has become one of the telecommunications industry's most promising growth areas.[48] The growth of automotive telematics – the use of information technologies in cars – has also proven to be a lucrative market for GPS manufacturers. From missiles and airplanes, to farmers and hikers, the potential applications of GPS grow exponentially.

CONCLUSIONS: PIONEERING SPACE COMMERCE

Technological innovation – such as conceiving of GEO for telecommunications, observing a Doppler effect and reversing it to calculate ground location for navigation – has been remarkable in the history of commercial space. Yet these advances have been insufficient as conditions for commercial success. As important has been the willingness of commercial and policy visionaries to push the frontier of public policy, given that space has been an industry dominated by government as gatekeeper in access to space and as regulator in use of space.

Government subsidies, whether of payload launches by the Space Shuttle or of Landsat data, have at times made it difficult for the commercial sector to compete. The timing of transforming sectors of space industry from government dominance to market competition has been driven largely by events, rather than careful policy planning. Initial restrictions on the resolution of commercial remote sensing data and on the quality of GPS signals, based on national security concerns slowed the development of nascent markets. More recently, over the last decade, the situation has changed and now the US Government more actively and effectively fosters space commerce development, often times driven by individual actors and firms who push the technological envelope. This is exemplified by the lessening of restrictions on remote sensing spatial resolutions, with the free access to the high-precision GPS signal and liberalization of space launch procurement policies. At the same time, the space commerce sector is not yet a mature industry, with government support at some level remaining necessary to sustain commercial activities (see Illustrations and Documentation: Figure 5.1 Risk to Commercial Return Relationship).

Transitioning from an era of government-dominated space to a balance between private- and public-sector interests is likely to remain a challenge driven more by events than by farsighted policy. It may well be that the next generation of space policy continues to rely not on lessons learned, but rather on individual champions expanding both regulatory and technological frontiers. The new space actors highlighted herein have begun to push these boundaries creating ever expanding room for commercial enterprise to flourish in the new frontier.

NOTES

1 George P. Osling, *The Story of Telecommunications* (Macon, GA: Mercer University Press, 1992), 390.
2 Andrew J. Butrica, ed., *Beyond the Ionosphere: Fifty Years of Satellite Communication* (Washington, DC: National Aeronautics and Space Administration (NASA), 1997), 4 and 293.
3 John L. McLucas, *Space Commerce* (Cambridge, MA: Harvard University Press, 1991), 18.
4 Walter A. McDougall, ... *The Heavens and the Earth: A Political History of the Space Age* (New York, NY: Basic Books, 1985), 353.
5 John L. McLucas, *Space Commerce* (Cambridge, MA: Harvard University Press, 1991), 23.
6 Ibid., 28.
7 Stojče Dimov Ilčev, *Global Mobile Satellite Communications: For Maritime, Land and Aeronautical Applications* (Dordrecht, The Netherlands: Springer, 2005), 85.
8 'Letter to the Vice President on the Need for Developing Operational Communications Satellites', The Public Papers of John F. Kennedy 1961, written 15 June 1961, released 24 June 1961, www.jfklink.com/speeches/jfk/publicpapers/1961/jfk254_61.html (accessed 3 February 2008).
9 John L. McLucas, *Space Commerce* (Cambridge, MA: Harvard University Press, 1991), 40.
10 Ibid., 36.
11 Walter A. McDougall, ... *The Heavens and the Earth: A Political History of the Space Age* (New York: Basic Books, 1985), 353.

12 John L. McLucas, *Space Commerce* (Cambridge, MA: Harvard University Press, 1991), 42.
13 Don I. Dagleish, *An Introduction to Satellite Communications* (London: Institute of Engineering and Technology, 1989), 23.
14 'The Top Ten', Space News International 15th Anniversary edition, www.space.com/news/space_news100–101.html (accessed 15 March 2008).
15 For additional discussion, see US General Accountability Office, 'Intelsat Privatization and the Implementation of the ORBIT Act', GAO-04-891, September 2004.
16 Bonnie Fought, 'Legal Aspects of the Commercialization of Space Transportation Systems', *Berkeley Technology Law Journal*, 3: 1 (1988), www.law.berkeley.edu/journals/btlj/articles/vol3/fought.html (accessed 5 November 2008).
17 Numerous analyses argued that Space Shuttle fees were well below the actual cost of operations and amortization; for example, see Michael A. Toman and Molly K. Macauley, 'No Free Launch: Efficient Space Transportation Pricing', *Land Economics*, 65: 2 (May 1989): 91–99; and US Congressional Budget Office, 'Pricing Options for the Space Shuttle', Washington, DC, March 1985.
18 Carl E. Behrens, 'Space Launch Vehicles: Government Activities, Commercial Competition, and Satellite Exports', *CRS Issue Brief for Congress*, updated 20 March 2006, www.fas.org/sgp/crs/space/IB93062.pdf (accessed 5 November 2008).
19 Fact Sheet: United States Launch Strategy, National Security Decision Directive 254, 27 December 1986, www.fas.org/irp/offdocs/nsdd/nsdd-254.htm (accessed 5 November 2008).
20 'Setting space transportation policy next term for the 1990s', *Space Policy*, 3: 1 (February 1987): 13–16.
21 Originally for five years, the indemnification provision has been renewed several times and next expires in December 2009. See also Carl E. Behrens, 'Space Launch Vehicles: Government Activities, Commercial Competition, and Satellite Exports', *CRS Issue Brief for Congress*, updated 20 March 2006, www.fas.org/sgp/crs/space/IB93062.pdf (accessed 5 November 2008) and Timothy J. Brennan, Carolyn Kousky, and Molly Macauley, "Public Private Co-Production of Risk: Government Indemnification of the Commercial Launch Industry," *Risk, Hazards and Crisis in Public Policy* 1(1), 2010, 7–24.
22 Leonard David, 'Rutan Blasts FAA's Suborbital Safety Regulations', *Space News*, 25 April 2005, www.space.com/spacenews/archive05/Rutan)042505.html (accessed 5 November 2008).
23 Erica Werner, 'FAA Issues Rules for Space Tour Operators', Associated Press, 15 December 2006, www.msnbc.msn.com/id/16229923 (accessed 24 February 2008).
24 'Starship Enterprise: The Next Generation', *The Economist*, 24 January 2008, www.economist.com/science/displaystory.cfm?story_id=10566293 (accessed 24 February 2008).
25 Carl Hoffman, 'The New Space Race', *Wired*, June 2007, 150.
26 Anton Gonsalves, 'Iridium Chooses SpaceX To Launch Satellites,' Information Week, 17 June 2010, www.informationweek.com/news/storage/data_protection/showArticle.jhtml?articleID=225700418&cid=RSSfeed_IWK_News (accessed June 17, 2010).
27 Pamela E. Mack, *Viewing the Earth: The Social Construction of the Landsat Satellite System* (Cambridge, MA: The MIT Press, 1990), 61.
28 Ibid., 115.
29 Nicholas M. Short, 'The Commercialization of Remote Sensing', *The Remote Sensing Tutorial*, EOS-Goddard Program Office, 2005, www.fas.org/irp/imint/docs/rst/Intro/Part2_26f.html (accessed 18 October 2008).
30 Jon Trux, 'Desert Storm: A Space-age War', *New Scientist*, 27 July 1991, 30, www.newscientist.com/article/mg13117794.900.html (accessed 18 October 2008).

31 T. A. Heppenheimer, 'Operational Remote Sensing Satellites', *US Centennial of Flight Commission*, www.centennialofflight.gov/essay/SPACEFLIGHT/remote_sensing/SP36.htm (accessed 18 October 2008).
32 Oliver Morton, 'Private Spy', *Wired*, Issue 5.08, August 1997, www.wired.com/wired/archive/5.08/spy_pr.html (accessed 18 October 2008).
33 Ibid.
34 Evan Ratliff, 'Google Maps is Changing the Way We See the World', *Wired*, Issue 15.07, June 2007, www.wired.com/techbiz/it/magazine/15-07/ff_maps?currentPage=all (accessed 18 October 2008).
35 Ibid.
36 Mike Gruntman, *Blazing the Trail: The Early History of Spacecraft and Rocketry* (Reston, VA: AIAA, 2004), 264.
37 Scott Pace et al., *The Global Positioning System: Assessing National Policies* (Santa Monica: RAND, 1995), 238.
38 Mike Gruntman, *Blazing the Trail: The Early History of Spacecraft and Rocketry* (Reston, VA: AIAA, 2004), 264.
39 Scott Pace et al., *The Global Positioning System: Assessing National Policies* (Santa Monica, CA: RAND, 1995), 239; and J. Rosalanka, 'Appendix A: The Modern History of Space', *The Remote Sensing Tutorial* (EOS-Goddard Program Office, 2005), ftp.fas.org/irp/imint/docs/rst/AppA/Part1_10.html (accessed 25 October 2008).
40 Scott Pace et al., *The Global Positioning System: Assessing National Policies* (Santa Monica, CA: RAND, 1995), 239.
41 Ibid., 240.
42 Ibid., 241.
43 Ibid.
44 'Statement by the President Regarding the United States' Decision to Stop Degrading Global Positioning System Accuracy', Office of Science and Technology Policy, Executive Office of the President, 1 May 2000, www.ostp.gov/html/0053_2.html (accessed 25 October 2008).
45 'United States Updates Global Positioning System Technology', US International Information Programs, US Department of State, usinfo.state.gov/xarchives/display.html?p=washfile-english&y=2006&m=February&x=20060203125928lcnirellep0.5061609 (accessed 25 October 2008).
46 Scott Pace et al., *The Global Positioning System: Assessing National Policies* (Santa Monica: RAND, 1995), 244.
47 'Statement by the President Regarding the United States' Decision to Stop Degrading Global Positioning System Accuracy', Office of Science and Technology Policy, Executive Office of the President, 1 May 2000, www.ostp.gov/html/0053_2.html (accessed 25 October 2008).
48 'Location, location, location', *The Economist*, 4 October 2007, www.economist.com/business/displaystory.cfm?story_id=9916519 (accessed 25 October 2008).

International Organizations in Civil Space Affairs

HENRY R. HERTZFELD*

International organizations (IOs) are important components of the political, economic and social environment of space activities and space policy. This chapter explores the extent and type of IOs with a significant impact on civil space affairs.[1] IOs reviewed here include the United Nations (UN), European Space Agency (ESA), European Union (EU), Asia-Pacific Space Cooperation Organization (APSCO), UN Specialized Agencies dealing with space, and other non-UN international space organizations that focus on space. For each organization covered, the discussion reviews mission, membership, general programme of activities and accomplishments in space policy and law.

UNITED NATIONS

The UN system is structured around a number of principal organs, which encompass the Trusteeship Council, Security Council, General Assembly, Economic and Social Council, International Court of Justice, and Secretariat.[2] The General Assembly is the main deliberative assembly of the UN. The first session met in 1946 and included representatives of 51 nations. Today, the General Assembly convenes all UN member states on an annual basis under a President elected from among the member states. When the General Assembly votes on issues such as budgetary matters, admission, suspension and expulsion of members, or questions of peace and security, a two-thirds majority of those present is required to pass the vote; all other issues are determined by majority vote. Each member state of the UN has one vote. The General Assembly is divided into the following categories: committees (30 total, six main); commissions (seven); boards (six); councils and panels (five); working groups; and 'other'. The main committees are: Disarmament and International Security (DISEC); Economic and Financial (ECOFIN); Social, Humanitarian and Cultural (SOCHUM); Special Political and Decolonization (SPECPOL); Administrative and Budgetary; and Legal.

United Nations in Space Affairs

The major bodies within the UN that have important functions related to outer space affairs are reviewed in this section (see Illustrations and Documentation: Table 6.1 Functions of Selected Bodies in the United Nations). Space-related organizations within the UN are located under the General

Assembly and the Economic and Social Council. The primary UN body that deals with space, the United Nations Committee on the Peaceful Uses of Outer Space (COPUOS), is a member of the SPECPOL committee.

The UN agencies that oversee specific space issues include: International Civil Aviation Organization (ICAO); International Telecommunication Union (ITU); and the World Meteorological Organization (WMO). UNESCO makes use of space applications and has become more involved in space issues in recent years.[3] These organizations are all Specialized Agencies, defined as autonomous organizations working with the UN and each other through the co-ordinating machinery of the Chief Executives Board for Coordination (CEB) at the Inter-Secretariat level and through the Economic and Social Council (ECOSOC) at the intergovernmental level.[4] ECOSOC assists the General Assembly in promoting international economic and social co-operation and development. ECOSOC has 54 members, all of whom are elected by the General Assembly for a three-year term. ECOSOC provides policy coherence to co-ordinate the overlapping functions of the UN's subsidiary bodies through its information gathering and advising of member nations.

United Nations Committee on the Peaceful Uses of Outer Space

COPUOS is the primary body in the UN that works on international space issues. It was formed as an ad hoc committee in December 1958, just over a year after the beginning of the space age with the launch of Sputnik 1. COPOUS had 18 members at the time of its formation. It became a permanent body of the UN in 1959. Since then, COPOUS has grown to 69 members and is one of the largest committees in the UN, with the UN Office for Outer Space Affairs serving as its Secretariat. One of its most important guiding principles, incorporated into the first sentence of the resolution that formally established COPOUS (adopted by the UN General Assembly on 2 December 1959), is that space should be used for peaceful purposes and for the benefits of mankind ((see Illustrations and Documentation: Documentation 6.1 International Cooperation in the Peaceful Uses of Outer Space).

COPUOS is the central forum to discuss the legal and scientific/technical aspects of maintaining peace in outer space, and to conceive and implement global regulation in this field.[5] The purpose of COPUOS is to review the scope of international co-operation in the peaceful uses of outer space, to devise programmes in this field to be undertaken under UN auspices, to encourage continued research and the dissemination of information on outer space matters, and to study legal problems arising from the exploration of outer space. The organization has two subcommittees, the Legal Subcommittee and the Scientific and Technical Subcommittee. Decisions are made by consensus to achieve the most acceptable decision for the entire group.[6]

COPUOS has had major achievements since its establishment and its work on universal space principles is the most far-reaching and effective treatment of the subject. As the only forum for the development of international space law, COPUOS has concluded five Space Treaties and a number of principles governing space activities adopted by the General Assembly ((see Illustrations and Documentation: Table 6.2 United Nations Treaties and Resolutions of Space Law).[7] The Treaties include: Treaty on Principles Governing the Activities of States in the Exploration and Use of Outer Space, including the Moon and other Celestial Bodies;[8] Agreement on the Rescue of Astronauts, the Return of Astronauts and the Return of Objects Launched into Outer Space;[9] Convention on International Liability for Damage Caused by Space Objects;[10] Convention on Registration of Objects Launched into Outer Space;[11] and Agreement Governing the Activities of States on the Moon and Other Celestial Bodies.[12]

The Principles entail: The Declaration of Legal Principles Governing the Activities of States in the Exploration and Uses of Outer Space;[13] The Principles Governing the Use by States of Artificial Earth Satellites for International Direct Television Broadcasting;[14] The Principles Relating to Remote Sensing of the Earth from Outer Space;[15] The Principles Relevant to the Use of Nuclear Power Sources in Outer Space;[16] The Declaration on International Cooperation in the Exploration and Use of Outer Space for the Benefit and in the Interest of All States, Taking into Particular Account the Needs of Developing Countries;[17] Application of the Concept of the 'Launching State';[18] and Recommendations on Enhancing the Practice of States and International Intergovernmental Organizations in Registering Space Objects.[19] COPOUS also adopted debris mitigation guidelines in 2007 (see Illustrations and Documentation: Documentation 6.10 on the Inter-Agency Space Debris Coordinating Committee—IADC).

A registry of launchings has been maintained by the UN Secretariat since 1962 in accordance with General Assembly resolutions. Since the Convention on Registration of Objects Launched into Outer Space came into force in 1976, another registry of launchings has been established for information received from Member States and intergovernmental organizations that are parties to the Convention. Finally, COPUOS has led three major conferences on the peaceful uses and exploration of outer space: UNISPACE I in 1968, UNISPACE II in 1982 and UNISPACE III in 1999. UNISPACE III established 12 Action Teams to address the recommendations of the Conference and be reviewed by the General Assembly on the basis of their progress and effectiveness in 2004.[20] These Action Teams focused on a number of different issues, including environmental monitoring strategy, management of natural resources, weather and climate monitoring, public health, disaster management, knowledge sharing, the Global Navigation Satellite System (GNSS), sustainable development, near Earth objects (NEOs), capacity building, awareness increase and innovative funding sources.[21]

The issues highlighted by the Action Teams establish the outline of the Committee's present and future agenda. COPUOS recently established both the International Committee on GNSS and the United Nations Platform for Space-based Information for Disaster Management and Emergency Response

International Organizations in Civil Space Affairs

(UN-SPIDER). The Scientific and Technical Subcommittee of COPUOS established a Working Group on the Use of Nuclear Power Sources in Outer Space to aid the International Atomic Emergency Agency with its goal of developing a set of safety standards for future nuclear power sources in outer space by the year 2010. The Scientific and Technical Subcommittee also established a Working Group on Space Debris[22] to collaborate with the IADC.

Currently (2008 and 2009), the Scientific and Technical Subcommittee is focused on a variety of topics, including space debris, space system-based disaster management support, recent developments in global navigation satellite systems, the use of nuclear power sources in outer space, NEOs and an examination of the physical nature and technical attributes of the geostationary Earth orbit (GEO), and utilization and applications. The Legal Subcommittee is considering a review and possible revision of the Principles Relevant to the Use of Nuclear Power Sources in Outer Space, an examination and review of the developments concerning the draft protocol on matters specific to space assets, capacity building in space, and the legal implications of space applications for climate change. For example, the 51st session of COPUOS, held in 2008, discussed topics that included maintaining outer space for peaceful purposes and ensuring that space technology applications continue to be used in many areas critical to all humanity, like disaster management, climate change, food security and education.[23]

MULTINATIONAL SPACE ORGANIZATIONS

There are three major multinational organizations that specifically focus on space activities: the ESA, EU and APSCO (see Illustrations and Documentation: Table 6.3 Regional Space Organizations). The ESA promotes co-operation among European states in space research and technology and their space applications, with a view to their being used for scientific purposes and for operational space applications systems. The EU plays a role in formulating overall space policy for European states that are members, though not all members of ESA are members of the EU, nor is there a one-to-one correspondence between the members of the EU and ESA. The EU, beyond its policy initiatives in space, funds research and development (R&D) in space applications through its R&D programmes. APSCO seeks to promote the peaceful uses of outer space in the Asia-Pacific region through attention to space science, technology, education, training and co-operative research.

European Space Agency

Since its founding in 1975, the ESA has served as Europe's 'gateway to space'. The ESA shapes the development of Europe's space capabilities to ensure that investment in space continues to deliver benefits to the citizens of Europe and the world. The ESA co-ordinates and uses the financial and intellectual resources of its member states[24] to design and undertake programmes and activities that would be beyond the scope of a single European country.[25] The preamble of the ESA Convention of 1975 emphasizes the ESA's role as co-ordinating and integrating a European space programme (see Illustrations and Documentation: Documentation 6.2 European Space Agency).

The ESA designs and implements programmes in a range of research areas to learn more about Earth, the Earth's immediate space environment, the solar system and the universe. The ESA also works to develop satellite-based technologies and services, and to promote European industries.[26] The specific programmes under ESA auspices cover the full spectrum of space activities.

- Advanced concepts – cutting-edge space R&D
- Aurora – human exploration of space, including the Moon and Mars
- Future Launchers – development of R&D for a next-generation launcher to be operational in around 2020
- Galileo – a global positioning and navigation satellite system
- General Studies – an internal think tank
- Living Planet Programme – Earth observations, Global Monitoring for Environment and Security (GMES) system
- Science – development of new space science instruments
- Technology R&D – systems engineering and technology demonstrations
- Telecommunications – new system concepts, standardization

From its headquarters in Paris, France,[27] the ESA works closely with space organizations outside Europe, and maintains liaison offices in Belgium, the USA and Russia, in addition to a launch base in French Guiana and ground and tracking stations throughout the world. The ESA is governed by the ESA Council, which provides the policy guidelines within which the Agency develops its space programme. Each Member State has one vote on the Council, regardless of its size or financial contribution.[28] The ESA's 'mandatory activities' – space science programmes and the general budget – are funded by financial contributions from each Member State, calculated on the basis of each country's gross national product (GNP).[29] The ESA operates on the principle of geographic return, by which it invests in each Member State, through industrial contracts for space programmes, an amount approximately equivalent to each country's contribution. The ESA also operates a number of optional programmes. Each Member State may decide in which optional programmes to become involved and how much to contribute.[30]

European Union

The EU is an economic and political partnership among democratic European countries, which aims to foster peace, prosperity and freedom for its citizens. The main agencies of the EU are: the European Parliament (representing the people of Europe); the Council of the European Union (representing national governments); and the European Commission (representing the common EU interest). Since the signing of the 'Treaty on European Union' in 1993, some of the goals towards which the EU has worked include: frontier-free travel and trade; the implementation of the euro (the single European currency); safer food; a 'greener' environment; better living standards in poorer regions of Europe; joint action on crime and terrorism; and less expensive phone calls and air travel between EU Member States.[31]

As the EU has expanded its domain from the economic to the scientific and cultural, it has become more involved in matters of European space policy. With this expansion of EU interest, a number of steps have been taken to strengthen the relationship between the ESA and the EU, including the establishment of the European Commission-ESA Framework Agreement (2003), and the launching of two joint European space projects: a global navigation system called Galileo and the GMES system.[32] Space has been included in both the proposed European Constitution and the European Treaty. Although the people of Europe have endorsed neither document as yet (2009), space remains a high priority for the EU. In addition to the R&D sponsored by the ESA, space applications have been funded annually as part of the EU's Framework Programme in R&D.

In 2007 29 European countries endorsed a new European Space Policy created to unify the space approaches of the ESA, EU and the individual Member States of each IO. The document was jointly drafted by the European Commission and the ESA (see Illustrations and Documentation: Documentation 6.3 European Space Policy). The goal of the document was to create a basic vision and strategy for the space sector to allow Europe to tackle together issues of security, defence, access to space and human space exploration. The document commits the EU, ESA and their respective Member States to increase co-ordination and co-operation between the two organizations.[33]

Although the future organization of space in Europe remains undecided, the EU has assumed a role in representing Europe in a number of international matters including antitrust, trade, telecommunications, safety and security. The ESA continues to be the primary space R&D organization, alongside the research laboratories of the nations of Europe. Each state in Europe, as signatories of the various UN Space Treaties, retains sovereign control of its territorial rights that cover many space related issues, such as property rights, taxes, overflight and national security. This multitiered system adds to the complexity of the politics of space in Europe.

Asia-Pacific Space Cooperation Organization

APSCO is a regional organization for multilateral space co-operation to enable Asia-Pacific countries to benefit from each others' strengths and help address the technological and financial challenges to their space causes. APSCO grew out of the Asia-Pacific Multilateral Cooperation in Space Technology and Applications, jointly proposed by the People's Republic of China, Pakistan and Thailand in 1992, which evolved over time to lead to the establishment of a more formal co-operative IO. The eight original signatories of the APSCO Convention (see Illustrations and Documentation: Documentation 6.4 Asia-Pacific Space Cooperation Organization), which came into force in 2005, are China, Bangladesh, Indonesia, Iran, Mongolia, Pakistan, Peru and Thailand. Turkey signed the Convention in 2006.[34] APSCO headquarters are located in Beijing, China and host nation China funded APSCO until the end of 2006; other Member States did not have to contribute financially until 2007.[35] Each Member State has one vote on the Council, the highest decision-making body of the organization.[36]

The fields of co-operation, as defined by the APSCO Convention include: space technology and programmes of its applications; Earth observation; disaster management; environmental protection; satellite communications and satellite navigation and positioning; space science research; education, training and exchange of scientists and technologists; establishment of a central data bank for the development of programmes of the organization and dissemination of technical and other information relating to the programmes and activities of the operation; and other co-operative programmes agreed upon by the members. To this end, the basic activities of the organization include: establishment of the organization's plans for space activities and development; carrying out fundamental research concerning space technology and its applications; extending the applications of matured space technology; conducting education and training activities concerning space science and technology and their applications; managing and maintaining the branch offices and relevant facilities, as well as the network system of the organization; and undertaking other necessary activities to achieve the objectives of the organization.[37]

APSCO has reported initial successes in the expansion of applications of the space technology in remote sensing, disaster mitigation, environmental protection and other fields. Beginning in 2000, APSCO in co-operation with the China National Space Administration (CNSA), has held five training courses on space technology and remote sensing application for government officials and technicians from Asia-Pacific countries. Approximately 200 trainees from over 30 countries have attended the courses, facilitating the improvement of these countries' capabilities in space technology and its applications.[38]

UNITED NATIONS SPECIALIZED AGENCIES

International Telecommunication Union

The ITU is the specialized UN agency responsible for information and communication technologies in three primary sectors: radiocommunication, standardization and development. Organized by the ITU, World Radiocommunication Conferences (WRC) are held every two to four years.[39] WRC is charged with reviewing and, if necessary, revising the Radio Regulations. The Radio Regulations is the international treaty governing the use of the radio frequency spectrum and GEO satellite use of that spectrum as well as

GEO orbital slot allocations. The ITU Council sets the agenda for the conference two years prior to it, with the concurrence of a majority of Member States. Under the terms of the ITU Constitution, a WRC can revise the Radio Regulations and any associated frequency assignment and allotment plans, address any radiocommunication matter of world-wide character, instruct the Radio Regulations Board and the Radiocommunication Bureau, review their activities, and determine questions for study by the Radiocommunication Assembly and its Study Groups in preparation for future WRC.[40]

Radio frequency spectrum is a limited natural resource that is shared among nations on a regional and global basis. The conferences are charged with managing the international use of radio frequency spectrum in a rational and equitable manner. The decisions of the WRC have a significant impact on US consumers and industry. As such, the US Federal Communications Commission (FCC) sets up an Advisory Committee before each WRC to voice US interests during the process.[41]

The most recent WRC (WRC-07) took place in 2007. Topics discussed at the 2007 conference include: mobile, aeronautical mobile, radionavigation and radiolocation services; space science services; low-frequency, middle-frequency and high-frequency bands and maritime mobile services; and regulatory procedures and associated technical criteria applicable to satellite networks. The next conference is planned for 2011.[42]

The ITU facilitates the relationship of the world's Information and Communication Technology (ICT) community through such events as the 2003-5 World Summit on the Information Society, for which the ITU was the lead organizing agency. From its base in Geneva, Switzerland, the ITU uses space-based technologies to connect its Member States and sector members and associates (see Illustrations and Documentation: Documentation 6.5 International Telecommunication Union).[43]

To carry out its mission, the members of the ITU have agreed to work towards the common purpose of: 'Enabling the growth and sustained development of telecommunications and information networks, and facilitating universal access so that people everywhere can participate in, and benefit from, the emerging information society and global economy'.[44] The ITU has a specific Space Services Department, which manages all procedures and data for space systems and Earth stations. The department is also responsible for 'examining frequency assignment notices submitted by administrations for inclusion in the form co-ordination procedures or recording in the Master International Frequency Register and managing the procedures for space related assignment or allotment plans of the ITU and for provision of assistance to administration on all of the above issues'.[45]

International Civil Aviation Organization

The ICAO is the specialized agency of the UN charged with regulating international air travel. It works to:

[...] insure the safe and orderly growth of international civil aviation throughout the world; encourage the arts of aircraft design and operation for peaceful purposes; encourage the development of airways, airports and air navigation facilities for international civil aviation; meet the needs of the peoples of the world for safe, regular, efficient and economical air transport; prevent economic waste caused by unreasonable competition; insure that the rights of Contracting States are fully respected and that every Contract State has a fair opportunity to operate international airlines; avoid discrimination between Contracting States; promote safety of flight in international air navigation; and promote generally the development of all aspects of international civil aeronautics.[46]

Formed in 1944 with the signing of the Convention on International Civil Aviation in the USA, the agency has 188 Member States, a membership that almost entirely reflects the membership of the UN. Representatives from all member states form the Assembly, the sovereign body of ICAO. The Assembly meets every three years. The Council, the governing body of ICAO, is elected by the Assembly for a three-year term and is composed of 36 states. The members of the Council are chosen to represent three areas: states of primary importance in air transport; states whose facilities are the most necessary for air navigation; and states that ensure that the Council's membership reflects all major areas of the world.[47]

The ICAO's objectives deal with aviation: 'ICAO works to achieve its vision of safe, secure, and sustainable development of civil aviation through co-operation amongst its member states'.[48] The ICAO's Strategic Objectives for the period 2005–10 reflect this focus on aviation: 'Safety (enhance global civil aviation safety); Security (enhance global civil aviation security); Environmental Protection (minimize the adverse effect of global civil aviation on the environment); Efficiency (enhance the efficiency of aviation operations); Continuity (maintain the continuity of aviation operations); and Rule of Law (strengthen law governing international civil aviation)'.[49]

Even though the ICAO has no formal role in space as yet, the advent of commercial space operations for tourists, and possibly transport, in the next few years may change this. Suggestions have been made to have ICAO, or a bureau based on the structure of ICAO, as the appropriate agency to tackle space regulation. In a 2007 report, 'An ICAO for Space?', the International Association for the Advancement of Space Safety (IASS) discussed the need for the establishment of space regulation guidelines.

> Though there is great enthusiasm about the commercial potentials of space, and the safety risks are very real and growing, there is no international safety regulatory framework to balance the multiple commercial interests in space with internationally enforced safety risks mitigation measures. Because of this, the IASS Legal and Regulatory Committee felt that to continue to fuel the commercial growth of space that it was

important to support a corresponding international safety regulatory regime that can enable the growth in the commercial sector while ensuring that people, property and goods are adequately protected [...] Probably the best analogy is ICAO, which was created towards the end of World War II. For this reason, and also because there is a wide commonality of interests, first of all the sharing of a crowded airspace, the white paper focuses particular attention on how ICAO could be a model for international commercial space safety regulations, thus, the title 'An ICAO for Space?'[50]

Although ICAO has been brought up as a model for space regulation several times, most recently at the Council of European Aerospace Societies (CEAS) European Air and Space Conference in 2007,[51] any near-term ICAO expansion into co-ordinating or regulating international space flight is unlikely. As private space activities develop, this topic will likely be revisited, given the number of issues in commercial space flight that parallel those in aviation.

United Nations Educational, Scientific and Cultural Organization

UNESCO aims to build peace through education, social and natural science, culture and communication.[52] Two intergovernmental bodies govern UNESCO: the General Conference and the Executive Board. The General Conference convenes the Member States and Associate Members, together with observers from non-member states, intergovernmental organizations and non-governmental organizations, once every two years. Every country has one vote in the General Conference, regardless of the size of either the country or its contribution to the budget. The General Conference directs the policies and primary work of the organization. It sets the programmes and budgets of UNESCO, in addition to electing the members of the Executive Board and the Director-General once every four years. The Executive Board governs the overall operations of UNESCO. The 58 members are elected by the General Conference in an effort to represent the cultural and geographic diversity of the organization as a whole. The Executive Board meets twice a year to carry out the specific assignments given to it by the General Conference.[53] Since its founding in 1945, UNESCO has grown to encompass 193 Member States and six Associate Members, from every area of the world.[54]

UNESCO makes use of spaceborne technologies and their applications to tackle a number of challenges, including natural resource management, environmental planning and Earth observation for global monitoring. UNESCO uses space-based technologies to implement distance learning and training, helping to break down educational isolation. UNESCO examines the ethics of space policy as well, the main objective being to 'keep in mind the place of human beings and answer the anxieties of public opinion through an objective, independent and transparent approach'.[55]

World Meteorological Organization

Created from the International Meteorological Organization (founded in 1873), the World Meteorological Organization (WMO) was established in 1950 as the specialized UN agency for meteorology, operational hydrology and related geophysical sciences (see Illustrations and Documentation: Documentation 6.6 World Meteorological Organization). The WMO is composed of 188 Member States and Territories and performs a leading role in international efforts to monitor and protect the environment.[56] The WMO includes a specific space programme dedicated to organizing environmental satellite affairs within the WMO, expanding the space-based Global Observing System (GOS), and advancing the use of satellite data for climate, water, weather and related purposes.

The four cornerstones of the WMO include: consolidate requirements for satellite observation; develop the space-based component of the GOS; enhance the availability of satellite data, products and services; and enhance users' capability to take advantage of satellite data.[57] The objectives of the WMO space programme are to: co-ordinate environmental satellite matters and activities throughout all WMO programmes and provide guidance on the potential of remote sensing techniques in meteorology, hydrology, and related disciplines and applications; and improve the provision of data, products and services from operation and R&D satellites contributing to GOS, as well as facilitating and promising the wider availability and meaningful use of these data, products and services around the globe.

Past achievements of the space programme include: a fully operational GEO and low Earth orbit (LEO) observing system complemented by advanced R&D satellites; the development and implementation of an Integrated Global Data Dissemination Service (IGDDS); and the establishment and initial implementation of the Global Space-Based Inter-Calibration System (GSICS).[58] The space programme has also published a long-term strategy plan for the period 2008–11, which outlines a number of goals, including contribution to the development of GOS, the promotion of high-quality, satellite-related continuing education to keep the knowledge and skills of members' operational and scientific staff up to date, and a review of the space-based components of a number of observing systems throughout the WMO programmes and support programmes.[59]

OTHER INTERNATIONAL SPACE ORGANIZATIONS

This section presents an overview of a number of important IOs that are not formally affiliated with the UN, although they closely co-ordinate their activities with the UN. Four of these organizations – Committee on Earth Observation Satellites (CEOS), Group on Earth Observations, International Charter 'Space and Major Disasters' and IADC – are specifically established to deal with space issues. Three other IOs reviewed – the International Institute for the

Unification of Private Law (UNIDROIT), the Organisation for Economic Co-operation and Development (OECD), and the Wassenaar Arrangement on Export Controls for Conventional Arms and Dual-Use Goods and Technologies – are organized for non-space purposes, but have over the past few years undertaken activities related to space (see Illustrations and Documentation: Table 6.4 Global Space Organizations). The growth of commercial interests in space, technology proliferation issues exacerbated by global space relations and, in general, international issues concerning unresolved political, legal and economic problems, have been the main stimuli for these organizations to become involved with space commerce,[60] economic development and space, and export controls of space technologies, respectively.

Committee on Earth Observation Satellites

CEOS is an independent organization that co-ordinates international civil spaceborne missions, which examine and analyse the Earth. The organization serves as a global focal point for Earth observation satellite programmes and the co-ordination of these programmes with satellite data users world-wide (see Illustrations and Documentation: Documentation 6.7 Committee on Earth Observation Satellites). CEOS has three primary objectives: (1) to optimize the benefits of spaceborne Earth observation through co-operation of its members in mission planning and in the development of compatible data products, formats, services, applications and policies; (2) to aid both its members and the international user community by inter alia, serving as the focal point for international co-ordination of space-related Earth observation activities, including the Group on Earth Observations and entities related to global change; and (3) to exchange policy and technical information to encourage compatibility among spaceborne Earth observation systems currently in service or development and the data received from them. CEOS addresses issues of common interest across the spectrum of Earth observation satellites.

CEOS was created in 1984 as a result of the Economic Summit of Industrialized Nations Working Group on Growth, Technology and Employment's Panel of Experts on Satellite Remote Sensing. Originally termed the International Earth Observations Satellite Committee, the organization represents the melding of the Coordination on Ocean Remote Sensing Satellites and the Coordination on Land Observing Satellites. Through its 26 members (primarily national space agencies) and 20 associates (primarily associated national and international organizations), CEOS constitutes an international framework for the co-ordination of all spaceborne Earth observation missions.[61]

CEOS defined 10 priority activities for the period 2002–7. The overarching goal is to strengthen the positioning of CEOS, specifically at the political level. CEOS strives to project a more coherent, defined sense of identity in the political and scientific arenas.[62] At the most recent CEOS plenary, held in November 2007, the organization committed to a series of detailed and comprehensive actions to support development of the Global Earth Observation System of

Systems (GEOSS) space segment. In addition, CEOS reviewed reports from the pilot phases of the four CEOS Virtual Constellations – Atmospheric Composition, Land Surface Imaging, Ocean Surface Topography, and Precipitation – to decide upon the next steps in their development. The agency also took steps to expand its consultation and co-ordination with the WMO space programme and Coordination Group for Meteorological Satellites.

Group on Earth Observations

The Group on Earth Observations is dedicated to constructing GEOSS. The group grew out of the 2002 World Summit on Sustainable Development as a result of the recognition that international collaboration is essential for exploiting the potential of Earth observations to support decision-making in an increasingly complex and environmentally stressed world. The emerging public infrastructure is interconnecting a diverse and growing array of instruments and systems for monitoring and forecasting changes in the global environment. This 'system of systems' supports policymakers, resource managers, science researchers and many other experts and decision-makers.[63] The Group's 10-Year Implementation Plan adopted in 2005 outlines its goals for building GEOSS in respect to nine 'Societal Benefit Areas': disasters, health, energy, climate, water, weather, ecosystems, agriculture and biodiversity (see Illustrations and Documentation: Documentation 6.8 Global Earth Observation System of Systems).[64]

The group is a voluntary partnership of 73 governments, the European Commission, and 51 intergovernmental, international and regional organizations with a mandate in Earth observation or related issues. The US Group on Earth Observations was established in March 2005 as a standing subcommittee of the Natural Science and Technology Council Committee on Environmental and Natural Resources.[65] The US contribution to GEOSS is the Integrated Earth Observation System. GEOSS and the Integrated System will facilitate the sharing and applied usage of global, regional and local data from satellites, ocean buoys, weather stations and other surface and airborne Earth observing instruments.[66] The Group on Earth Observations is governed by a Plenary comprising all members and participating organizations. It meets at least once a year at the level of senior officials and periodically at the ministerial level. An Executive Committee oversees the group's activities when the Plenary is not in session. The group has four committees and one working group to implement the 10-Year Plan. The four committees are: Architecture and Data; Science and Technology; User Interface; and Capacity Building; and the Working Group is dedicated to tsunami activities.[67]

International Charter: Space and Major Disasters

Following the UNISPACE III conference, the ESA and the French national space agency, Centre National d'Etudes Spatiales (CNES) developed the International charter 'Space and Major Disasters'. The charter aims to provide a unified system of space data acquisition and delivery to those affected by natural or technological disasters, through authorized users (see Illustrations and Documentation: Documentation 6.9 Space and Major Disasters). The agreement comes from the recognition that no single operator or satellite can meet the challenges of natural disaster management. Each member agency commits resources to support the provisions of the charter, helping to mitigate the effects of disasters on human life and property.[68]

The charter was deemed fully operational on 1 November 2000 and now includes the ESA, CNES, Canadian Space Agency (CSA), Indian Space Research Organization (ISRO), National Oceanic and Atmospheric Administration (NOAA) in the USA, Argentina's Comisión Nacional de Actividades Espaciales (CONAE), the Japan Aerospace Exploration Agency (JAXA), the US Geological Survey (USGS), DMC International Imaging (DMC), and the China National Space Administration (CNSA).[69] The UN Office for Outer Space Affairs acts as a 'co-operating agency' supporting the charter. A framework co-operation agreement was signed with the office in July 2003. In case of emergencies falling within the scope of the charter, the bodies and organizations of the UN system that are in a position to provide support, are involved in disaster management and have identified the need for data and information, may forward a request for assistance to the charter through the Office for Outer Space Affairs.[70]

From its inception in 2000 to 2008, the charter has been activated more than 200 times, and the satellites of the charter constellation have successfully provided both wide swath and metric precision images, day and night and in all weather conditions. Activations thus far have involved natural disasters, such as floods, hurricanes, forest fires, earthquakes, volcanoes and landslides, and technological disasters, like oil spills and industrial accidents. Specific recent instances of note include: the floods in Iowa, USA in June 2008; the earthquake in Rwanda in February 2008; the volcanic operation in Ecuador in January 2008; and the oil spill in the North Sea in December 2007.[71] When the charter is activated, a series of commands are put into place to ensure the timely delivery of necessary information.

International Institute for the Unification of Private Law

UNIDROIT is an independent, intergovernmental organization located in Rome, Italy, the purpose of which is to 'examine ways to harmonize and coordinate private law of States and of groups of States'.[72] To accomplish this, UNIDROIT drafts international commercial agreements and serves as a forum for international law. Established in 1926 as an auxiliary organ of the League of Nations, the organization was re-established independently in 1950 on the basis of the multilateral UNIDROIT Statute.[73]

The Statute states:

> The purposes of the International Institute for the Unification of Private Law are to examine ways of harmonizing and coordinating the private

law of States and of groups of States, and to prepare gradually for the adoption by the various States of uniform rules of private law. To this end, the Institute shall: (a) prepare drafts of laws and conventions with the object of establishing uniform internal law; (b) prepare drafts of agreements with a view to facilitating international relations in the field of private law; (c) undertake studies in comparative private law; (d) take an interest in projects already undertaken in any of these fields by other institutions with which it may maintain relations as necessary; and (e) organize conferences and publish works, which the Institute considers worthy of wide circulation.[74]

UNIDROIT is composed of 61 Member States from five continents and has a three-tiered structure comprising a Secretariat, a Governing Council and a General Assembly. UNIDROIT's primary connection to space policy is its work to facilitate commercial transactions regarding 'space assets'.[75] The UNIDROIT Convention on International Interests in Mobile Equipment of 2001 addresses the international legal complexities involved in the use and movement of mobile equipment in three categories: (1) airframes aircraft engines and helicopters; (2) railway rolling stock; and (3) space assets.[76] The Protocol to the Convention on International Interests in Mobile Equipment on Matters Specific to Aircraft Equipment came into force in 2001, but the protocols on railway rolling stock and space assets are draft formats and they continue to be debated.[77]

In the matter of space assets, UNIDROIT is acting on the need for a method to protect secured interests and creditors' rights as more money is invested in space commerce. Identification of assets would include: a report of the name of the debtor and creditor; the address; a general description of the asset, including manufacturer and serial numbers and the intended location; the data and location of the launch; and a description of separately identified components. The choice of law would be determined by contract.

As of 2009, there are a number of issues surrounding UNIDROIT's work with space assets that still need to be resolved. These problems include: the definitions of space property and assets; the potential liability of a launching state for space assets no longer owned by a citizen of that state; the housing of this new registry; and associated rights (for example, in aircraft and landing rights). In addition, any new law would need to be integrated with UN treaties and national laws.[78] Finally, the UNIDROIT proposal is actually a mix of public and private law, as many governments are satellite owners themselves. UNDROIT must overcome these significant issues before any formal agreements will be made regarding mobile space assets.[79]

Organisation for Economic Co-operation and Development

The Organisation for European Economic Co-operation (OEEC) was established in 1948 with support from the USA and Canada to help co-ordinate the implementation of the Marshall Plan for the reconstruction of Europe after the Second World War. The OECD, the renamed and more comprehensive successor organization, took over from the OEEC in 1961 with the purpose of bringing together the governments of democratic market economy countries to forge a better world economy.[80]

Unlike many IOs, the OECD strictly reviews all applicant countries before they are allowed to join. The members of the organization meet in the Council to decide whether a country should be invited to join the OECD and on what conditions. The OECD currently has 30 members from around the world.[81] In 2007 the OECD invited Chile, Estonia, Israel, Russia and Slovenia to open discussions for membership in the organization, while offering 'enhanced engagement', with the possibility of membership, to Brazil, China, India, Indonesia and South Africa.[82] The OECD is run by a Council, a Secretariat and individual committees.

The OECD's primary connection to space affairs and space policy dates back to 2002, when the International Futures Program began to analyse space applications and the ability of space applications to address economic concerns facing all nations of the world. Their work on space has produced several publications and a set of recommendations for encouraging the growth of commercial and civil space applications. Of note are the following space-related OECD reports of the International Futures Program: Space 2030: Exploring the Future of Space Applications (2004); Space 2030: Tackling Society's Challenges (2005); The Space Economy at a Glance 2007 (2007); and Space Technologies and Climate Change: Implications for Water Management, Marine Resources and Maritime Transport (2008).[83]

The OECD plans to continue its work on space issues, emphasizing several important areas: the development of more reliable international economic statistics and indicators of space activity; specific analyses of the importance of space applications to the economy of the world; and a continuous monitoring of national economic and legal activities related to the recommendations developed in 2005 to enhance commercial and civil space investments and benefits.

Inter-Agency Space Debris Coordination Committee

IADC is an independent, intergovernmental agency that acts as an international focal point for the global co-ordination of actions connected to the problem of human-made and natural orbital debris in space (see Illustrations and Documentation: Documentation 6.10 Inter-Agency Space Debris Coordination Committee). To that end, the IADC strives 'to exchange information on space debris research activities between member space agencies, to facilitate opportunities for co-operation in space debris research, to review the progress of ongoing co-operative activities, and to identify debris mitigation options'.[84] The IADC is composed of a Steering Group and four specialized Working Groups: Measurements; Environment and Data; Protection; and

Mitigation. Eleven national and regional space agencies are members of the IADC, representing most of the major spacefaring nations.[85]

Space debris in LEO has increased significantly since the COPUOS Scientific and Technical Subcommittee published its 1999 Technical Report on Space Debris. The test of a Chinese anti-satellite (ASAT) system on 11 January 2007, which successfully targeted and destroyed a Chinese satellite, and the collision between an Iridium communications commercial satellite and a dysfunctional Russian satellite (defunct Soviet-era Cosmos spacecraft) on 10 February 2009 created the most severe orbital debris clouds in the history of the space age.[86]

In 2002 the IADC published a set of Space Debris Mitigation Guidelines based on the debris mitigation standards of several national and international space organizations. These guidelines are centred on three major principles: preventing on-orbit breakups; removing spacecraft and orbital stages that have reached the end of their mission operations from densely populated orbit regions; and limiting the objects released during normal operations.[87]

The 2006 International Academy of Astronautics Cosmic Study on Space Traffic Management recommended that COPUOS endorse the guidelines as a UN legal document with the goal of universal acceptance and implementation.[88] The Scientific and Technical Subcommittee of COPUOS adopted the guidelines by consensus in 2007 and the committee endorsed the guidelines at its 572nd meeting.[89] The IADC frequently works with the COPUOS Subcommittees and presents an update at its annual meetings.[90] The IADC is currently focusing on a number of on-going studies, including research on entry criteria and procedures, long-term presence of objects in the GEO region, sensor systems to detect impacts on spacecraft, re-entry predictions, and the comparison of meteoroid models.[91]

Wassenaar Arrangement and Export Controls

With the goal of using export control as a means to combat terrorism, the Wassenaar Arrangement: contributes to regional and international security and stability; promotes transparency and greater responsibility in transfers of conventional arms and dual-use goods and technologies; and complements and reinforces the existing control regimes for weapons of mass destruction and the delivery systems.[92] Export controls are implemented by each individual Participating State. Although the scope of export controls in Participating States is determined by Wassenaar Arrangement lists, practical implementation varies from country to country in accordance with national procedures (see Illustrations and Documentation: Documentation 6.11 Wassenaar Arrangement).[93]

The Wassenaar Arrangement has grown from its original 33 members to 40 Participating States.[94] All Participating States have agreed to maintain national export controls, implemented through national legislation, on listed items and to report on transfers and denials of specified controlled items to destinations outside the Arrangement. The states exchange information on sensitive dual-use goods and technologies in accordance with the agreed Best Practices, Guidelines or Elements. The Plenary, the decision-making body of the Wassenaar Arrangement, meets once a year; subsidiary bodies meet periodically. All decisions are made by consensus and the deliberations are kept in confidence.[95]

Space is not a separate category in the structure of the Wassenaar Arrangement. However, following the US lead in export controls, most space objects, including commercial communications satellites, are considered as part of the 'missiles' category of controlled goods. Although this classification has been severely criticized, it has been in effect for over 10 years and, as described below, appears to have had a significant international effect on the manufacture and distribution of space hardware and services, not only in the USA, but in other nations as well.

The US export control system and the Wassenaar Arrangement are cited as reasons for the difficulties the USA faces in competing in world satellite markets:

> The present situation in the space industry is one in which US companies, mainly those that produce satellites, have great difficulty competing in the world market. The most serious barrier to US competitiveness in this field is government policy on export controls. The outcome of 20 years of cooperation and joint ventures with partners, such as Chinese and Russian space firms, to produce launch vehicles, is a system with a rigid interpretation of ambiguous statutory requirements and a confusing licensing process that leads to long delays and uncertain results. The main problem is identified in the length of time it takes to obtain an International Traffic in Arms Regulations (ITAR) approval. According to reports from US manufacturers, the time taken to get a license has increased to 150 days. In addition, the situation is complicated by the uneven application of international agreements. The Wassenaar Arrangement, for instance, binds US commercial satellite companies with restrictions that several companies in Europe, Canada, Russia, and Japan, not being members of the agreement, are not subjected to.[96]

According to Satellite Industry Associate data, the US share of global satellite sales has dropped from 64% of the US $12.4bn market in 1998, to 36% in 2002. Current export-control policy for dual-use items has increased the cost of doing business with US satellite manufacturers while simultaneously decreasing their ability to compete in the global marketplace. This decrease has allowed other companies, particularly European manufacturers, to increase their grip on the world satellite market. For example, the space business Alcatel announced that it would create an 'ITAR-free' spacecraft in the early 2000s. By 2004 Alcatel had been able to double its market share from around 10% in 1998 to over 20% in 2004.[97]

CONCLUSION

The role and scope of IOs vary from those at the global level to those that have more regional impacts and, finally, to those that focus on specific functional areas of concern. The UN system is the largest and most important world-wide forum for international discussion about space and for addressing specific space-related issues. A second type of IO includes the multinational space organizations that span specific regions of the world – the ESA, EU and APSCO. Following these are UN Specialized Agencies that deal with space, and other non-UN international space organizations that focus on space-related topics dealing with Earth observations, natural and human-made disaster management, commerce and economic development, Earth orbital debris and export controls of technology.

NOTES

* The author is indebted to Jordan Bock, research assistant at the Space Policy Institute, George Washington University during the summer of 2008, for her help in preparing this chapter. Ms Bock is currently an astrophysics major at Harvard University.
1. Military organizations, such as the North Atlantic Treaty Organization (NATO), which use space applications primarily for security and defence are not discussed in this chapter.
2. 'The Structure of the United Nations', www.historylearningsite.co.uk/structure_of_the_united_nations.htm (accessed 24 June 2008).
3. Because space issues span a number of disciplines and areas, other agencies in the UN also involve activities or issues that relate to space. The Office for Disarmament Affairs and the International Maritime Organization are just two of many that in some way engage with outer space. These entities will not be addressed in this review because their primary function does not directly involve space.
4. 'The United Nations System', The United Nations, www.un.org/aboutun/chart_en.pdf (accessed 12 June 2008).
5. Kai-Uwe Schrogl, 'Basic space science in a future COPUOS: Emphasizing the role of developing countries', 6th United Nations/European Space Agency Workshop on Basic Space Science, 1996, The United Nations, www.seas.columbia.edu/~ah297/un-esa/paper-schrogl.html (accessed 23 June 2008).
6. 'United Nations Committee on the Peaceful Uses of Outer Space', United Nations Office for Outer Space Affairs, 2006, www.unoosa.org/oosa/COPUOS/copuos.html (accessed 10 June 2008).
7. See 'United Nations Treaties and Principles on Space Law', United Nations Office for Outer Space Affairs, 2006, www.unoosa.org/oosa/en/SpaceLaw/treaties.html (accessed 24 June 2008).
8. General Assembly resolution 2222 (XXI), adopted on 19 December 1966, opened for signature on 27 January 1967, entered into force on 10 October 1967.
9. General Assembly resolution 2345 (XXII), adopted on 19 December 1967, opened for signature on 22 April 1968, entered into force on 3 December 1968.
10. General Assembly resolution 2777 (XXVI), adopted on 29 November 1971, opened for signature on 29 March 1972, entered into force on 1 September 1972.
11. General Assembly resolution 3235 (XXIX), adopted on 12 November 1974, opened for signature on 14 January 1975, entered into force on 15 September 1976.
12. General Assembly resolution 34/68, adopted on 5 December 1979, opened for signature on 18 December 1979, entered into force on 11 July 1984.

13 Adopted on 13 December 1963 by General Assembly resolution 1962 (XVIII).
14 Adopted on 10 December 1982 by General Assembly resolution 37/92.
15 Adopted on 3 December 1986 by General Assembly resolution 41/65.
16 Adopted on 14 December 1992 by General Assembly resolution 47/68.
17 Adopted on 13 December 1996 by General Assembly resolution 51/122.
18 Adopted on 10 December 2004 by General Assembly resolution 59/115.
19 Adopted on 17 December 2007 by General Assembly resolution 62/101.
20 UN/BSS Science Organizing Committee Planning Meeting, 'UN Programme on Space Applications: A Brief Overview', *United Nations Office for Outer Space Affairs*, 19–21 October 2004, ihy2007.org/img/PSA_2004–10–13_final.ppt (accessed 24 June 2008).
21 'International Committee on Global Navigation Satellite Systems', Inter-Agency Meeting on Outer Space Activities, 19 January 2007, www.uncosa.unvienna.org/pdf/iamos/2007/ois-06.pdf (accessed 25 June 2008).
22 Space debris is defined as all human-made objects, including fragments and elements thereof, in Earth orbit or re-entering the atmosphere, that are non functional.
23 'Committee on the Peaceful Uses of Outer Space: 2008, Fifty-first session, 11–20 June 2008', United Nations Office for Outer Space Affairs, 2008, www.unoosa.org/oosa/en/COPUOS/index.html (accessed 24 June 2008). The Fifty-second session was planned for 3–12 June 2009.
24 Member States are: Austria, Belgium, Czech Republic, Denmark, Finland, France, Germany, Greece, Ireland, Italy, Luxembourg, the Netherlands, Norway, Portugal, Spain, Sweden, Switzerland and the United Kingdom. Canada takes part in some projects under a Co-operation agreement. Hungary, Poland and Romania are formal European Co-operating States. Turkey, Estonia, Ukraine and Slovenia have co-operation agreements with the ESA.
25 'About ESA', 19 June 2008, European Space Agency, www.esa.int/SPECIALS/About_ESA/index.html (accessed 2 July 2008).
26 ESA clearly has a function to promote jobs and income in Europe. Its role in industry policy is clearly stated in the ESA Convention.
27 ESA maintains a number of centres throughout Europe with different responsibilities: EAC, the European Astronauts Centre in Cologne, Germany; ESAC, the European Space Astronomy Centre, in Villafranca del Castillo, Madrid, Spain; ESOC, the European Space Operations Centre in Darmstadt, Germany; ESRIN, the ESA Centre for Earth Observation, in Frascati, near Rome, Italy; and ESTEC, the European Space Research and Technology Centre, Noordwijk, the Netherlands.
28 The head of ESA is a Director-General who is elected by the Council every four years. Each individual research sector has its own Directorate and reports directly to the Director-General.
29 For example, ESA's 2007 budget was €2,975m.
30 'About ESA', 19 June 2008, European Space Agency, www.esa.int/SPECIALS/About_ESA/index.html (accessed 2 July 2008).
31 'Panorama of the European Union', European Union, europa.eu/abc/panorama/index_en.htm (accessed 2 July 2008).
32 'European Space Policy', European Commission – Enterprise and Industry, European Union, ec.europa.eu/enterprise/space/index_en.html (accessed 2 July 2008).
33 'ESA and the EU', European Space Agency, www.esa.int/SPECIALS/About_ESA/SEMFEPYV1SD_0.html (accessed 2 July 2008).
34 'APSCO', Asia-Pacific Multilateral Cooperation in Space Technology and Applications, www.intelligence.gov/3-student-opportunities.shtml (accessed 2 June 2008).
35 James Lowe, 'APSCO set to be born', *NASA Spaceflight*, 2 July 2008.
36 Asia-Pacific Space Cooperation Organization, Convention of the Asia-Pacific Space Cooperation Organization (APSCO), Beijing: 2005.
37 Ibid.

38 'APSCO', Asia-Pacific Multilateral Cooperation in Space Technology and Applications, www.intelligence.gov/3-student-opportunities.shtml (accessed 2 June 2008).
39 Prior to 1993 the conference was known as the World Administrative Radio Conference, or WARC. In 1992 the Additional Plenipotentiary Conference in Geneva dramatically remodeled ITU with the goal of giving it great flexibility to adapt to the increasingly complex, interactive and competitive environment of telecommunications. The ITU was streamlined into the three sectors of which it is composed today: Standardization, Radiocommunication and Development. WARC came under the purview of ITU-R and was reworked as the World Radiocommunication Conference.
40 'World Radiocommunication Conferences (WRC)', ITU Radiocommunication Sector, 15 July 2008, International Telecommunication Union, www.itu.int/ITU-R/index.asp?category=conferences&rlink=wrc&lang=en (accessed 17 July 2008).
41 'US Proposals to the ITU', 8 February 2007, Federal Communication Commission, www.fcc.gov/wrc-07/itu/welcome.html (accessed 17 July 2008).
42 Yvon Henri, 'Agenda for the WRC-2007', *International Telecommunication Union Radiocommunication Bureau,* Geneva, Switzerland, 2004.
43 'About ITU', International Telecommunication Union (accessed 11 June 2008) www.itu.int/net/about/index.aspx.
44 'The ITU Mission', 11 June 2008, International Telecommunication Union, www.itu.int/net/about/mission.aspx (accessed 11 June 2008).
45 'Space Services Department', 6 May 2008, International Telecommunication Union, www.itu.int/ITU-R/space (accessed 11 June 2008).
46 'Memorandum on ICAO', International Civil Aviation Organization, www.icao.int/cgi/goto_m.pl?icao/en/pub/memo.pdf (accessed 14 June 2008).
47 'About ICAO', International Civil Aviation Organization, www.icao.int/icao/en/m_about.html (accessed 14 June 2008).
48 'Strategic Objectives of ICAO for 2005–10', 17 December 2004, International Civil Aviation Organization, www.icao.int/icao/en/strategic_objectives.htm (accessed 23 June 2008).
49 Ibid.
50 'An ICAO for Space?', 29 May 2007, International Association for Advancement of Space Safety, www.iaass.org/Publications.htm (accessed 15 June 2008).
51 'ESPI at the First CEAS Conference', 13 September 2007, European Space Policy Institute, www.espi.or.at/index.php?option=com_content&task=view&id=93&Itemid=37 (accessed 16 February 2009).
52 'About UNESCO', 10 August 2007, United Nations Educational, Scientific and Cultural Organization, portal.unesco.org/en/ev.php-URL_ID=3328&URL_DO=DO_TOPIC&URL_SECTION=201.html (accessed 27 June 2008).
53 'Governing Bodies', 10 December 2007, United Nations Educational, Scientific and Cultural Organization, portal.unesco.org/en/ev.php-URL_ID=3973&URL_DO=DO_TOPIC&URL_SECTION=201.html (accessed 27 June 2008).
54 'About UNESCO', 10 August 2007, United Nations Educational, Scientific and Cultural Organization, portal.unesco.org/en/ev.php-URL_ID=3328&URL_DO=DO_TOPIC&URL_SECTION=201.html (accessed 27 June 2008).
55 Yolanda Berenguer and Malcolm Hadley, *Space Activities in UNESCO* (Paris, France: Menager Imprimeurs, 2002).
56 'WMO in brief', World Meteorological Organization, www.wmo.int/pages/about/index_en.html (accessed 12 June 2008).
57 'WMO Space Program Goals and Objectives', www.wmo.int/pages/prog/sat/Goalsandobjectives.html (accessed 27 June 2008).
58 Donald Hinsman, 'The WMO Space Programme', World Meteorological Organization, 12 May 2005 www.wmo.ch/pages/prog/gcos/scXV/13_WWW_Space_Program.pres.pdf (accessed 16 February 2009).

59 'Outline of the WMO Space Program Implementation Plan for 2008–11', 1 August 2007, WMO Space Program, www.wmo.int/pages/prog/sat/documents/SAT-ST-06_v1_WSPImplementationPlanoutline2008–11.pdf (accessed 24 June 2008).
60 The World Trade Organization and the World Intellectual Property Organization also affect space commerce activities, but are not covered in this review.
61 'Overview', 23 March 2007, Committee on Earth Observation Satellites, www.ceos.org/pages/overview.html (accessed 24 June 2008).
62 'CEOS Five Year Plan (2002–7)', 22 January 2008, Committee on Earth Observation Satellites, www.ceos.org/pages/pub.html##plan (accessed 23 June 2008).
63 'What are GEO and GEOSS', Group on Earth Observations, earthobservations.org/#WhatIsGEO (accessed 14 July 2008).
64 'About GEO', Group on Earth Observations, earthobservations.org/about_geo.shtml (accessed 14 July 2008).
65 Ibid.
66 'About USGEO', United States Group on Earth Observations, usgeo.gov (accessed 14 July 2008).
67 'About GEO', Group on Earth Observations, earthobservations.org/about_geo.shtml (accessed 14 July 2008).
68 'Charter on Cooperation to Achieve the Coordinated Use of Space Facilities in the Event of Natural or Technological Disasters', 11 April 2008, International Charter 'Space and Major Disasters', www.disasterscharter.org/charter_e.html (accessed 27 June 2008).
69 'About the Charter'. 11 April 2008, International Charter 'Space and Major Disasters', www.disasterscharter.org/about_e.html (accessed 27 June 2008).
70 'Space and Major Disasters', International Charter 'Space and Major Disasters', www.disasterscharter.org/downloadable/CharterBrochure.pdf (accessed 14 July 2008).
71 'Recent Activations', 13 June 2008, International Charter 'Space and Major Disasters', www.disasterscharter.org/new_e.html (accessed 27 June 2008).
72 'UNIDROIT: An overview', 2008, UNIDROIT, www.unidroit.org/dynasite.cfm?dsmid=84219 (accessed 13 June 2008).
73 Ibid.
74 'UNIDROIT: International Institute for the Unification of Private Law Statute', 26 March 1993, UNIDROIT, www.unidroit.org/english/presentation/statute.pdf (accessed 12 June 2008).
75 Space assets are defined as any separately identifiable asset that is in space, any separately identifiable component forming a part of an asset, any separately identifiable asset or component assembled in space, or any launch vehicle that is expendable or can be reused.
76 'Convention on International Interests in Mobile Equipment', UNIDROIT DCME Document Number 74, article 2.
77 See www.unidroit.org/english/conventions/mobile-equipment/main.htm#NR1 (accessed 16 February 2009)
78 Both UNIDROIT and COPUOS have signalled their willingness to work together to avoid conflicts in the registration process and resolve this issue.
79 'Preliminary Draft Protocol on Matters Specific to Space Assets – Study LXXII J', UNIDROIT, 28 October 2004, www.unidroit.org/english/documents/2004/study72j/s-72j-13rev-e.pdf (accessed 16 February 2009).
80 'About the OECD', Organisation for Economic Co-operation and Development, www.oecd.org/pages/0,3417,en_36734052_36734103_1_1_1_1_1,00.html (accessed 27 June 2008).
81 Members of the OECD are: Australia, Austria, Belgium, Canada, Czech Republic, Denmark, Finland, France, Germany, Greece, Hungary, Iceland, Ireland, Italy, Japan, Korea, Luxembourg, Mexico, the Netherlands, New Zealand, Norway,

Poland, Portugal, Slovak Republic, Spain, Sweden, Switzerland, Turkey, the United Kingdom and the USA.
82 'Members and partners', Organisation for Economic Co-operation and Development, www.oecd.org/pages/0,3417,en_36734052_36761800_1_1_1_1_1,00.html (accessed 27 June 2008).
83 For publications of the OECD International Futures Program, see www.oecd.org/document/8/0,3343,en_2649_34815_40862152_1_1_1_1,00.html (accessed 16 February 2009).
84 'Welcome to the Inter-Agency Space Debris Coordination Committee', The Inter-Agency Space Debris Coordination Committee, www.iadc-online.org/index.cgi?item=home (accessed 11 June 2008).
85 Members of the IADC include: Italian Space Agency, British National Space Centre (BNSC), Centre National d'Etudes Spatiales (CNES), China National Space Administration (CNSA), German Aerospace Centre (DLR), European Space Agency (ESA), Indian Space Research Organization (ISRO), Japan Aerospace Exploration Agency (JAXA), National Aeronautics and Space Administration (NASA), National Space Agency of Ukraine (NSAU), and Russian Federal Space Agency (ROSCOSMOS). See www.iadc-online.org/index.cgi?item=members (accessed 11 June 2008).
86 'Chinese Anti-satellite Test Creates Most Severe Orbital Debris Cloud in History', *Orbital Debris Quarterly News* 11: 2 (2007), orbitaldebris.jsc.nasa.gov/newsletter/newsletter.html (accessed 18 June 2008); for the satellite collision, see www.agi.com/corporate/mediaCenter/news/iridium-cosmos (accessed 16 February 2009).
87 'IADC Space Debris Mitigation Guidelines', 15 October 2002, Inter-Agency Space Debris Coordination Committee.
88 Corinne Contant-Jorgenson, Petr Lála and Kai-Uwe Schrogl, 'The IAA Cosmic Study on Space Traffic Management', *Space Policy* 22: 4 (2002): 283–88.
89 'Report of the Committee on the Peaceful Uses of Outer Space' (General Assembly Official Records: Sixty-Second Session, 2007).
90 'IADC Document Registration List', May 2008, Inter-Agency Space Debris Coordination Committee, www.iadc-online.org/index.cgi?item=docs_pub (accessed 14 July 2008).
91 Inter-Agency Space Debris Coordination Committee, 'Long Term Sustainability of Space Activities: International Informal Working Meeting' (Paris, France, 7 February 2008).
92 'Overview', Wassenaar Arrangement, www.wassenaar.org/introduction/overview.html (accessed 10 July 2008).
93 'Frequently Asked Questions', Wassenaar Arrangement, www.wassenaar.org/faq/index.html (accessed 10 July 2008).
94 Participating countries include: Argentina, Australia, Austria, Belgium, Bulgaria, Canada, Croatia, Czech Republic, Denmark, Estonia, Finland, France, Germany, Greece, Hungary, Ireland, Italy, Japan, Republic of Korea, Latvia, Lithuania, Luxembourg, Malta, Netherlands, New Zealand, Norway, Poland, Portugal, Romania, Russian Federation, Slovakia, Slovenia, South Africa, Spain, Sweden, Switzerland, Turkey, Ukraine, the United Kingdom and the USA.
95 'How does the Wassenaar Arrangement work', Wassenaar Arrangement, www.wassenaar.org/introduction/howitworks.html (accessed 10 July 2008).
96 Antonella Bini, 'Export control of space items: Preserving Europe's advantage', *Space Policy* 23: 2 (2007): 70–72.
97 Ibid.

Non-governmental Space Organizations

JAMES A. VEDDA

Space-related non-governmental organizations (NGOs) can range from small and local to large and global; from clubs of interested amateurs to collectives of seasoned professionals; and from politically insignificant to fairly influential. Some are relatively new, while others predate the age of space flight. In general, new groups formed and established groups increasingly turned their attention to public policy concerns in the aftermath of the Apollo programme. At that time, reduced funding and disappearing programmes put the future of space exploration and development in doubt, and there was a perceived need for action to stave off decline. Since then, organizations that have established their credibility have become valued information sources for policy-makers and, in some cases, for members of the general public as well.

The broadest definition of NGOs for space would include, as the name indicates, every space-related organization that is not part of a government. While this is conceptually simple, the immense size and diversity of the resulting community is too unwieldy for consideration unless it is divided into components of similar characteristics. Even this can be challenging, as the categories cannot avoid overlap in their goals, membership and types of activities. With this in mind, and focusing on not-for-profit organizations that are designed, at least in part, to influence space policy and strategy, this chapter frames its discussion using the following categories: non-governmental international organizations (IOs); industry and professional organizations; think tanks; space advocacy groups; and opposition groups.

Each of these categories is presented in general terms and illustrated through profiles of selected organizations in the community.[1] Organizational characteristics of interest, such as a group's purpose, history, target audience, size, geographic reach, major activities and publications, are discussed. Indications of effectiveness in influencing policy and strategy, for individual groups and for NGOs as a whole, are examined.

MODELLING NON-GOVERNMENTAL ORGANIZATIONAL BEHAVIOUR IN THE POLICY PROCESS

Political scientists and sociologists have studied interest group behaviour for decades, attempting to understand the internal dynamics that drive their actions. They also have sought to determine how the groups fit into the larger policy-making community and whether they are effective in this role. This interaction with the larger community is the focus of this chapter.

In the realm of domestic politics, the traditional 'iron triangle' has been used to describe the arrangement that existed in the early US space programme, with the three points of the triangle being: the executive agencies involved in space (the National Aeronautics and Space Administration, NASA and the US Department of Defense, DoD); the relevant congressional committees; and the space interest groups that are primarily large contractors to the US Government. While this may have been a compelling way of looking at early space policy-making, it describes a policy monopoly that could not be sustained for very long due to the tremendous expansion and diversification of space interest groups. The iron triangle model lacks the flexibility to accommodate the interaction of a large, multi-faceted NGO community whose members vary significantly in their capacity and willingness to act on each of the key issues facing the space community at any given time. A more accurate depiction of the NGO role in the policy process can be found in the issue network model,[2] which consists of a large number of participants with varying degrees of commitment. Participants move in and out of the network continuously, depending on their interest in particular issues, making it difficult to determine where the network ends. No one person or group appears to be in control of the policies and issues.

While the iron triangles of the past were small, autonomous operations, issue networks have many players at several levels of involvement. In the case of the space community, this includes older organizations that originally were content to serve their target audiences without direct involvement in public policy, as well as new organizations specifically designed to directly or indirectly influence policy. This chapter examines some of those players, starting with a category that features a number of well-established entities on the world stage.

NON-GOVERNMENTAL INTERNATIONAL ORGANIZATIONS

Many aspects of space activities are multinational and even global in nature. This was evident from the beginning in space science and has become accepted for technical and operational aspects as international co-operation has grown and globalization of space capabilities has progressed. A logical outgrowth of these circumstances was the emergence of IOs serving space constituencies in scientific, technical, legal and commercial disciplines.

The UN Department of Public Information/NGO Directory lists over 1,600 NGOs in 46 categories.[3] Only some of these appear in the Outer Space category,[4] but that does not mean space applications are excluded from the work of the others. Among the other NGO categories that involve space-related activities are: Agriculture; Climate Change; Environment; International Law; and Science and Technology.[5]

The largest of the organizations in the UN directory's Outer Space category, as well as the oldest, is the International Astronautical Federation (IAF).

Non-governmental Space Organizations

Headquartered in Paris, France, it was formed in London in 1951 by interest groups from 10 states: Argentina, Austria, France, Germany, Italy, Spain, Sweden, Switzerland, the United Kingdom (UK) and the USA. Its major activities include: promoting awareness of space activities world-wide; aiding exchange of information on space programme developments; helping to develop highly motivated and knowledgeable workforces; recognizing achievements in space activities; and promoting the uses of space systems for human development.[6]

The IAF, in co-operation with its associates, the International Academy of Astronautics (IAA) and the International Institute of Space Law (IISL), organizes an annual International Astronautical Congress, which provides a forum for the exchange of news and ideas and a blending of cultures from spacefaring nations. The IAF also sponsors focused symposia on specific topics of current interest to complement the larger annual meetings.

The IAF membership consists of 165 organizations from 45 countries. There is no way to accurately determine how many individuals this represents, but they certainly number in the hundreds of thousands given the size of the astronautical and other professional societies, space agencies, IOs, space companies, universities, research institutes and non-profit organizations that constitute the membership.

The IAF established the IISL[7] in 1960, replacing the Permanent Committee on Space Law that had been created two years earlier. The IISL has members from over 40 countries and holds an annual colloquium on space law in conjunction with the International Astronautical Congress. Although a component of the IAF, the IISL functions independently. In co-operation with IOs and national legal institutions, it assists in the development of space law and related studies through meetings and competitions on legal and other social science aspects of space activities.

A particularly interesting IISL activity is the annual space law competition, which helps to nurture upcoming generations of space law professionals.[8] The Manfred Lachs Space Law Moot Court Competition was started in 1992 to challenge law school and other graduate level students with hypothetical space law problems, requiring competitors to argue the case in a courtroom setting. National and regional competitions are held in Europe, North America and the Asia-Pacific region, and the three regional winners compete in the world finals at the International Astronautical Congress, judged by three sitting members of the International Court of Justice. At all levels of the competition, this experience serves to cultivate an international perspective in the next generation of space lawyers.

The 1,600 NGOs registered with the UN are just a small fraction of the total number of such organizations around the world. Similarly, the small Outer Space category in the UN NGO directory is not representative of the number and variety of space-related organizations that can be found worldwide. Some other well-established and well-known space NGOs are noted below.

Since 1958, the Committee on Space Research (COSPAR) has provided a forum to promote international co-operation and exchange of information on space science research. Based in Paris, France, COSPAR sought to stay above Cold War geopolitics and build bridges between East and West for co-operation in space. In the post-Cold War era, COSPAR has continued to focus on stimulating progress in all kinds of space research. COSPAR organizes meetings such as its biennial scientific assemblies and publishes a journal called *Advances in Space Research*.[9]

The International Academy of Astronautics (IAA) was founded in 1960 in Stockholm, Sweden, and today is based in Paris, France. Established to foster the development of astronautics for peaceful purposes, IAA provides avenues for international co-operation and recognizes individuals who have distinguished themselves in space fields. The organization holds scientific meetings and publishes the journal *Acta Astronautica*.[10]

These international societies and others specifically tied to scientific disciplines, such as astronomy, physics and planetary geology, undoubtedly have encouraged international information sharing and collaboration that would not have existed otherwise. It is impossible to determine how many government space projects, especially in the sciences, owe their existence to interactions stimulated by these forums. Nor is it possible to accurately judge whether national space programme priorities and budgets were influenced in any significant way by the international meetings, publications and personal contacts spawned by these organizations. None the less, their longevity, prestige and determination to stretch the limits of their disciplines speak volumes about their importance in sustaining global space efforts and guiding them into the future.

INDUSTRY AND PROFESSIONAL ASSOCIATIONS

Among the oldest space-related NGOs are the professional associations that emerged in the early days of aeronautics and astronautics. Today, they have grown in number and can be found in spacefaring states around the world. They serve their membership by providing venues for exchange of technical information, supporting education and generally looking after the interests of their respective professions. This can include taking positions on government programmes or courses of action, often handled by a sub-group specializing in public policy. Examples of such organizations are described here.

The American Institute of Aeronautics and Astronautics (AIAA) states its mission as follows: 'To advance the arts, sciences and technology of aeronautics and astronautics, and to promote the professionalism of those engaged in these pursuits'.[11] The organization was established in 1963 by the merger of the American Rocket Society and the Institute of Aerospace Science, both of which originated in the early 1930s. As of 2007, the AIAA claimed over 31,000 members, including engineering and science professionals, and university/college student members in aviation, space and defence

Non-governmental Space Organizations

fields. The group acts as an industry advocate, information source, conference sponsor and education supporter. It is also the publisher of a vast number of books and conference papers, and the monthly magazine *Aerospace America*.

In addition to its more than 70 technical committees focused on a wide variety of aeronautics and space activities, AIAA has a public policy committee and an international activities committee, often working together, to formulate organizational positions on current and emerging issues and manage communications with US executive agencies and the US Congress. This includes actions such as publishing policy papers and press releases, holding seminars and workshops on prominent policy issues and testifying at congressional hearings.

AIAA can draw on a large pool of expertise, which includes all of the technical areas relevant to the space enterprise as well as social science areas such as political science and economics. From one perspective, AIAA can be viewed as just one of the multitude of special interest advocates, but its reputation for authoritative input to the policy process in a complex technical area makes it a valuable player among executive and legislative offices that are seeking useful, reliable information.

While the AIAA is built on individual members from aerospace professions, the Aerospace Industries Association (AIA) represents the leading manufacturers and suppliers in the USA of civil, military and commercial space systems and components, as well as those for aeronautical applications.[12] AIA can trace its roots as least as far back as 1919, when the Aeronautical Chamber of Commerce was founded by 100 charter members 'to foster, advance, promulgate and promote aeronautics, and generally, to do every act and thing which may be necessary and proper for the advancement' of American aviation. Early members included such aviation pioneers as Orville Wright and Glen H. Curtiss, as well as representatives of major aircraft manufacturing units in the USA. Its name was changed to the Aircraft Industries Association of America after the Second World War, and to the Aerospace Industries Association in 1959 to encompass the emerging space industry.

As of 2007, the AIA included over 100 member companies and 170 associate member organizations. Its mission, stated in simple terms, is to shape policy. It does this by regularly publishing newsletters, providing testimony and background briefings to relevant committees of the US Congress, speaking at conferences and other industry events, and interacting with news and trade media. The AIA can be very effective in getting the industry position heard, but as the organization operates by consensus, not every issue will find sufficient agreement among all members.

Another long-lived association is the American Astronautical Society (AAS), which was formed in 1954 and draws its membership from both individual space professionals and corporate and academic organizations.[13] Like the groups mentioned previously, the AAS also seeks to influence public policy, but is best known for its two annual conferences, in March and

November, and its specialized technical conferences on topics such as space flight mechanics, and guidance and control. The AAS is well respected for its publications, including *The Journal of the Astronautical Sciences* (a peer-reviewed quarterly), the bi-monthly magazine *Space Times*, and book series on space science, technology and history.

THINK TANKS

Public policy think tanks are created for the purpose of researching policy options and making recommendations to government and private-sector entities. Many of them touch on space policy issues as part of their science and technology portfolio, especially as they relate to defence and international security. Think tanks that deal exclusively, or even routinely, with space policy issues are relatively few in number. Those that do are primarily concerned with foreign policy, arms control and disarmament, and science and technology funding priorities.

The oldest organization that qualifies as a space-related think tank is the RAND Corporation, which started out as a component of the Douglas Aircraft Company in 1946.[14] Its first study, released in May of that year, was *Preliminary Design of an Experimental World Circling Spaceship*.[15] Produced in just three weeks, the study was more than 300 pages of detailed engineering analyses, which foresaw all of the major space applications that later became vital to the world: communications, weather monitoring, surveillance, reconnaissance and navigation.

In the decades that followed, RAND studies were highly influential not only in satellite design and applications, but also in nuclear weapons, missiles, information technology and other areas of interest to government customers. By the 1970s, RAND's list of technical disciplines had expanded to encompass many more issues of interest in public policy, business and economics.[16] RAND's online archive of publications shows a large output of publications on a vast array of topics. A detailed examination reveals that space-related studies have declined in number and prominence in recent years. This may be a result of changing interests among RAND's customers as well as the departure of key personnel with space expertise. RAND has moved away from its space roots as it has diversified.

In contrast, another think tank that traditionally has not been associated with space is expanding its portfolio in this area. The Center for Strategic and International Studies (CSIS) was established in Washington, DC in 1962 to study challenges to national and global security and suggest new insights and possible policy solutions. Technology issues have always played a role in CSIS studies, but space became a more prominent subject in 2003 when the organization began its Human Space Exploration Initiative programme to examine the global implications of humanity's movement into space.[17]

CSIS has published papers on space policy issues and has conducted a series of seminars under the heading 'The Global Space Agenda'.[18] Presenters have been international experts addressing questions such as the emergence of the China and India as spacefaring nations, the role of the UN in space development, global co-operation in Earth observation and international regulation of space tourism. The impact of this attention remains to be seen. CSIS is a respected non-partisan think tank with no track record of advocating a particular course of action in the exploration and development of space, so its work could be welcomed by a space community looking for fresh, independent perspectives.

Some think tanks take clearly partisan positions on space issues. Examples include the Center for Defense Information,[19] the Federation of American Scientists,[20] the Henry L. Stimson Center[21] and the Union of Concerned Scientists,[22] all of which are sceptics of space control and space weapons proposals; and the George C. Marshall Institute[23] and the Heritage Foundation,[24] conservative organizations supporting missile defence and expanded military space applications.

SPACE ADVOCACY GROUPS

In addition to groups designed to represent the professional and economic interests of their membership, there are those that include members who typically derive no direct benefits from advocacy of space activities other than satisfying their personal interest in the subject. Most of these 'grassroots' groups appeared in the aftermath of the Apollo programme, when the budgets of civil space exploration and development programmes declined dramatically. However, space flight enthusiasts began to organize much earlier.

One noteworthy advocacy group that originated prior to the space age and still exists today is the British Interplanetary Society (BIS). Formed in 1933, it was a founding member organization of the IAF in 1951 and continues to serve the interests of its international membership, which consists of both space professionals and nonprofessionals. It publishes the monthly *Journal of the British Interplanetary Society*, a peer-reviewed technical journal, and the monthly magazine *Spaceflight*, which caters to a wider audience.[25]

The BIS took the long-term view of progress in astronautics. In the 1930s, when rocket societies in Germany, Russia and the USA were striving to reach the upper atmosphere and the edge of space, the Society was already working on design concepts for multi-stage rockets that would reach the Moon. In the 1970s, after Apollo had ended and the Space Shuttle had yet to fly, the BIS produced its Project Daedalus study on the use of a fusion-powered rocket to send a science probe on a 50-year journey to Barnard's Star.[26] Although the Society never built the ambitious hardware it designed, these projects provided technically grounded inspiration to existing and aspiring space professionals well beyond the group's membership.

An assortment of post-Apollo pro-space groups have come and gone over the years, a few of which have demonstrated staying power to survive beyond the dedicated efforts of their originators.[27] As with other types of NGOs, advocacy groups feature a variety of interests that sometimes provide opportunities for collaboration between groups and other times generate friction. Areas of interest include: public education at all levels, including 'teaching the teachers'; economic ambitions, including employment opportunities, industrial growth and the development of new forms of space commerce; construction, test and application of amateur space hardware (e.g. rocket clubs or ham radio operators); space sciences, especially astronomy and planetary geology; space tourism and settlement; space defence, including space weapons; and space arms control.

Most advocacy groups have mixed purposes, although some have focused on specific goals. An outside observer may see these groups as sharing a common bond – support for the exploration and development of space – which should compel them to routinely band together to address pressing issues, as environmental advocacy groups often do, or merge into a smaller number of larger, more powerful groups. However, a variety of forces have prevented this from happening, often to the detriment of efforts to influence governments and educate citizens.

- Some groups are local or regional clubs with no plans to expand beyond their immediate geographic area and, therefore, no prospects for growing to a large membership.
- Some groups were created by one or two strong personalities who do not wish to co-operate with other organizations, which would require them to share leadership or, in their view, dilute their message.
- Larger groups are hesitant to team with smaller groups, which they believe will do little more than drain their funding.
- Groups disagree on three divisive issues: the relative value of human and robotic space flight; the relative importance and potential of civil government and commercial space efforts; and the role of military activities in space.

Some examples illustrate the persistent problems. The National Space Society (NSS)[28] and the Space Frontier Foundation (SFF)[29] both emphasize the opening of space for all of humanity and the development of its resources. However, their approaches to achieving this can sometimes conflict. Traditionally, the NSS has given across-the-board support to NASA programmes, while the SFF sees the government in general, and NASA in particular, as a big part of the problem and seeks ways to bypass the space agency. In other words, the NSS, headquartered in Washington, DC, has positioned itself as an insider, while the SFF, headquartered in Nyack, NY, has assumed the role of the vocal outsider. Unless one of these organizations changes its approach, the two are unlikely to find common ground on civil space policy, although they may be in agreement on commercial space and public education issues.

Tension between groups already was evident early in the post-Apollo years, as exemplified by the early 1980s attempt to unite space advocacy groups under the banner of the National Coordinating Committee for Space. In terms of membership, the largest of the groups at the time was the Planetary Society with approximately 120,000 members.[30] The unwillingness of the Planetary Society to team with smaller, typically underfunded, groups was a showstopper for collaborative efforts.[31] Combined with other difficulties, such as disagreement over how to treat military space activities, this prevented the National Coordinating Committee from achieving its unification goal during its three years of existence.

The Planetary Society was also a vocal opponent of weapons in space. This stance, shared by other advocacy groups and some think tanks, runs afoul of those who believe aggression in space is inevitable, increased military activity in space is desirable and programmes in this area benefit space development in general. Such disagreements came to a head during the Strategic Defense Initiative programme in the US Reagan Administration, and were re-energized during the George W. Bush Administration. An advocacy group more supportive of military space projects, such as the Space Foundation,[32] may find it impossible to reconcile differences with a group like the Planetary Society, foreclosing collaboration even in areas where both have a strong interest, such as public education.

A closer look at some of the groups already mentioned is instructive in assessing their influence on public policy. The Planetary Society, the NSS and the SFF differ in some of their goals and in their approach to achieving them, but each has been around long enough to establish a track record of interests and group behaviour.

The Planetary Society was formed in 1980 by astronomer Carl Sagan, geologist Bruce Murray, then Director of the Jet Propulsion Laboratory, and engineer Louis Friedman. It quickly became the most prominent group endorsing a higher priority for robotic exploration over human space flight, although in the late 1980s it began to support human missions to Mars. It does not publish its current membership numbers, but claims to have members in more than 125 states.

In addition to being an advocate for space science programmes, the Planetary Society is a participant in several projects with partners around the world, including investigations of Mars and near Earth objects, development of solar sails, the search for extra-solar planets, and optical and radio searches for extraterrestrial intelligence. Despite this apparent diversity and its large membership compared to other space advocacy groups, the Society has never achieved the broad appeal that might have made it the nexus of the advocacy movement. Nor has it attempted to establish a lobbying arm to the US Government. Early in its existence, the Society relied on its high-profile founders to be its spokesmen. Various attempts at political intervention, including letter-writing campaigns to the US executive branch and Congress, proved ineffective and in some cases may have been counterproductive.[33]

The group's website claims that 'There is no doubt that without The Planetary Society, Earth's efforts to explore other worlds would not have reached as far into our solar system and beyond as they have today'.[34] This bold assertion, based on the Society's contention that it is at least partially responsible for the rovers on Mars and the New Horizons mission to Pluto, would be difficult to prove. For example, the Society staged letter-writing and petition campaigns to the US Congress on behalf of the troubled Pluto mission, which finally ascended from its launch pad in January 2006.[35] Although the desired outcome eventually was achieved, the effectiveness of the campaigns must be questioned, as they had to be repeated four times in four years.

Although the Planetary Society has been around since 1980, it has not yet demonstrated its ability to survive and flourish after a generational transition beyond its founders – a key indicator for long-term sustainability. As of 2007, Friedman was still the executive director and Murray was the chairman of the board of directors (Carl Sagan died in 1996). When these two gentlemen retire, a smooth transition to the next generation of leaders will be critical to the Society's longevity and effectiveness.

The NSS was formed by the 1987 merger of the National Space Institute and the L5 Society, both of which had their roots in the post-Apollo advocacy movement of the mid-1970s. Both groups sought to rally grassroots support for space development and build their membership to a politically significant size, but they differed in style and approach. The Institute had the reputation of being the pro-NASA 'establishment' group; government-centric, it sought big corporate sponsors and well-known board members, and defended NASA programme budgets, which pleased both the space agency and its contractors. In contrast, L5 was seen as a nonconformist, even renegade organization that wanted to build its political power through a large network of local chapters. As memberships declined and operating costs went up for both groups in the mid-1980s, they joined forces, and the NSS retains some characteristics of each. For example, it still has a Washington, DC headquarters, but has inherited the L5 chapter network.

On its website, NSS claims to be 'the preeminent citizen's voice on space'. The size of its membership is simply described as 'thousands of members and over 50 chapters in the United States and around the world'. The combined membership of the National Space Institute and the L5 Society at the time the two organizations began merger discussions was estimated at nearly 20,000,[36] but the number today is likely much lower. One possible indication of reduced membership is that the Society's publication, *Ad Astra* started as a monthly magazine, later became bi-monthly and more recently was reduced to quarterly.

NSS representatives frequently have made appearances before the US Congress over the years, usually by testifying before the US House subcommittee in charge of NASA's budget authorization. Like other advocacy organizations that periodically enjoy high-level access to the policy-making process, NSS has portrayed such events as evidence that it is helping to shape space policy and has been a critical factor in saving specific NASA programmes, such as the International Space Station (ISS) and the Galileo probe

Non-governmental Space Organizations

to Jupiter. However, there is no evidence that NSS or any other space advocacy group has wielded significant influence over major programmatic or policy decisions.[37]

To date, the biggest policy issue in which an advocacy group has had a reasonable claim that its efforts resulted in victory was the 1979 L5 Society campaign to stop US Senate ratification of the Moon Treaty,[38] which critics saw as detrimental to future commercial development of space. However, even this claim is weak, based solely on the fact that L5 paid a lobbyist to fight the Treaty and the desired outcome was achieved. Upon closer examination, one finds that the Treaty had no organized support and faced much opposition, primarily due to the fear of the precedents it would set in other areas such as seabed mining.[39] Treaty ratification was doomed to fail with or without help from L5 lobbying.

A noteworthy strength of the NSS compared to many other space advocacy groups is that it has survived a generational change in its leadership. The NSS and its predecessor organizations have been around for over three decades, proving that it is not a personality cult that collapses when its founders move on, nor is it too fragile to weather the storms of political transitions, fluctuations in issue salience, organizational evolution and economic challenges. This does not automatically equate to policy influence, even for a group embedded in the Washington, DC scene, but it does signal the potential for positive achievements in the future.

After the L5 Society merged with the National Space Institute in 1987, it appeared that the space advocacy movement's only prominent rebel organization had all but disappeared. Just a year later, however, a new rebel emerged: the SFF. Focused on enabling the human settlement of space as quickly as possible, the group's founders articulated what they termed 'three truths': large-scale industrialization and settlement of the inner solar system is technically possible within one or two generations; this is not happening and it cannot happen under a centrally planned and exclusive US government space programme; and the existing bureaucratic programme must be replaced with an inclusive, entrepreneurial, 'frontier-opening' enterprise, primarily by working on the outside to promote radical reform of US space policy.[40]

Concluding that no existing organization was appropriate to this task, the group's creators decided that the responsibility fell to them and that the Internet would allow them to get their message out to the world undiluted. The SFF approach has included strongly worded and often antagonistic statements that blame NASA and its major contractors for stifling entrepreneurial space commerce and human settlement goals.[41] Like other space advocacy groups before it, the Foundation has produced its own publication, sponsored conferences, created a lobbying organization, ProSpace, and highlighted its perceived influence on space policy and programmes. The group's website displays a list of accomplishments, although most of these are space-related activities in which its members have participated, rather than Foundation initiatives. Items on the list that represent the group's own efforts include defining the goal of cheap access to space, saving the funding for the DC-X experimental rocket, which was later cancelled, and 'stopping NASA's pre-1994 plans to maintain its monopoly on human space flight well into the 21st century'.[42]

At this point it is premature to say whether or not the SFF has made any long-term contribution to space development. If it succeeds in its goal of popularizing the visions of Konstantin Tsiolkovsky, Gerard K. O'Neill and Arthur C. Clarke, and facilitates collaboration between entrepreneurs, policy-makers and others who will turn these visions into reality, then it will have performed a service of immeasurable value. However, this work remains to be done. The Foundation has yet to prove that it can survive beyond its creators and flourish under a new generation of leadership.

Overall, the impact of US space advocacy groups on major space programme and policy decisions has been minimal. They have lent their voices in support of government space projects conceived without their input, sometimes claiming success and sometimes suffering failure in their attempts to boost the fortunes of these projects. In no case can it be clearly demonstrated that decisions would have gone differently in the absence of their efforts.

While some groups have claimed that they 'saved' troubled programmes or were instrumental in initiating new starts, there is no conclusive evidence that such claims are anything more than a desire to be associated with successful space projects, or rhetoric to enlist new members. Testifying at US congressional hearings or providing information to policy-makers can be a valuable service, but it may amount to no more than reinforcing existing positions, rather than altering them or inspiring new initiatives. The groups have made little headway in promoting new directions for policies or programmes. As far back as the 1979 episode in which the L5 Society participated in the defeat of the Moon Treaty, the space movement discovered that it is far easier to oppose something than it is to create something new.

Government decision-making on space continues to be dominated by internal processes, supplemented by information and lobbying from industry and other direct stakeholders. Although political influence has spread within the expanding space community, citizen advocates have not shared in this expansion. For example, there was no involvement of space advocacy groups in the formulation of the 2006 US National Space Policy[43] or the 2004 US Space Exploration Policy,[44] despite three decades of efforts by these groups to play a role in this process.

The space advocacy movement is a diverse, eclectic, fragmented phenomenon, without a dominant leader or an agreed-on agenda. In the USA, the space advocacy groups representing a few tens of thousands of citizens are very small in membership, resources and visibility compared to influential lobbying organizations in other areas, such as the environmental group Sierra Club of 1.3m. members,[45] the National Rifle Association of 3m. members,[46] and the American Association of Retired Persons of 39m. members.[47] The space groups compete with each other for membership and often are in conflict over basic issues. Such conflicts may cause groups to ignore public opinion polls on space if their views are contradicted, choosing instead to use polling data only to the extent that they can extract supportive results. Clearly, those groups that claim to be the 'voice of the people' on space issues are really speaking for a very small subset of the population.

OPPOSITION GROUPS

Although there is no evidence of active NGOs, the sole purpose of which is to oppose the exploration and development of space, there are groups that find certain space programmes to be in direct or indirect conflict with their interests, causing them to focus part of their efforts on preventing certain types of activities in space. Some interest groups may choose to lobby against space projects simply because they are competitors for scarce government funds. The most popular substantive objections are in response to existing or potential military space operations and the use of nuclear power in space. The dual-use nature of almost all space technologies fuels suspicion in some communities that all space projects have a military purpose, and may help enlist the efforts of a variety of groups, e.g. environmental, anti-nuclear or arms control, in protests over missions conducted by all sectors of the space community.

Environmental groups have traditionally opposed space projects for various reasons. For example, in the late 1970s there was opposition to proposals to build space-based solar power satellites that would beam energy down to the terrestrial power grid. This may seem counterintuitive, as solar energy from space was being proposed as an alternative to fossil fuels. However, many environmentalists became convinced that the satellites' transmission of energy through microwave beams would damage the Earth's atmosphere and endanger airplanes, birds and possibly life on the ground near the beam's receiving antenna. The potential for these negative side-effects was being studied at the time.[48] Nevertheless, when solar power satellite funding was cancelled in the USA in 1980, the decision was driven primarily by concerns about cost and technological maturity, with environmental concerns playing only a minor role.

More generally, there have been concerns about the pollution caused by rocket fuel during its manufacture and use.[49] On the other hand, there also has been recognition by many environmentalists that space-based capabilities, particularly imaging of the Earth, are vital tools serving their interests in tracking deforestation, loss of wildlife habitat, air and water pollution, ozone depletion, urban growth and climate change.

In a different area of interest group concern, an animal rights group, People for the Ethical Treatment of Animals (PETA), has protested the use of animals for experiments in space. At one point, the group took its objections to the NASA Administrator's office. In 1996 PETA representatives protested NASA's involvement in the Russian Bion mission, which was scheduled to carry monkeys into space. In part, the protesters may have been motivated by erroneous press reports that the monkeys would die as a result of the mission.[50] PETA has taken credit for ending the Bion programme and for reductions in other NASA projects that included animal experiments,[51] but it is unclear what influence, if any, the group had on these decisions.

The most widespread and consistent interest group hostility to space activities has been directed against US and allied national security space policy. An example of one small, but persistent opposition is the Global Network Against Weapons and Nuclear Power in Space. This group is a UN registered NGO that arranges protests by aligning itself with anti-military, anti-nuclear and peace activist groups in North America and other parts of the world. The group's website critiques the George W. Bush Administration's 2006 National Space Policy by making linkages between civil and military space programmes, interpreting the policy as part of a strategic plan to dominate outer space and to deploy missile defences and space-based weapon systems.[52] Additionally, the group has attempted without success to prevent the launch of deep space probes with nuclear power sources, such as NASA's Galileo, Cassini and New Horizons missions.

To date (2009), special interest groups opposed to space activities have been small, fragmented in their approach to space issues and, therefore, not influential. For the past two decades, the most serious manifestation of resistance to civil space programmes has been the use of the court system by anti-nuclear groups. Opposition to the launch of nuclear-powered space science missions in each case involved at most a few dozen pounds of plutonium to provide modest power levels to spacecraft systems for journeys to the outer planets. Despite strict safety protocols and assessments showing minimal risk, anti-nuclear groups have responded with protests and have sought the intervention of the courts to prevent launches. Even though the courts have denied requests for restraining orders on NASA planetary launches, it cannot be assumed that this favourable treatment will continue. Only one successful challenge is needed to set a legal precedent, which could have implications for all nuclear-powered missions that follow.

In the future, it is possible that anti-globalization sentiments may become anti-space as well, especially if there is an anti-technology component to the movement. The same entities that dominate space activities – government institutions and large corporations – are seen by critics as orchestrating globalization to serve the wealthy at the expense of the poor. Space technology could be seen by globalization critics as a tool of transnational corporations that exploit workers, of foreign investors who undermine local businesses, or of wealthy, i.e. spacefaring, states that economically take advantage of developing nations. So far, space activities have not been directly targeted by globalization opponents.

CONCLUSIONS: SPACE NON-GOVERNMENTAL ORGANIZATIONS AND PUBLIC POLICY

There is no doubt that NGOs have an effect on public policy related to space, but it is more subtle than many organizations would like to believe. Overt actions by NGOs to alter policy significantly or initiate new programmes in the short term have proven ineffective. When national or international space agencies and their associated government leadership formulate space policy, NGOs do not have a seat at the table and only rarely are they formally asked for input.[53] Groups made up of a few thousand individual members that advertise themselves as the most influential voice in the space community, or the 'saviors' of major space programmes, are engaging in the type of self-promotion that is to be expected in membership organizations. In short, there is little evidence that direct interventions by NGOs, particularly advocacy groups, have been necessary or sufficient to ensure positive outcomes in public policy decisions on space.

Taking a broader view, it is clear that space-related NGOs – the types discussed in this chapter, plus those in business and academia – constitute the majority of the active space community world-wide, vastly outnumbering their government counterparts. As a result, NGOs collectively are the keepers of the 'culture'. They shape it and preserve it through a multitude of conferences, workshops, publications and other formal and informal contacts across programmes, disciplines, national borders and generations. They are the primary sources of expert information and new ideas. In other words, public policy on space would be bankrupt without them. This role requires the patient development of networks and credibility over an extended period of time.

Space tourism is an example of an idea that needed to be cultivated in the NGO community before being accepted in public policy. Only recently have governments recognized that this is an endeavour to be nurtured, requiring their near-term attention on licensing and regulation. It took three decades of evolution in technologies, economics and attitudes to make space tourism an accepted activity that governments could no longer ignore. Other examples include global satellite navigation and the Hubble Space Telescope, both of which eventually became reality as a result of 30 years of conceptual, technical, bureaucratic and budgetary evolution, which had to overcome perceptions that they would not be worth the investment. Direct intervention by NGOs did not produce immediate policy or programmatic action in these areas, but patience and persistence allowed them to become mainstream activities.

Significant and sustainable developments in space policy and law have been produced through many years of improving the state of the art, while communicating to policy-makers the possibilities and pitfalls of such developments. This communication takes place in countless impromptu meetings outside conference sessions and in behind-the-scenes briefings to legislative staffers. Results take time and visibility is low, but the long-run productivity of this approach can exceed efforts that are more visible, like aggressive lobbying and letter-writing campaigns. The groups most successful at affecting public policy on space have been, and will continue to be, those that demonstrate credibility, longevity and a process for routinely educating the broader community, especially policy-makers, about the evolution and insights of the global space community.

NOTES

1 This chapter does not address academic or research organizations. The extent to which these groups are able to influence space policy, or desire to do so, varies greatly. Space-related academic and research organizations exist to increase the body of knowledge in relevant areas, advance the state of the art in space technologies, and convey this knowledge to users and upcoming generations. Typically, they are not involved in formulating or lobbying for particular policies, although individuals connected with these organizations may participate in political activities due to personal interest or through special circumstances such as service on government advisory committees or involvement in community efforts to respond to a particular policy dilemma.
2 Hugh Heclo, 'Issue Networks and the Executive Establishment' in A. King (ed.), *The New American Political System* (Washington, DC: American Enterprise Institute, 1978).
3 United Nations Department of Public Information/NGO Directory, www.un.org/dpi/ngosection/dpingo-directory.asp (accessed 9 March 2009).
4 The International Astronautical Federation (IAF) in Paris, France, www.iafastro.com (accessed 9 March 2009); the National Space Society (NSS) in Washington, DC, www.nss.org (accessed 9 March 2009); the Institute for Cooperation in Space (ICIS) in Vancouver, Canada and Loja, Ecuador, www.peaceinspace.com; and the Global Network Against Weapons and Nuclear Power in Space in Brunswick, ME, USA, www.space4peace.org (accessed 9 March 2009).
5 An example of a space-related NGO in the science and technology category is the International Society for Photogrammetry and Remote Sensing (ISPRS), www.isprs.org (accessed 9 March 2009).
6 International Astronautical Federation, 'About IAF', www.iafastro.com/index.php?id=60 (accessed 15 November 2008).
7 International Institute of Space Law, www.iafastro-iisl.com (accessed 15 September 2008).
8 Manfred Lachs Space Law Moot Court Competition, www.spacemoot.org (accessed 20 October 2008).
9 Committee on Space Research, cosparhq.cnes.fr (accessed 10 September 2008).
10 International Academy of Astronautics, www.iaaweb.org (accessed 10 September 2008).
11 American Institute of Aeronautics and Astronautics, www.aiaa.org (accessed 10 September 2008).
12 Aerospace Industries Association, www.aia-aerospace.org (accessed 15 September 2008).
13 American Astronautical Society, www.astronautical.org (accessed 10 October 2008).
14 RAND Corporation, www.rand.org (accessed 10 October 2008).
15 For this historic report, see www.rand.org/pubs/special_memoranda/SM11827 (accessed 20 October 2008).
16 Virginia Campbell, 'How RAND Invented the Postwar World', *Invention & Technology*, Summer 2004.
17 Center for Strategic and International Studies, 'Human Space Exploration Initiative', www.csis.org/space (accessed 9 March 2009).
18 For example, see: 'The Still Untrodden Heights: Global Imperatives for Space Exploration in the 21st Century', 2005; 'US Leadership and Space Exploration', April 2006; 'NASA FY 2007 Budget Proposal – An Analysis', April 2006; Vincent Sabathier and G. Ryan Faith, 'Minding the Gap: Keeping Exploration Alive', October 2007.
19 Center for Defense Information, www.cdi.org (accessed October 2007).

Non-governmental Space Organizations

20 Federation of American Scientists, 'Space Policy Project', www.fas.org/spp/index.html (accessed 10 October 2008).
21 See Henry L. Stimson Center, Space Security Program, www.stimson.org/space/programhome.cfm (accessed 20 October 2008).
22 Union of Concerned Scientists, Nuclear Weapons and Global Security, 'Space Weapons Policy Issues', www.ucsusa.org/global_security/space_weapons (accessed 10 October 2008).
23 George C. Marshall Institute, 'Space Security and National Defense', marshall.org/category.php?id=8 (accessed 20 October 2008).
24 The Heritage Foundation, 'Space Issues', www.heritage.org/research/space (accessed 10 October 2008).
25 British Interplanetary Society, www.bis-spaceflight.com (accessed 5 August 2008).
26 A. Bond, A. R. Martin, R. A. Buckland, T. J. Grant, A. T. Lawton, et al., 'Project Daedalus – The Final Report on the BIS Starship Study', *Journal of the British Interplanetary Society*, 31 (Supplement), 1978.
27 For a chronicle of the emergence of these groups in the USA, see Michael Michaud, *Reaching for the High Frontier: The American Pro-Space Movement, 1972–84* (New York: Praeger Publishers, 1986).
28 National Space Society, www.nss.org (accessed October 2007).
29 Space Frontier Foundation, www.space-frontier.org (accessed 5 October 2008).
30 The Planetary Society, www.planetary.org (accessed 5 October 2008).
31 Personal Correspondence, David Webb, former chairman of the National Coordinating Committee for Space, 9 September 2007.
32 The Space Foundation, www.spacefoundation.org (accessed 5 October 2008).
33 Michael Michaud, *Reaching for the High Frontier: The American Pro-Space Movement, 1972–84* (New York: Praeger Publishers, 1986): 211–13.
34 The Planetary Society, 'History', planetary.org/about/history.html (accessed 10 October 2008).
35 The Planetary Society, 'Pluto Campaign', www.planetary.org/programs/projects/pluto (accessed 10 October 2008).
36 Michael Michaud, *Reaching for the High Frontier: The American Pro-Space Movement, 1972–84* (New York: Praeger Publishers, 1986): 343–44.
37 Ibid., 308–9.
38 UN Agreement Governing the Activities of States on the Moon and Other Celestial Bodies (Moon Treaty or Moon Agreement), 1979.
39 Michael Michaud, *Reaching for the High Frontier: The American Pro-Space Movement, 1972–84* (New York: Praeger Publishers, 1986): 92–93.
40 Space Frontier Foundation, www.space-frontier.org/History (accessed 5 October 2008).
41 Opinions of the Space Frontier Foundation can be accessed at spacefrontier.org/blog (accessed 5 March 2009).
42 Space Frontier Foundation, spacefrontier.org/about (accessed 5 March 2009).
43 National Security Presidential Directive (NSPD) 49, 'US National Space Policy', 31 August 2006.
44 National Security Presidential Directive (NSPD) 31, 'US Space Exploration Policy', 14 January 2004.
45 Sierra Club, www.sierraclub.org (accessed October 2007).
46 National Rifle Association, 'A Brief History of the NRA', www.nra.org/aboutus.aspx (accessed 25 October 2008).
47 American Association of Retired Persons, www.aarp.org/about_aarp/aarp_overview/a2003-01-13-aarphistory.html (accessed 25 October 2008).
48 US Office of Technology Assessment, *Solar Power Satellites*, August 1981, 179–224, 275–88, www.princeton.edu/~ota/ns20/alpha_f.html (accessed 25 October 2008).
49 See Environmental Working Group, 'Rocket Fuel in Lettuce', 29 April 2003, www.ewg.org/reports/rocketlettuce (accessed 25 October 2008); Steven Aftergood,

'Poisoned plumes: Across the US, environmentalists are protesting against rocket launches. Toxic exhaust fumes from rockets packed with solid propellant attract the greatest concern'. *New Scientist*, 7 September 1991, space.newscientist.com/channel/space-tech/space-shuttle/mg13117854.400 (accessed 15 October 2008).
50 'Money Woes Delay Russian Launches', seds.org/spaceviews/9611/news.html (accessed 25 October 2008).
51 Examples can be found at www.peta.org by searching the website for 'NASA' (accessed 25 October 2008).
52 Karl Grossman, 'Bush Opens Outer Space to Combat', www.space4peace.org/articles/bush_opens_space_to_combat.htm (accessed 20 October 2008); and Bruce Gagnon, 'NASA Plans Moon Base to Control Pathway to Space', www.space4peace.org/articles/nasa_moon_base.htm (accessed 20 October 2008).
53 An example of a case in which non-governmental organizations were formally solicited for space policy input can be found in the report of the Ronald Reagan Administration's National Commission on Space, *Pioneering the Space Frontier* (New York, NY: Bantam Books, 1986). Input was taken in written form and at the Commission's numerous town hall meetings. However, this activity was a couple of steps removed from policy-makers, and the Commission's recommendations were never implemented. A more recent example of broad input to a high-level policy commission can be found in the report of the President's Commission on Implementation of United States Space Exploration Policy (Aldridge Commission), 'A Journey to Inspire, Innovate, and Discover', US Government Printing Office, June 2004. A website was set up to accept input from individuals and organizations. However, space exploration policy already had been established by this time, making the outreach effort an exercise in solicitation of public support and collection of implementation ideas.

Public Sector Actors

Roger B. Handberg

At the very dawn of the space age, the governments pursuing various activities in space were driven by security and prestige to the effective exclusion of other considerations. Other facets of space activity emerged gradually from those initial efforts, but their growth in importance and activity were always bounded by the earlier and more established national space activities. The experience of the USA saw the initiation of civil and private activities, or at least allowed others to do so, while the Soviet programme remained more focused on a narrower socialist and government view of what was required. For example, telecommunications companies in the USA sought to move quickly to exploit the obvious potential for space-based communications. Those efforts forced a response by the US Government to keep such endeavours under control, so that they would not interfere with security priorities, while also facilitating other social and economic priorities or goals. Here, the focus is public sector organizations whose primary mission is not space oriented, but the activities of which impact what is accomplished in the area of space.

One important facet of activity by these newer participants, whether public or private, is the ability to change expectations first and then policies. The traditional or long-standing space participants usually at first resist such changes, but over time new attitudes and perspectives arise regarding what is deemed acceptable, leading to dramatic changes in policy. As this political and bureaucratic process continues, it brings private players into the policy process providing necessary political push to achieve the changes sought. This change process remains incremental rather than dramatic, but the outcomes appear enduring.

The analysis herein encompasses diverse levels of engagement in space activities, including the international and regional levels, but primarily focuses upon national and sub-national public space actors, their private de facto partners and their impact on the space policy process. This chapter begins with a brief survey of the larger historical context within which to place the activities of these more recent players, the existence of which was not considered in the early central government-driven scenarios. Other participants were found in science fiction, the bold entrepreneurs in space. This was long considered unlikely due to the cost of travelling into orbit and working there, but now these new players are becoming engaged without, in many cases, physically entering Earth orbit or expecting to do so.

EARLY DAYS OF THE SPACE AGE

The roots of the space age lie among isolated individuals and groups who pursued the dream of humans and their machines entering outer space and moving off to explore other celestial bodies.[1] The jump start that moved those dreams from aspirations to the beginnings of the human adventure in outer space came from the military, at first not directly, but through the pursuit of ballistic missile weapons. A Faustian bargain was made in which the military supported their efforts to build missiles, prototype space launch vehicles from another perspective.[2] The complex and expensive problems inherent in building such rockets was well beyond the financial capabilities of the space enthusiasts. At first, there existed no immediate economic prospects, although many saw their potential. After the Second World War, military efforts accelerated as the defence potential of ballistic missiles became ever clearer. Entering Earth orbit was more problematic, as the assumption was that all energies should be solely focused on achieving a successful missile launch and target impact. However, the military potential for Earth orbiting satellites for communications and Earth observation was already clear even in the absence of achieving access to orbit.[3]

Given this national security focus and the fact that the military in the first two space states, the USSR (Union of Soviet Socialist Republics) and the USA, funded development of what became the first lift vehicles, the opening of the space age was one characterized by government-controlled and -dominated activities.[4] These activities were justified in some manner by their links to national security and international prestige. Initially, there was comparatively little discussion of other facets, although their potential was acknowledged as future goals. This perspective dominated space activities generally, not just for the two first spacefaring states, and meant that all decisions made regarding space activities took place in a policy environment premised on national government control and pursuit of national advantage. This mindset continues to dominate the field, despite the growing independence of other actors, both public non-military and private. Development of private endeavours is not prevented by the existing international legal regime, but they operate only with the agreement of national governments. Without government licensing, regulation and economic support, in most cases independent private space activities would not exist.

National advantage as a motivator has shifted in terms of its content over time.[5] Initially, pursuit of national military advantage trumped all competitors with national prestige considerations; as long as the national security aspect dominated international space policy discussions, the other factors were subordinate to that goal. By the early 1970s, and even earlier among the Western industrial states, space activities began to be perceived as possible agents for facilitating national economic growth and enhanced economic competitiveness. The demonstrated capacity to pursue space activities successfully signalled the state's technological capabilities to its competitors in an enormously visible manner. As has occurred over the last decade, the People's Republic of China and India have both signalled their arrival on the international stage as major technological players through their demonstrations of space capabilities and plans of expansive space goals.[6] In order to independently launch and recover payloads, including humans in the case of China, each state demonstrates a technological capability that attracts international investment and economic partners from all levels of the economic development spectrum.

Along with this escalating interest in economic competitiveness, global interest in the social applications of space technologies also took hold. These attracted more states to the field, including those that were otherwise reluctant for political and ideological reasons, or for reasons of economic capacity, to become involved. The expanding availability of ready access to space applications and their products empowers states, otherwise excluded from participation, economically and militarily. For example, military-grade imagery can be obtained from commercially available remote sensing satellites.[7]

In the early days of their space programmes, other states outside the USA and the USSR defined outer space as a realm for peaceful activities, excluding military spacecraft. This definition did not exclude further expanding space applications to include military purposes, as occurred both in Europe and Japan. The shift of Japan from a peaceful space programme to one encompassing at least some military space applications was the most profound change, given the severe limits in their national constitution regarding military matters. The practical distinction accepted by all states was that military satellites were acceptable as long as no actual weapons were placed in orbit. Space became a sanctuary that fostered the development of commercial and other applications.[8]

The shift in emphasis from security and prestige to a mix also including economic competitiveness and social utility, brought more diverse public players into the space policy milieu despite their earlier reluctance in some cases to become involved, or their initial lack of institutional capacity to intervene. Agencies that dealt with normal regulatory matters, at first including telecommunications, were largely marginal to space activities. Over time, the growing importance of space-based telecommunications had the effect of focusing attention on telecommunications regulatory matters; communication satellites (comsats) used the radio spectrum, which meant that the spectrum was unavailable for terrestrial carriers – a matter of great economic and competitive concern.[9]

For national space programmes, becoming socially useful was critical in order to gain support from other public agencies, owing to the value of space-based applications as social goods. This economic and social dimension is critical for fostering programme growth. Societies plagued by a lack of reliable domestic communications, especially in remote areas, now saw comsats as cost-efficient and-effective options for resolving such disconnects in social communications. Remote sensing satellites likewise added great social value, allowing governments and later commercial interests the ability to assess weather conditions, flooding, drought, insect infestations, fires and the like, which allowed earlier mobilization of assistance. Obviously, as after the Katrina hurricane disaster in 2005 in the USA, effective assistance does not necessarily follow the receipt of such information, but the potential for effective response now exists where it previously did not. For many governments that are marginally effective, these informational aids are particularly important for natural disaster mitigation.

INTERNATIONAL AND REGIONAL PLAYERS

When national security and prestige considerations lost momentum due to the high costs, developing space applications for social and economic purposes began in earnest. The initial developments tended to reflect political and strategic considerations, with the USA and the USSR as the polar forces around which the other states aligned. Evidence of this can be seen in the development of various international organizations (IOs) aimed at rationalizing and facilitating space activities, such as Intelsat and Inmarsat, in order to conduct telecommunications through comsats. Intersputnik International Organization of Space Communications was one Soviet response to Intelsat.

Equivalent arrangements, such as the World Meteorological Organization (WMO), were developed for handling meteorological data collection and dissemination across member states. While some states kept their climatic information outside such international arrangements, most in time joined the WMO based on their needs for enhancing their national economic and technological development using such data. The reciprocal nature of the WMO made co-operation mandatory by those states interested in participating.

On a regional basis, organizations arose around specific functional areas, like the European Space Agency (ESA), the European Telecommunications Satellite Organization (Eutelsat), and the Arab Satellite Communications Organization (Arabsat). The ESA represents the European effort to maximize opportunities in space despite domination by the USA. Eutelsat and Arabsat, meanwhile, are examples of organizations focused on a single application: communications using Earth orbiting satellites. Their purpose was to create larger entities, allowing some inherent efficiencies of scale and allow the merging or pooling of scarce technological resources. This allowed smaller or less technologically developed states, otherwise excluded from the space realm, to become involved at a significant level beyond their national capabilities.

More recently, regional organizations, such as the Asia-Pacific Space Cooperation Organization (APSCO), have arisen where the implicit model is the ESA one, but the resources employed are national rather than collective in nature. In the case of APSCO, the organizational problem is that China is technologically advanced compared to other members, especially given Japan's non-participation until now. The Asia-Pacific Regional Space Agency Forum (APRSAF) and the Asia-Pacific Satellite Communications Council are other efforts at developing co-operation and assistance among their members. The actual work of these organizations remains largely conducted through a series of bilateral agreements, rather than the more integrated multilateral approach embodied by the ESA.

By the 1970s and 1980s, the original Cold War-driven political, economic and security divisions declined in intensity. Market-driven approaches to space activity became more prominent with the result that successful international governmental arrangements – Intelsat, Inmarsat and Eutelsat – were privatized. In the case of Intelsat, the organization was split into a commercial side, Intelsat Ltd, and an attenuated public entity known as the International Telecommunications Satellite Organization (ITSO). The latter survives as the last vestige of the 1960s global social and economic development model that underlay Intelsat's original public justification as an international monopoly; the US interest in controlling the development of the communications satellite sector through Intelsat for security and economic reasons went unspoken, but was understood by all the states that joined Intelsat. As US space dominance declined, other possibilities arose, including an expanding private-sector role and the privatization of Intelsat.

The rise of the private sector altered the landscape of international space commerce by removing or reducing the public, government sector's dominant role and influence. At a national level, private and quasi-private space-sector efforts changed the policy dynamics as government institutions became more engaged and influential in how and for what purposes space policy was pursued. Such interventions, despite their merely national jurisdictions, have influenced the development of international space commerce in unexpected ways.

THE EMERGING SPACE POLICY ENVIRONMENT

Government sector intervention to both promote and regulate international space commerce was manifested in a mixture of political events and policy decisions. As various international and regional space organizations were put into play, powerful competitors at national and sub-national level also became more publicly engaged in considering the role and scope of space activities, especially in democratic industrial states (see Illustrations and Documentation: Table 8.1 Government Sector Interventions in Space). As space applications became accessible to individuals and groups across society, national and sub-national organizations were drawn into the field, or created to exploit it. The emergence of these players reflected the growing usefulness of space applications for science and commerce.

The focus here is first on the bureaucratic competitors that emerged from the national level with different frames of reference as to what it was they desired in terms of policy outcomes. Secondly, the analysis shifts to consider the impact of private actors on how space activities were conducted. Private-sector presence and interest in space applications cannot be ignored because they are proven adroit political players. Traditional space players increasingly find that they no longer control their policy environment. In fact, private entities, due to their political connections, are often more effective because their range of political allies is wider. They are able to mobilize influential associates unavailable to government agencies.

The first and most obvious examples of intervention were, in fact, the least competitive or disruptive, but demanded the attention of the space agencies. These were the functional governmental organizations that arose from traditional government functions, which, for example, pursued remote sensing and space-based meteorological activities. Such agencies, however, remained dependent upon national space agencies.

To illustrate, the US National Oceanic and Atmospheric Administration (NOAA) relies upon the National Aeronautics and Space Administration (NASA) to build and launch weather and other environmental monitoring satellites. Although private contractors build the satellites, NASA designs and monitors construction. Acquisition of such expertise is a burden to the requesting agency's budget, which NOAA feels could be more productively used elsewhere; therefore, NASA is still expected to support them. Additionally, the responding agency does so initially because to not respond is to create a potential competitor with its own technical expertise. What causes difficulties for the national space agency is that the outside agency has inconvenient demands; demands that do not fit the space agency's vision of future directions or present ones, which for reasons of delay and technological challenges drain resources from activities deemed more important by the space agency.

The impact of outsiders upon space agency operations is also exemplified by the multi-decade dispute over funding for the US Land Remote Sensing Satellite Program (Landsat) successor spacecraft. The first Landsat satellite entered orbit in 1972 and began collecting environmental and ecological data. NASA's focus on human space flight, planetary sciences and astronomy has left this remote sensing programme to seek continuation funding from wherever it can. The loss of data continuity was essentially irrelevant to NASA, which had moved on to more important (in its view) projects. The difficulty was that there was no obvious successor agency to pick up the cost and technological challenge to sustain Landsat. The 'free rider' problem was endemic to the programme – many data users, especially scientists, wanted the programme to continue, but lacked the fiscal resources, and without a lead agency there was no ability to pool resources to retain the programme. Efforts to commercialize Landsat failed due to high costs and technical factors.[10]

Complete programme termination of Landsat has been threatened a number of times, but the satellite technology has proven robust enough to keep functioning long after the satellites should have failed. Landsat's survival has extended the ability to collect environmental data over additional years, while pressuring NASA to help broker a solution to the satellite replacement issue. Ironically, if NASA had not designed the technology to be as robust as the satellites proved to be, the problem would have solved itself many years ago with satellite failures in orbit, effectively ending the data collection process. Combining Landsat data with recently released intelligence imagery collected first by the US Corona spy satellite programme has extended some data sets out over five decades; this is valuable for assessing environmental changes.[11] The result is that the Landsat Data Continuity Mission (LDCM), the proposed successor to Landsat, is moving towards implementation.

Given that nonmilitary space programmes are expensive and always seeking political justifications for their continuation and expanded budgets, responding to such pressures is useful for advancing commercial development. Commercial space-related applications have become critical for developing states with much tighter budget parameters, such as Algeria, Brazil, India, Nigeria, Turkey and China, and provide a powerful justification for pursuing space activities beyond the realm of military applications. Those countries that lack the requisite technology or expertise can outsource to contractors, such as Surrey Satellite Technology Ltd (in the UK) for satellites and the Great Wall Industry Corporation (in China) for launch services.[12]

One area of recent development outside the military and government space sectors is commercial remote sensing. The French with their Spot Image commercial programme in the 1970s directly challenged the American monopoly on readily available civilian remote sensing. It offered an alternative system with much more detailed ground resolution, military grade in some estimates. With the end of the Cold War, remote sensing entered the global space marketplace. The resolution being offered by new competitors became much more accurate, to less than 1m, allowing for detailed imagery. The rise of international competitors also ended US and French dominance of the civilian and commercial markets, respectively. Changes in US policy allowed private vendors to enter the field, reflecting both US market ideology and pragmatism, as the private imaging vendors eventually supplemented US Department of Defense (DoD) reconnaissance satellites with government data-purchase contracts.

Space applications previously had remained somewhat abstract in nature due to their exotic 'out of this world' nature and the ever-present spectre of national government secrecy. Moving such applications to a private level demystified the activities and brought the more normal processes of commerce to bear. One of the most interesting examples of this demystification process occurred with the development of the individual level satellite disk receiver and its proliferation, first to the rural countryside and then, in time, to urban areas. These dishes, shrinking in physical size and improving in functionality, eliminated the 'middle man', as they operated off the traditional communications grid. The emergence of direct broadcast satellites (DBS) explicitly built for this function expanded the field, providing a further challenge to traditional communications carriers. Governments, such as Iran, that wished to control access to certain programming content found such devices subversive to their efforts. In the Western case, the rise of DBS helped reinforce an already existing regulatory challenge. The US case provides an instructive example of when a non-space national regulatory agency came to have great impact upon the field's future direction.

The newly involved government players enter the policy fray with their own institutional and political perspectives, rather than simply acquiring existing space policy perspectives. The range of national and sub-national activities that can end up being relevant to space applications is often unanticipated. This change process is only just beginning and will accelerate if decentralization of national government control over space activities continues globally.

The international space policy and legal regime is now being impacted by agents formerly believed to be restricted or irrelevant, due to the security dimensions of space policy. This appears most clearly in regulatory activities, in particular the case of the US Federal Communications Commission (FCC) regarding Intelsat and its effective monopoly over international space telecommunications. The expanded regulatory activities engaged in by the FCC only became possible when the heavy security focus of earlier years was lessened, removing the psychological and legal block that space activities were mostly outside the realm of domestic regulation. This subtle shift allowed the negation of traditional US commercial values of competition and general resistance to monopolies, striking at the heart of the then-existing US space communications policy, as Intelsat was for years used to maintain US domination of the satellite-based communications field.

The rise of potential competitors in satellite communications, initially in Europe, was resisted by the USA, which sought to deny Intelsat control over access to the US market. Nevertheless, some regional adjustments allowing competition were made by the 1980s, when the pressure for significant change became overwhelming. A private company, PanAmSat, helped to break Intelsat's monopoly over space telecommunications by posing a private corporate threat stronger than the earlier government-led efforts, such as Telesat Canada and Eutelsat. Breaking that barrier led the US Congress to again become engaged in discussing Intelsat's future in relation to the USA – an earlier debate that had been hotly contested when the Communications Satellite Corporation (Comsat), the then-public-private majority owner of Intelsat, was first established in 1962 as a monopoly. Eventually, Intelsat was dissolved into two entities, as mentioned earlier: the ITSO to perform some of the organization's public functions, and Intelsat Ltd as a purely commercial operation.

The US Congress took a more assertive market-based approach in support of the FCC.[13] The result was that policy decisions by a national regulatory commission not intrinsically involved in space applications had a profound effect upon them. These changes allowed for private and other national telecommunications organizations to enter the field. Their efforts expanded the range of services provided. The FCC's importance was not exhausted by its initial independent foray into space applications. Later, in an FCC decision facilitating DBS receiver use, the agency pre-empted local regulations prohibiting satellite dishes or otherwise restricting their usage. For the FCC, this was a straightforward application of a long-standing US Federal constitutional law and statute prohibiting states and local government from interfering in the free flow of commerce, in this case, the signal from a satellite.

These new space policy participants, i.e. private companies and government regulatory agencies, which are adverse to the space regime with its rules and understandings, did not set out to become competitors. Yet, their different organizational goals and policy values led to conflict with existing policies and expectations. In fact, the relationships often started as partnerships, the nonspace participant facilitating the work of the space players, as with the FCC; but over time that relationship shifted to a more distant, often adversarial one, usually driven by specific issues and value conflicts. These newly emergent actors could be government, private or some combination of public-private. The FCC, for example, continues to distribute or auction off the electromagnetic spectrum authorized for sole use by the USA by the International Telecommunication Union (ITU).

There exist arrangements that are a shifting mixture of public and private. One must remember that the existing legal space regime is one based upon a 1950s and 1960s statist regime; one most space participating states were comfortable with ideologically and politically. Private-sector involvement is channelled through national-level agents, either public or quasi-public. The latter include entities that are commercial, but are either owned by the national government or operate within clearly understood governmental parameters. Signs of change in the public and private mix in space occurred in the 1970s with challenges to the statist regime arising in the areas of communications and remote sensing, as noted above, and in space transportation. In the latter case, the initial commercial enterprises were derivatives of military space programmes, such as Atlas, Delta and Pegasus in the USA. As such, the actual challenges were fairly muted until the 1990s when a plethora of space launchers were proposed world-wide.[14]

The US Government context provides some interesting examples of intrusion by sub-national agents, due to its relatively robust federal and capitalist culture. The end of the Cold War saw a general decline in national security-based restrictions on most space applications. Yet the reality was that nothing changed quickly because access to Earth orbit was through national spaceports controlled by the military. The military was not intrinsically hostile, but commercial activities beyond supporting the usual contractors were not seen as their remit. With the growth of the economic possibilities of space applications, the security restrictions or indifference of the military to the commercial aspects of space were seen as stunting growth in the field. The US Air Force's (USAF) tight control of federal spaceports hampered the private sector's flexibility in conducting operations. By definition, space applications have to be sent to Earth orbit in order to function, so this choke point or bottleneck was restrictive.

The question became: from whence are these new spaceports, that is facilities capable of launching payloads to outer space on a routine basis, going to come? Initially, the field's development was constricted by the lack of available launch vehicle options. The major launch options, the Atlas and Delta systems in the USA, were owned by large government contractors who were comfortable remaining so. The rise of purely private launch options was deemed the prerequisite for escaping the military yoke. These launch vehicles were generally smaller than traditional ones, which meant spaceports could be less elaborate, with smaller infrastructure, in principle cheaper to operate and use.

US state governments became the most logical funders of new spaceports, because they were able to fund what were highly speculative ventures, considered unlikely to find private investment. Some states have long pursued economic development initiatives as one of their major goals.[15] States such as California, Florida, Alaska and Virginia were the original pioneers and represented a variety of approaches. The California, Florida and Virginia used launch pads on the existing federal spaceports, while Alaska built an entirely state-owned spaceport, although one heavily subsidized by the Federal Government. Their success has been checkered, largely because most of the proposed private launch vehicles failed to fly at all, or on time. Development of new flight options proved more difficult than the announced projects proclaimed. In fact, most spaceports and private space launch initiatives hit a dead end through lack of funding, technical issues and market questions.

The parameters of what could be done in the area of private space activities changed after the flight of SpaceShipOne in 2004, when it clinched the Ansari X-Prize. The prize was a competition that awarded US $10m. to anyone able to fly to 100km above the Earth's surface twice within a 14-day period, using the same flight vehicle. The vehicle carried sufficient weight to simulate passengers on board. The goal of the prize was to jump-start the space tourism industry, which was seen by many analysts as the only way that a high flight rate could be created, thus drastically lowering the cost to orbit. To sustain the flight rate established by the prize requires a reusable launch vehicle (RLV). NASA cancelled its RLV development programmes, the X-33 and X-34, due to technical issues. The difficulty is that SpaceShipOne was a suborbital flight, rather than to Earth orbit, for which the energy required is significantly higher.

However, the well-publicized success of the Ansari X-Prize led the state of New Mexico to invest $250m. in building a spaceport to support space tourism, indicating the clear economic value many states attach to space activities; 15 other states have indicated some efforts in the direction of spaceport development. This activity is unprecedented, in that space activities were thought to be the sole province of national government. Likewise, US states systematically support state-level institutes to support commercial research and space science, although their emphasis is upon the former rather than the latter given their economic development potential. States are starting to impact traditional agencies with their potential as alternatives to existing national spaceports and the launch vendors who use them. The change process will be slow, but change is coming as the private launch sector grows more diverse and robust.[16]

Lowering the cost of space lift is one of the aims of private space activities. Following the collapse of the USSR in the early 1990s, the resulting decline in the Russian and other former Soviet states' economies forced the government bureaus, which had been the organizational basis of the Soviet space programme, to seek new funding sources. Prior to the Soviet collapse, their launch vehicles had been suggested as an alternative for Western satellite builders, who were discouraged by the high cost of Western, especially US, launch vehicles.

One solution pursued to resolve this cost problem was the linking of Western corporations with the former Soviet bureaus in a series of launch alliances. Major examples of these alliances include: Sea Launch, with four partners (Boeing Commercial Space Company of the USA, S.P. Korolev Rocket and Space Corporation Energia of Russia, also known as RSC Energia, SDO Yuzhnoye/PO Yuzhmash of Ukraine and Aker ASA Group of Norway); International Launch Services, with three partners (Khrunichev State Research and Production Space Centre and RSC Energia of Russia and the former Lockheed Martin, now Space Transport, Inc.); and Eurockot, with two partners (European Aeronautic Defence and Space Company (EADS) Astrium of Europe and Khrunichev State Research and Production Space Centre). All these players are important political participants, the businesses of which are considered important by their respective governments. Their existence, inconceivable prior to the breakup of the USSR, adds another layer of complexity, due to their diverse international and multinational partnerships. National space policies have had to adjust to their presence and the competitive challenge they pose for national launch companies.

An additional example of this growing impact of formerly excluded or restricted private parties upon existing space players can be readily seen in the changes occurring in space-based navigation policy and practice. Since the dawn of the space age, it was apparent that satellites in Earth orbit could be used as reference points from which those operating within the atmosphere and on the Earth's surface could locate their exact position. Such precision in determining location was obviously militarily significant in lifting the 'fog of war' from the battlefield and allowing ships at sea to operate confidently regardless of the weather.

Both the USSR and the USA pursued suitable programmes with the Soviet Global Navigation Satellite System (GLONASS) and the US Navstar Global Positioning System (GPS) systems. These systems provide the requisite precision to allow successful operations. In time, GLONASS became operationally constrained, as replacement satellites were not orbited and this loss of functionality decreased its global usefulness. Replenishing GLONASS has become a national priority, given Russian political and economic aspirations internationally. Upgrading GLONASS (planned by 2010) keeps Russia as a player in the marketplace both militarily and commercially.

The US GPS system remained global in support of its naval units. The GPS was completely operational by 1994 and by default became the global navigation system employed by all users, both military and nonmilitary. What proved an unanticipated consequence of those space-based navigation aids were their commercial and civil uses. In its pursuit of military advantage, the US DoD designed and built signals that proved extremely useful for business purposes. The navigation signal was the first exploited, but the timing function provided by the atomic clocks on board GPS satellites has also proven valuable. However, the DoD retained absolute control over the system by means of Selective Availability (SA), including the capacity to deny the signal to nonmilitary users. In addition, the GPS signal was encrypted and the signal available to nonmilitary users was distorted, denying the more precise military-quality signal.

Despite the deliberate distortion introduced, commercial devices were developed, which effectively eliminated most of the measurement error built into the signal. US and international government agencies, notably the FAA, and vendors pressured the DoD to loosen its complete control until in 2000 the US Government turned-off the SA function; in 2007 it was established that the SA technology would no longer be built into replacement GPS satellites.[17] The decision to eliminate SA was one response to the competitive threat posed by the European Union's (EU) global navigation system, known as Galileo, Russia's previously mentioned decision to upgrade GLONASS, China's development of the Beidou/Compass system, the Indian Regional Navigation Satellite System, and Japan's Quasi-Zenith Satellite System. All these endeavours represent a decline in GPS commercial dominance.

Within the field of global navigation systems, nontraditional space participants are forcing significant policy changes by reluctant traditional space players, such as the US DoD and the EU. The more established players are being forced to respond in ways contrary to their deeply held views of how various space applications should be regulated and operated. What is occurring is a gradual pressure to change by nontraditional users, such as the FAA, which incorporates GPS, for example, for flight control management. The Galileo challenge, along with incipient other competitors forced a response from a space player that a decade ago was extremely unlikely to respond under the best of circumstances. These changes in policy are particularly interesting because the USA, through the DoD, has become extremely conscious of emerging space power threats, particularly China with its demonstrated interest in anti-satellite (ASAT) weaponry.[18]

CONCLUSIONS

Space policy is clearly moving beyond the strictures of the early legal space regime, with its focus on state-centric (statist) activities; change, however, comes more dramatically and quickly when other, nonspace public agencies responding to growing commercial pressures become engaged in the policy process. Within many states, more capitalist in outlook, those pressures are difficult to resist, barring a return to a Cold War setting. States with vestiges of the old government-dominated space order, including the USA, are finding that sustaining their control is becoming more difficult. Adjustments are being made to accommodate pressures from new players, which are often strongly linked to important political elements. The USA has been the most obvious example, with calls for privatization and commercialization since the beginning of the space age. The statist model of space exploration and exploitation continues in place, but increasing pressures are coming from nontraditional space players inside and outside government, which are forcing policy adjustments. National space agencies are no longer able to ignore the demands and needs of other players in the space policy arena.

When the legal space regime will be reformed to accommodate these new players and the changes they are forcing is unknown, but the need for reform is becoming important. Further complicating the situation is the rise of multinational economic alliances in space commerce, as exemplified in telecommunications satellites and space launch facilities. These alliances are similar to traditional multinational corporations, wherein distinctive national identification becomes blurred. International space law forces national, and does not allow for multinational, identification in terms of launch and operation of spacecraft. Also, denying access to space assets by one state in conflict with another becomes more problematic when the operator is multinational and commercial. This is compounded by the fact that the military space sector makes use of commercial space assets. Ad-hoc arrangements to accommodate complaints and demands for change will not suffice in the long term because further inconsistencies will arise, creating a morass of regulatory rules, which will adversely impact future growth in space applications.

NOTES

1 The works of Frank Winter have explored the history of those early efforts in *The First Golden Age of Rocketry* (Washington, DC: Smithsonian Institution Press, 1990), and in the immediate run up to the space age in *Prelude to the Space Age: The Rocket Societies, 1924–1940* (Washington, DC: Smithsonian Institution Press, 1983).
2 Michael J. Newfeld, *The Rocket and the Reich* (New York: Free Press, 1995). Slave labour built the V-2 and thousands died in that effort due to abuse, starvation and execution.
3 William E. Burrows, *This New Ocean: The Story of the First Space Age* (New York: Modern Library, 1999); Walter A. McDougall, *... The Heavens and the Earth: A Political History of the Space Age* (New York, NY: Basic Books, 1985, reprinted in 1999 by Johns Hopkins University Press); and Dwayne A. Day, John M. Logsdon and Brian Latell, eds, *Eyes in the Sky: The Story of the Corona Spy Satellites* (Washington, DC: Smithsonian Institution Press, 1998).
4 Roger D. Launius and Dennis R. Jenkins, eds, *To Reach the High Frontier: A History of US Launch Vehicles* (Lexington, KY: University Press of Kentucky, 2002).
5 Roger Handberg, *The Future of Space Industry* (Westport, CT: Quorum Books, 1995). The early trends were clearly visible by the end of the Cold War and have

only accelerated since that time. See Roger Handberg, *International Space Commerce: Building from Scratch* (Gainesville, FL: University Press of Florida, 2006), 227–28.
6 Roger Handberg and Zhen Li, *Chinese Space Policy: A Study in Domestic and International Politics* (London: Routledge, 2007); Brian Harvey, *China's Space Program: From Conception to Manned Spaceflight* (New York: Springer, 2007); and Brian Harvey, *The Japanese and Indian Space Programmes: Two Roads to Space* (New York: Springer, 2000).
7 John C. Baker, Kevin M. O'Connell and Ray A. Williams, *Commercial Observation Satellites: At the Leading Edge of Global Transparency* (Santa Monica, CA: RAND, 2001); and Pamela E. Mack, *Viewing the Earth: The Social Construction of the Landsat Satellite System* (Cambridge, MA: MIT Press, 1990).
8 Bruce M. Dublois, 'Space Sanctuary: A Viable National Strategy', *Airpower Journal* (Winter 1998), www.airpower.maxwell.af.mil/airchronicles/apj/apj98/win98/deblois.html (accessed 20 June 1999).
9 Roger Handberg, *International Space Commerce: Building from Scratch* (Gainesville, FL: University Press of Florida, 2006); and Joseph N. Pelton, 'The History of Satellite Communications', in John M. Logsdon, ed., *Exploring the Unknown, Volume III: Using Space* (Washington, DC: NASA, Government Printing Office, 2003).
10 Molly K. Macauley, 'Rethinking Space Policy: The Need to Unearth the Economics of Space', in Radford Byerly Jr, ed., *Space Policy Reconsidered* (Boulder, CO: Westview Press, 1989).
11 Dwayne A. Day, John M. Logsdon and Brian Latell, eds, *Eyes in the Sky: The Story of the Corona Spy Satellites* (Washington, DC: Smithsonian Institution Press, 1998).
12 See chapter 7 in Roger Handberg, *International Space Commerce: Building from Scratch* (Gainesville, FL: University Press of Florida, 2006).
13 Congressional Public Law 106–80, Open Market Reorganization for the Betterment of International Telecommunications Act, signed 27 March 2000.
14 See Steven J. Isakowitz, Joshua B. Hopkins and Joseph P. Hopkins Jr, *International Reference Guide to Space Launch Systems* (Reston, VA: American Institute of Aeronautics and Astronautics, 2004, 4th edition) for a list of launch vehicles globally, including new starts.
15 Roger Handberg and Joan Johnson-Freese, 'State Spaceport Initiatives: Economic and Political Innovations in an Intergovernmental Context', *Publius* 28 (Winter 1998): 1–20.
16 Leonard David, 'State Spaceports Grow in Number', *Space.com*, 16 June 2006, www.space.com/news/060616_spaceports_update.html (accessed 1 October 2007).
17 See Removal of GPS Selective Availability (SA), www.ngs.noaa.gov/FGCS/info/sans_SA (accessed 27 September 2007).
18 See Larry M. Wortzel, 'The Chinese People's Liberation Army and Space Warfare', *Astropolitics* 6:2 (2008).

A–Z Glossary of Space Organizations

A survey of major space organizations in the civil, commercial and military space sectors

Eligar Sadeh and Cris Sadeh

There are thousands of space organizations worldwide. This glossary focuses on a selection of them, including national space agencies, governmental organizations with a role in space, intergovernmental organizations, space and aerospace companies, satellite communications service companies, and non-governmental organizations including research institutes, laboratories, grassroot groups and various types of private non-profit entities.

A

Aerojet-General Corporation, USA, www.aerojet.com

Aerojet-General Corporation was founded in 1942 by a group of scientists from the California Institute of Technology. One of the company's products—Jet Assist Take Off rocket motors—provided boosting power for US military planes during the Second World War. Growth in products and technologies in the 1950s and 1960s led Aerojet to build a site for rocket engine development, testing and production in California. In 2010 this facility served as Aerojet's headquarters and site of missile and space propulsion operations.

During the 1960s Aerojet engines sent Gemini missions into space and Aerojet's Apollo Space Propulsion System placed astronauts in orbit around the Moon. In the 1970s and 1980s Aerojet developed space electronics, including satellite sensors for weather forecasting and missile detection. Aerojet also continued to design and build propulsion systems, such as the Space Shuttle Orbital Manoeuvring System engines.

After divesting its space electronics division in 2001, Aerojet worked in the development of diversified propulsion systems, culminating in the acquisition of the General Dynamics Space Systems and the propulsion business of Atlantic Research Corporation—a developer and manufacturer of advanced solid rocket propulsion systems and auxiliary rocket motors for space and defence applications. Aerojet provides in-space propulsion for the **National Aeronautics and Space Administration** (NASA) Discovery missions, plays a significant role in the US missile defence programme, and supplies liquid engines for the US launch vehicles, Titan and Delta.

Aeronautics and Space Research and Dissemination Centre—*see* **Centro de Investigación y Difusión Aeronautico Espacial (CIDA-E), Uruguay.**

AeroSpace and Defence Industries Association of Europe (ASD), www.asd-europe.org

The AeroSpace and Defence Industries Association of Europe represents aeronautics, space, defence and security industries in Europe in all matters of common interest, with the objective of promoting and supporting the competitive development of the sector. ASD pursues joint industry actions on a

European level and generates common industry positions. It has 28 member associations in 20 countries across Europe and represents more than 2,000 companies with a further 80,000 suppliers. The aerospace and defence industry sectors in Europe employed 676,000 people with a turnover of more than US $180,000m. in 2009.

The European Association of Aerospace Industries, the European Defence Industries Group, and Eurospace, a non-profit professional organization incorporated in 1961 to foster the development of space activities in Europe and promote understanding of space industry issues and problems, merged to form ASD in 2004. Eurospace became the Space Group of ASD representing space interests. The main focus of the Space Group is space policy and strategy, and it formulates recommendations based on the identification of issues affecting the space industry. The Space Group maintains policy, programmatic and technology working groups to assess issues.

Aerospace Corporation, USA, www.aero.org

In 1960, the US Air Force Ballistic Missile Division headquarters announced the formation of the Aerospace Corporation, a non-profit corporation to serve the Air Force in the scientific and technical planning and management of its missile space programmes. The Aerospace Corporation operates a federally funded research and development centre (FFRDC) for the US Air Force and the **National Reconnaissance Office**, and supports all national security space programmes.

The Aerospace Corporation also works with projects for civil agencies, such as the National Aeronautics and Space Administration (NASA) and the National Oceanic and Atmospheric Administration (NOAA), commercial companies, universities and international organizations in the national interest. Aerospace Corporation involvement spans space systems, including systems engineering, testing, analysis and development; acquisition support; launch readiness and certification; anomaly resolution; and the application of new technologies for existing and next-generation space systems.

The Aerospace Corporation is involved in numerous programmes and projects, including: Advanced Launch Vehicles Aerospace; the Atlas II and III programmes; Centaur rockets; Delta II launch vehicles; Evolved Expendable Launch Vehicle (EELV); Inertial Upper Stage (IUS) general systems engineering; Space Shuttle mission planning and integration support; Titan II and IV launch vehicles; Defense Satellite Communications System (DSCS); Defense Meteorological Satellite Program (DMSP); Defense Support Program (DSP); Global Broadcast Service (GBS); Global Positioning System (GPS); Milstar satellite; Navy Space Systems support of the Ultrahigh Frequency Follow-On (UFO) Mobile User Objective System (MUOS); National Reconnaissance Office (NRO) programmes; Space-Based Space Surveillance (SBSS); Space Test Program (STP); weather satellites; Wideband Global Satcom System; Air Force Satellite Control Network (AFSCN); Air Force

Agence Spatiale Algérienne (ASAL)

Satellite Communications (AFSATCOM) System; Consolidated Satellite Operations Center (CSOC); the Fleet Satellite Communications (FLTSATCOM) System; the Military Satellite Communications (MILSATCOM) System; Rapid Attack Identification and Detection System (RAIDRS); and Space-Based Infrared System (SBIRS) Ground System.

Aerospace Industries Association (AIA), USA, www.aia-aerospace.org

The Aerospace Industries Association represents US manufacturers and suppliers of civil, military and business aircraft; helicopters; robotic aircraft systems; space systems; aircraft engines; missiles and related components; equipment; services; and information technology. AIA was founded in 1919. It receives its policy guidance from the direct involvement of CEO-level officers of more than 100 aerospace companies and 170 associate member companies. The Government seeks advice from AIA on issues, and AIA provides a forum for government and industry representatives to exchange views and resolve problems on matters related to the aerospace industry.

Aerospace Research Institute (ARI), Iran, www.ari.ac.ir

The Aerospace Research Institute works with space science and technology applications on the aerodynamic design and analysis of launch vehicles, conducting wind tunnel tests and designing payload and equipment. An ARI orbital debris team works on the categorization, characteristics and tracking of space debris. It is involved in the design and production of sounding rockets, which are designed to perform scientific experiments, and includes work on rocket motors and staging systems. ARI conducts test launches of Iran's exploratory rocket, Kavoshgar (Explorer). The Kavoshgar rocket is a two-stage solid fuel rocket capable of carrying a small payload and re-entering the lower atmosphere.

Agence Spatiale Algérienne (ASAL), Algeria, www.asal-dz.org

The Agence Spatiale Algérienne (Algerian Space Agency) was established by presidential decree in 2002. ASAL is responsible for designing and implementing Algeria's space programme to meet national needs. This entails the promotion and operation of the peaceful uses of space and strengthening national capacities to ensure the safety and well-being of the national community and to contribute to economic, social and cultural development, protection of the environment and the knowledge and management of natural resources of the country. To do this, ASAL established an infrastructure involving national institutions of higher education and research, industrial development and space technology. The agency advocates co-operation adapted to national concerns.

From 2002 to 2005 ASAL developed and launched a series of five microsatellites as part of an international Disaster Monitoring Constellation

(DMC). The objective of the mission was to provide multispectral images of medium resolution for monitoring natural disasters and other thematic applications of remote sensing. A number of institutions co-operated in this endeavour, including the Center for Research in Astronomy, Astrophysics, and Geophysics; the Center National des Techniques Spatiales; the Institut National de Cartographie et de Télédétection; Central Engineering Parasismique; and the Center for Development of Advanced Technologies. International partners in the project included the Institut de Physique du Globe de Strasbourg of France, Abdus Salam International Center for Theoretical Physics of Italy, and Atlantis Scientific of Canada. Using images acquired from the DMC, ASAL, in co-operation with the Ministry of Land and Environment, advanced sustainable land-use planning. See also **DMC International Imaging**.

Agence Spatiale Canadienne (ASC)—*see* Canadian Space Agency (CSA).

Agencia Bolivariana para Actividades Espaciales (ABAE), Venezuela, www.abae.gob.ve

The Agencia Bolivariana para Actividades Espaciales (Bolivarian Agency for Space Activities) is an agency of the Venezuelan Ministry of Popular Power for Science and Technology responsible for developing and carrying out the policies of the national executive of Venezuela regarding the peaceful use of space. Initially, it was called the Centro Espacial Venezolano (Venezuelan Space Center). ABAE designs, co-ordinates and implements the policies of the national executive and acts as a decentralized agency specializing in aerospace.

Objectives of ABAE are to propose space policy in the short term; to develop and implement activities and programmes in space; to ensure compliance with international treaties governing space activities; and to establish technical criteria to reconcile national initiatives in the field of space technology. Of note, is ABAE's Simón Bolívar satellite programme. The satellite, which was launched by the People's Republic of China, started operation in 2008. The mission is to facilitate access and transmission of internet data services, telephony, television, telemedicine and tele-education.

Agencia Chilena del Espacio, Chile, www.agenciaespacial.cl

The Agencia Chilena del Espacio (Chilean Space Agency) was created in 2001 as a governmental organization. The agency proposes national space policy and the measures and plans required for space programme execution; advises the President of Chile to ensure that Chile's foreign policy in space affairs reflects national space policy; promotes and advances international agreements; proposes criteria for allocating national resources and resources

Agencia Espacial Mexicana (AEXA)

obtained in the context of international co-operation for space development; promotes space activities and the use of space for peaceful purposes; maintains information on space activities at the national and international levels; and studies national legislation on space affairs and proposes relevant improvements and reforms at the institutional and operational levels. Chile partners with international agencies in the space arena for the benefit of technology transfer and investment funds for space projects. In 2009, Chile allocated US $1m. for civil space.

Agência Espacial Brasileira (AEB), Brazil, www.aeb.gov.br

The Brazilian space programme began in 1979 with the development of the Brazilian Complete Space Mission satellites, which were launched in 1993 and 1998. The Agência Espacial Brasileira (Brazilian Space Agency) was created in 1994. AEB is responsible for formulating and co-ordinating Brazilian space policy, and for the implementation, co-ordination and supervision of projects and activities related to satellites and their applications, together with ground-based platforms for satellite data collection.

As part of the Ministry of Science and Technology, AEB provides continuity to the efforts undertaken by the Brazilian Government to promote development of the space sector. Key organizations of the Brazilian space programme supported by AEB include the Instituto Nacional de Pesquisas Espaciais (INPE), the Departamento de Ciência e Tecnologia Aeroespacial (DCTA) of the Brazilian Air Force, and Brazsat Commercial Space Services. Civil space programmes in Brazil were funded at US $175m. as of 2009, and the 2010 authorization stands at close to US $200m.

Agencia Espacial Civil Ecuatoriana (EXA), Ecuador, www.exa.ec

The Agencia Espacial Civil Ecuatoriana (Ecuadorian Civilian Space Agency), founded in 2007, is responsible for conducting scientific research on planetary and space sciences and promoting development of science in the educational system of Ecuador. EXA manages and executes the civil space programme and its phases of space flight, including an Ecuadorian astronaut in space and an Ecuadorian satellite. EXA jointly operates a microgravity flight programme with the Ecuadorian Air Force and develops and operates, through its planetary sciences division, an ultraviolet (UV) radiation observatory and the National Radiation Monitor programme.

Agencia Espacial Mexicana (AEXA), Mexico, www.aexa.tv

The Agencia Espacial Mexicana (Mexican Space Agency) was created in 2010 to promote the development and exploration of space with a focus on technological, economic and industrial applications. AEXA is charged with the promotion, co-ordination and encouragement of research, exploration and use of space for Mexico. AEXA intends to promote the development of space-related technologies, increase competitiveness among Mexican companies and resume research performed by the former Comisión Nacional del Espacio Exterior (CONEE—National Commission for Outer Space), which existed from 1962 to 1977.

Agensi Angkasa Negara (ANGKASA), Malaysia, www.angkasawan.gov.my

The Agensi Angkasa Negara (Malaysian National Space Agency) is responsible for leading and observing the development of space science in Malaysia through providing leadership in the education and research of space science, assisting the government in formulating and executing the national space programmes and providing services to customers. ANGKASA's mission is to develop the country's potential in the space sector to support the development of the economy, generate knowledge and strengthen the national security infrastructure. Malaysia is involved with the TiungSAT, MeaSAT and RazakSAT satellite programmes. ANGKASA consists of four divisions: operations and space system division; technology development and space applications division; space science and education division; and administration and human resource division.

Agenţia Spaţială Română (ASR), Romania, web.rosa.ro

The Agenţia Spaţială Română (or Romanian Space Agency—ROSA) has been the Romanian national co-ordinating body for space technology, activities and programmes since 1991. It was reorganized by a government decision in 1995 as an independent public institution under the auspices of the Ministry of Research and Technology. The mission of ROSA is to promote and co-ordinate development and national efforts in the field of space, and to promote international co-operation. ROSA is authorized by the Romanian Government to establish research and development centres to realize objectives of the Romanian space programme. ROSA co-ordinates projects toward basic space science, space structures, space technologies, microgravity sciences, communications, education, Earth observations, remote sensing applications and life sciences. ROSA is the national representative for Romania for co-operative agreements with international partners.

Agenzia Spaziale Italiana (ASI), Italy, www.asi.it

The Agenzia Spaziale Italiana (Italian Space Agency) was created in 1988 to co-ordinate Italy's efforts and investments in the space sector, which began in the 1960s. ASI is involved in space science, satellite technologies and the development of robotic systems for exploring the universe. The agency works with the **European Space Agency** (ESA) and the **National Aeronautics and Space Administration** (NASA) in scientific missions and in the development of the International Space Station (ISS). ASI has developed scientific instruments that are on-board ESA and NASA robotic probes to Mars, Jupiter and Saturn. ASI also developed the multipurpose logistic modules—Leonardo, Raffaello and Donatello—to transport materials to the ISS on-board the Space Shuttle and Node 2 of the ISS, an interconnecting element between ISS laboratories.

Mission areas of ASI include space habitability, medicine and biotechnology, satellite navigation, observation of Earth, high-energy astrophysics, cosmology and fundamental physics, solar system exploration, telecommunications and space transportation.

ASI is involved with satellite navigation for vehicles, satellite tracking of dangerous goods across Europe; the Safety in Sea Traffic project to help manage navigation and sea traffic support services; the Italian National Satellite Navigation Program for Civil Aviation (ASI-ENAV); testing of Galileo; the Navigation for Disability Applications (NADIA) project to support disabled people; the Constellation of Small Satellites for the Mediterranean Basin Observation (COSMO-SKYMed), **Agenţia Spaţială Română** (ASR) radio occultation sounder for atmosphere, Probing Rotation and Interior of Stars: Microvariability and Activity (PRISMA) Earth observation mission; and the Missione Ottica su Microsatellite (MIOSAT) optical light mission. Other projects that involve ASI contributions include the Herschel Infrared Astronomy Observatory, LISA Pathfinder, Planck, BepiColombo, Cassini-Huygens, Dawn, EXOMARS missions, Mars Express, ROSETTA, SHARAD, Venus Express and the Vega space launch vehicle. See also **Avio**.

Algerian Space Agency—*see* **Agence Spatiale Algérienne (ASAL), Algeria.**

Alliant Techsystems (ATK), USA, www.atk.com

Alliant Techsystems was launched as an independent company from Honeywell in 1990. ATK expanded into the aerospace sector with the acquisitions of Hercules Aerospace Company in 1995 and Thiokol Propulsion in 2001, which transformed the company into a supplier of solid propellant rocket propulsion systems, missile defence systems, lightweight space deployable structures, solar arrays, satellite thermal management systems and technology development for small satellites. ATK's revenue for 2009 was US $4,600m.

ATK manufactures the Reusable Solid Rocket Motors of the US Space Shuttle. This same system was proposed with the National Aeronautics and Space Administration (NASA) Ares I crew launch vehicle. ATK was also slated to develop the launch abort and attitude control motors for the Ares I Orion crew capsule and the ultra-light solar arrays that would power the capsule; however, the Ares I launch system, under development by NASA, was proposed for termination by the Barack Obama Administration in April 2010.

In addition to human space flight systems, ATK is involved with Operationally Responsive Space for the US military space programme in the development of launch vehicles for small satellites. The company's small satellite capabilities, include responsive development, constellation solutions, and single-string and redundant avionics architectures. Of note, in the area of small satellites, is the development by ATK of a constellation of five small satellite buses for NASA's Time History of Events and Macroscale Interactions during Substorms (THEMIS) project. Furthermore, ATK's solar arrays powered NASA's Phoenix Mars Lander, and its composite structures are to provide the backbone for NASA's James Webb Space Telescope.

American Astronautical Society (AAS), USA, astronautical.org

The American Astronautical Society was established in 1954. AAS is focused on strengthening space programmes and advancing astronautics through cooperative efforts with other national and international space organizations and scientific societies. In 2010, AAS membership consisted of 1,400 individuals and more than 40 corporate and institutional sponsors.

The objectives of AAS are to present to its members, and to all people and institutions interested and concerned, creditable proposals, theories, facts and plans relating to the exploration and utilization of space; promote and support scientific research, engineering and exploration activities of the varied sciences required for the exploration and utilization of space; collect, publish and disseminate the conclusions of physical and theoretical research and their relationship to the astronautical sciences; provide a forum for the presentation, discussion and investigation of matters relating to space technology and the astronautical sciences; provide a scientific and practical approach to the problems of space exploration and utilization as a guide to planning by military, scientific and commercial organizations; develop and assume a position of authority in its chosen field; and bring astronautics to the attention of young people.

American Institute of Aeronautics and Astronautics (AIAA), USA, www.aiaa.org

The American Institute of Aeronautics and Astronautics was established in 1963 through the merging of the then American Rocket Society and the Institute of Aerospace Science. Both of these predecessor organizations brought a history to the relationship that formed AIAA—going back to the 1930s, when rocketry was science fiction and the aviation business was still young. AIAA has more than 30,000 members and is devoted to the progress of engineering and science in aviation, space and defence. AIAA provides information and resources for aerospace engineers, scientists, managers, policy-makers, students and educators. AIAA is also a resource for development in aerospace for corporations and government organizations world-wide.

Andøya Rocket Range (ARR)

The mission of AIAA is to enhance and support the viability of the future aerospace professional, practising aerospace professionals and the organizations and institutions involved in aerospace. The purposes of AIAA are to enable the global movement of people and goods within aerospace, assist with the global acquisition and dissemination of information and data related to aerospace, advance national security interests of the USA and provide for scientific progress and inspiration.

AIAA mission and activities are implemented through a number of programme and technical committees, which participate in numerous activities: they develop and administer technical conferences, conduct professional development training and courses, produce publications, and work with K-12 students to promote an interest in engineering and science education.

Analytical Graphics, Inc. (AGI), USA, www.agi.com

Analytical Graphics, a software development company, was founded in 1989 to provide commercial off-the-shelf software to national security and space professionals for integrated analysis of land, sea, air and space assets. AGI's Satellite Tool Kit (STK) software can propagate satellites, simulate missile trajectories, generate debris from impact through dispersion and analyse and visualize these capabilities in three dimensions.

AGI technology provides a geodynamic layer to standard geographic information system analysis and helps to support analysis of space situational awareness data. The STK software suite also provides support for space missions. A division of AGI, the Center for Space Standards and Innovation (CSSI), focuses on astrodynamic research and operates CelesTrak, a satellite tracking website. CSSI offers analysis of how likely satellite conjunctions are to occur through the use of the Satellite Orbital Conjunction Reports Assessing Threatening Encounters in Space (SOCRATES) programme.

Andøya Rocket Range (ARR), Norway, www.rocketrange.no

Andøya Rocket Range, a civilian-owned rocket launch site facility on Andøya Island in northern Norway, is a commercial provider of cost-effective access to space for polar orbits and a launch facility for sounding rockets and balloons. ARR was established in the 1960s with the first sounding rocket launch in 1962. Since then, 1,200 launch vehicles have lifted off from the range. Customers are national space agencies from the USA, Germany, Japan and France.

In 1994, ARR established the Arctic Lidar Observatory for Middle Atmosphere Research (ALOMAR) and, in 1997, a launch facility for sounding rockets and long duration balloons on Svalbard, Ny-Alesund, Norway. In 1994, ARR became a limited liability company, owned 90% by the Norwegian Ministry of Trade and Industry and 10% by Kongsberg Defence Systems of Norway. ARR is involved in building Norway's student satellites and it hosts activities, such as the European Space Camp organized by the Norwegian

Association of Young Scientists and the Norwegian Centre for Space-Related Education.

Angola Space Programme, Angola

Angola formalized a project in 2008 to produce, launch and operate an Angolan satellite named AngoSat. The Angolan Government has approved the contracts for the building, placing in orbit and operations of the satellite project estimated at more than US $300m. The Angolan Satellite Communications System project involves a contract with Russia that provides, in addition to technical resources, the training of Angolans in space technology to manage the satellite.

Rocket and Space Corporation (RSC) Energia of Russia is responsible for production of AngoSat, which will support the national infrastructure of telecommunications and digital terrestrial television across the country. The project takes into consideration the characteristics of Angola's national territory to meet the needs of economic growth, even in the remote areas. It will also take into account the growing demands for developing capability to acquire satellite data, as the Angolan Government aims to turn the country into an active member of the information society through the growing use of information technologies.

Ångström Aerospace Corporation Microtec (ÅAC), Sweden, www.aaerospace.com

Ångström Aerospace Corporation Microtec focuses on the development of small satellites that weigh 10 kg or less. In 1996 ÅAC started a research group at Uppsala University, which in 2000 became Ångström Space Technology Centre (ÅSTC). The centre was dedicated to developing microsatellites and attracted researchers and financial support from Uppsala University, the **European Space Agency** (ESA), and the **Swedish National Space Board** (SNSB). ÅSTC began to commercialize its knowledge and Ångström Aerospace was established as a spin-off in 2005. In 2008 the company changed its name to ÅAC Microtec.

Projects include the development of satellites for space research and microsystems for space applications; the SpriteSat mission; In-Orbit Verification of ÅAC Technologies on Rubin (INOVATOR), a nanosatellite avionics bus architecture being tested on RUBIN-9, an OHB System developed in-orbit verification platform; development of a miniaturized motion control chip for ESA; development of a miniaturized Point-of-Load (POL) converter for aerospace; and development of a miniaturized Software Defined Radio (SDR) platform.

Antrix Corporation Ltd, India, www.antrix.gov.in

Antrix Corporation Ltd, the commercial arm of the Indian Department of Space (DOS) and the **Indian Space Research Organisation** (ISRO), was incorporated in 1992 for the promotion and commercial exploitation of products and services for the Indian Space Programme.

Antrix markets space products and services in a number of areas, including global marketing of Indian Remote Sensing (IRS) satellite data products; supply of reliable satellite systems and subsystems; commercial satellite launches; execution of support services to international space agencies; and a telemedicine network in various states of India.

Argentine Association for Space Technology—*see* **Asociación Argentina de Tecnología Espacial (AATE), Argentina.**

Arianespace, France, www.arianespace.com

Arianespace was founded in 1980 for commercial space launch. In 2010 it had 23 shareholders from 10 European countries, including the **Centre National d'Etudes Spatiales** (CNES) with 32.5% and the **European Aeronautic Defence and Space Company (EADS) Astrium** with 30%. Since its creation, Arianespace has launched more than 275 payloads for more than 70 customers. The company generated revenue of US $1,400m. in 2009. Arianespace employs more than 300 people across its facilities, which include headquarters in France; the Guiana Space Centre (GSC) launch site in Kourou, French Guiana; and international offices in the USA, Japan and Singapore.

Arianespace develops and launches the Ariane space launch vehicle. The current operational vehicle, the Ariane 5 heavy-lift launcher, has been in commercial operation with Arianespace since 1999. In addition to this vehicle, Arianespace markets and launches the Russian Soyuz medium-lift launcher and Vega light-to-medium launcher since 2010, both from GSC. Vega was jointly developed by the Agenzia Spaziale Italiana (ASI) and the European Space Agency (ESA) for Arianespace.

Asia-Pacific Regional Space Agency Forum (APRSAF), Japan, www.aprsaf.org

The Asia-Pacific Regional Space Agency Forum was established in 1993, in response to the declaration adopted by the Asia-Pacific International Space Year Conference (APIC) in 1992, to enhance the development of Asia-Pacific space programmes and to exchange views toward future co-operation in space activities in the Asia-Pacific region. APRSAF holds annual meetings initiated jointly by the Ministry of Education, Culture, Sports, Science and Technology, Japan (MEXT), and the **Japan Aerospace Exploration Agency** (JAXA). APRSAF co-hosts organizations to discuss space-related issues and possible co-operation among countries mainly from the Asia-Pacific region. APRSAF intends to ensure wider participation of space agencies, government officials, regional and international organizations and institutions responsible for applying space technology and also space agencies from outside the region and private sectors as observers.

APRSAF provides opportunities to gather representatives from space agencies and international organizations in the Asia-Pacific region; to seek measures to contribute to socioeconomic development to the Asia-Pacific region and the preservation of the global environment, through space technology and its applications; to exchange views, opinions and information on national space programmes and space resources; to discuss possibilities for future co-operation among space technology developers and space technology users to bring mutual benefits of the countries in the Asia-Pacific region, to identify areas of common interest and assign priorities thereto; to review the progress of the implementation of the plans and programmes for further co-operation within the Asia-Pacific region; and to consider and recognize the importance of co-operating with space agencies and organizations outside the Asia-Pacific region that support APRSAF objectives.

Asia-Pacific Space Cooperation Organization (APSCO), People's Republic of China, www.suparco.gov.pk

The Asia-Pacific Space Cooperation Organization evolved from Asia-Pacific Multilateral Cooperation in Space Technology and Applications (AP-MCSTA), which came into being when a Memorandum of Understanding (MOU) was signed among China, Pakistan and Thailand in 1992. The objectives of APSCO are to focus on space science technology and its applications, education, training and co-operative research to promote peaceful uses of space. Workshops and international conferences were organized by AP-MCSTA from 1994 to 2003 in Bahrain, China, Iran, Pakistan, the Republic of Korea and Thailand. As a result, it was decided to formalize this initiative into a permanent organization for co-operation in, and promotion of, activities in space in Asia, with 14 member states.

The mission of APSCO is to undertake research and conduct pilot studies based on the applications of Satellite Remote Sensing (SRS) data and Geographic Information System (GIS) technology to natural resources surveying, mapping and environmental monitoring; undertake research studies in space and atmospheric sciences, including satellite meteorology, satellite radiance, troposphere and stratosphere studies, atmospheric pollution, satellite geodesy and astronomy; undertake research studies relating to the ionosphere and associated radio wave propagation and geomagnetism; development, design, fabrication, assembly and launching of satellites and rockets; establishment and operation of ground receiving stations for acquisition of SRS data and data for atmospheric and meteorological studies; transmitting and receiving signals from communications satellites; reception of signals from ships, boats and vehicles in distress under the satellite-aided search and rescue programme—Cosmicheskaya Sistema Poiska Avariynyh Sudov (Space System for the Search of Vessels in Distress) Search and Rescue Satellite-Aided

Tracking (COSPAS-SARSAT); establishment and operation of facilities for tracking satellites and rockets to determine their orbital parameters and trajectories; and development of instruments and software for various scientific and technological experiments.

Asociación Argentina de Tecnología Espacial (AATE), Argentina, www.aate.org

The Asociación Argentina de Tecnología Espacial (Argentine Association for Space Technology) is a civil non-profit, non-governmental organization, created to support technical and scientific activities related to space in Argentina. The association was formally founded in 1987. A number of projects are supported by the association, including hybrid, solid/liquid, propulsion engines for rockets and development of a reusable space vehicle. Past projects have involved the Argentine Experiments Package that flew on the US Space Shuttle in 2001. The Experiments Package was designed and constructed by Argentine universities and scientific institutions under the supervision of AATE.

The association, in collaboration with the Universidad Nacional del Comahue, supported the development of the Argentinean satellite, Pehuensat 1, launched by India in 2007 using the Polar Satellite Launch Vehicle (PSLV). Among other objectives, the satellite supported the advancement of space technology and education in Argentina.

Asociaţia Română pentru Cosmonautică şi Aeronautică (ARCA), Romania, www.arcaspace.ro

The Asociaţia Română pentru Cosmonautică şi Aeronautică (Romanian Cosmonautics and Aeronautics Association), created in 1999, is a non-governmental organization that promotes aerospace projects. One of ARCA's aerospace projects is a high-performance rocket engine. The purpose of this project is to develop capabilities and skills needed for future activities by using indigenous technology and materials. In 1995, ARCA entered into a contract with the **Agenţia Spaţială Română** (ASR) for the development of a rocket system with military applications. A second contract in 2007 with ASR was for a High Altitude Commercial Solar Balloon for Scientific Equipment (BASMATES), which consisted of designing, constructing, launching and recovering a high altitude solar balloon and pressurized capsule, built for transporting scientific and commercial payloads. A third contract with ASR, begun in 2007, is for the development of a vehicle for commercial operations is suborbital space.

Association Aéronautique et Astronautique de France (AAAF), France, www.aaafasso.fr

The Association Aéronautique et Astronautique de France (Association of Aeronautics and Astronautics of France) was formally approved by France in

1972. The association aims to bring together people involved in the science and technology of aeronautics and astronautics, encourage contact among members, develop a source of specialized information and establish a forum that allows members to express their views and their work. The activities of AAAF are related to supporting and advancing the French aeronautics industry, including space and technology development.

AAAF has supervised students who are working on Projet Etudiant de Recherche Spatial Européen Universitaire et Scientifique (PERSEUS), which is a technology development programme undertaken in 2005 as part of the research and innovation policy of the **Centre National d'Etudes Spatiales** (CNES) Launcher Directorate. The PERSEUS project has three objectives: the search for innovation and the development of promising technology applicable to space transportation systems; the undertaking of this work by young people within a university or association context to encourage them to choose a career in space; and the development of a set of ground-based and flight demonstrators for launching small satellites.

Association of Aeronautics and Astronautics of France—*see* **Association Aéronautique et Astronautique de France (AAAF), France.**

Association of Specialist Technical Organisations for Space (ASTOS), United Kingdom, www.astos.org.uk

In 1988, the Association of Specialist Technical Organisations for Space was created to serve the needs of small and medium-sized companies operating in the space sector of the United Kingdom (UK). ASTOS provides services to companies offering a range of technical expertise and products for space applications. The type of expertise that ASTOS members can provide, include feasibility and design, frequency co-ordination and access planning, risk assessment and management, component and subsystem manufacture, small satellite design and manufacture, software design and development and mission planning and support software and products. Members include manufacturers and consultants that are working in the civil and defence sectors, mobile and fixed applications, within the UK, Europe and elsewhere.

Association Tunisienne de la Communication et des Sciences Spatiales (ATUCOM), Tunisia, www.atucom.org.tn

The Association Tunisienne de la Communication et des Sciences Spatiales (Tunisian Association for Communication and Space Sciences) is a non-governmental organization governed by Tunisian law. The organization was registered in 1985 and was classified as a scientific association by the Tunisian Ministry of the Interior in 1992. The objectives of ATUCOM are to contribute to the development of science and communications technologies; to harness new technologies; to develop co-operation with similar institutions; to contribute to the development of legal research in the fields of information, communications and informatics; to contribute to the spread of digital culture; to disseminate space science and publicize the various peaceful uses of space; and to strengthen links with institutions and associations involved in space science.

Associazione Italiana di Aeronautica e Astronautica (AIDAA), Italy, www.aidaa.it

The Associazione Italiana di Aeronautica e Astronautica (Italian Association of Aeronautics and Astronautics) is a national non-profit organization that aims to bring together researchers, students, space enthusiasts, industries and institutions that are involved in scientific and technical aviation and space. AIDAA was founded in Rome, Italy in 1920, by a group of pioneers who were interested in aeronautics, as the Italian Association of Aerotecnica; it assumed its present name in 1969 when it merged with the Associazione Italiana Razzi. It is operated by a president and a group of shareholders from regional sections of Italy.

AIDAA promotes its activities through the organization of conferences, lectures, panel discussions, competitions, awards for studies and research, and publications. AIDAA maintains relationships with national and foreign institutions and is a member of the **International Astronautical Federation** (IAF), International Council of Aeronautical Sciences (www.icas.org), and the Council of European Aerospace Societies (www.ceas.org). It has an interest to connect with students interested in aeronautical studies. It co-ordinates a biennial conference, located in various Italian cities, where students and professionals can lecture on the results of their studies and network with others committed to the fields of aeronautics and space.

Associazione Italiana per l'Aerospazio (AIAD), Italy, www.aiad.it

The Associazione Italiana per l'Aerospazio (Italian Association for Aerospace) is the national industrial organization representing high-technology companies involved in the production, applications and services of aerospace, ground, naval and electronics systems, and equipment for customers. AIAD activities include monitoring and proposing actions in relation to legislative aspects of space policy; dissemination of information; co-ordination of suitable actions aimed at guaranteeing fair and equal treatment for the national industry in other **European Union** (EU) states and in non-EU states; statistical analysis of economic indicators relating to activities represented and the dissemination of results; co-ordination of promotional actions; and meetings, conferences and debates on topics of interest.

Astronaute Club Européen (ACE), France, www.ace-asso.eu

The Astronaute Club Européen (European Astronaut Club) was created in 2005 to engage institutions, universities and industry in Europe to develop suborbital space transportation. ACE informs the public on the techniques and technologies used in space flight through conferences, exhibitions and the publication of books.

ACE, with headquarters in Paris, France, promotes space tourism, drives the development of parabolic flights and private suborbital space travel to make them publicly available and promotes the design and development of a suborbital human spacecraft called Véhicule Hypersonique Réutilisable Aéroporté Suborbital Habité, VEHRA-SH (VSH). ACE participates in the VSH project, part of the Aerospace Student Challenge, which allows teams of European students to participate in the development of the project by addressing various aspects of the VSH system, such as propulsion, avionics, flight simulation, maintenance, management and legal aspects. The VSH vehicle launches from an aircraft and the vehicle requirements reach Mach 3.5 and an altitude of 100 km.

Astronautic Technology Sdn Bhd (ATSB), Malaysia, www.atsb.my

Astronautic Technology Sdn Bhd, created in 1997 and wholly owned by the Malaysian Minister of Finance under the supervision of the Malaysian Ministry of Science, Technology and Innovation (MOSTI), is a small satellite business in Malaysia. The company provides services and products, performs R&D of space and related technologies, leverages strengths into new products and services, and develops space engineering projects and associated technologies.

ATSB develops small satellite payloads and flight instruments. The company developed a Malaysian remote sensing microsatellite—TiungSAT-1—which was successfully launched in 2000. In 2010, the company developed a Malaysian Earth observation satellite named RazakSAT. In addition, ATSB provides launch services directed at the integration of Malaysian science, technology and education payloads for the Falcon 1 launch vehicle of **Space Exploration Technologies Corporation**. ATSB is also involved in navigation, tracking and communications systems and in ground-based systems for receiving and processing images and data from Earth observation satellites, including RazakSAT.

Astronautical Society of India (ASI), India, www.asindia.org/default.aspx

The Astronautical Society of India was established in 1990 to foster the development of astronautics in the country. ASI is engaged in the dissemination of technical and other information related to astronautics through seminars, workshops, publications and media. The society plays an active role in promoting the interests of other developing countries in the field of astronautics through the International Astronautical Federation. Initiatives of the society include a workshop on space instrumentation and payload engineering, support for university initiatives on microsatellite development, and a project dealing with satellite communications.

ASI objectives are to promote the development of astronautics; encourage dissemination of technical astronautical information; stimulate public interest and support for astronautics; encourage participation in research and applications; support activities related to professional growth; foster co-operation with international institutions; create subsidiary associations and institutions; organize meetings and conferences; and secure and administer funds, grants and endowments to realize objectives.

Astrotech Corporation (ASTC), USA, www.spacehab.com

Astrotech Space Operations, a wholly owned subsidiary of Astrotech Corporation, opened in 1984 as a division of Northrop Grumman overseeing satellite final assembly and checkout, solid rocket motor installation, liquid propellant loading, encapsulation, transport to launch pad and command and control of satellites during launch. Astrotech Space Operations has supported the processing of more than 260 major payloads.

Spacehab, renamed Astrotech Corporation in 2009, was incorporated in 1984 and made its initial public offering in 1995. The initial Spacehab mission was in 1993 on-board the Space Shuttle. Since then, Spacehab modules and cargo carriers have served as payloads on Space Shuttle missions and missions to the Russian space station, Mir, and the International Space Station.

ASTC services include preparing and sending satellites, cargo and science into space; satellite and spacecraft prelaunch processing; payload processing and integration; space hardware design and manufacturing; third-party space access acquisition and integration; microgravity commercial drug development; and space technology product commercialization for Earth applications.

Australian Space Research Institute (ASRI), Australia, www.asri.org.au

The Australian Space Research Institute is a non-profit research organization operated by volunteer members. ASRI has no employees and limited resources with which to undertake its development programmes. ASRI came about in the early 1990s as the result of a merger between the Launch Vehicle Development Group at Monash University in Melbourne and the Australian Space Engineering Research Association (ASERA). ASRI's vision is for the future of Australia's space community, including industry.

ASRI is involved in hybrid fuel, low-cost payload launch service and launch vehicle programmes. Small sounding rocket programme (SSRP) launch campaigns are usually conducted twice a year from ASRI's launch facility in the Woomera Test Facility, South Australia. Launch operations are conducted under the auspices of a Range Usage Agreement between ASRI and the Commonwealth of Australia and in accordance with the SSRP Safety and Operations Plan.

Austrian Research Promotion Agency (FFG), Austria, www.ffg.at/content.php

The Austrian Research Promotion Agency was founded in 2004 as a result of consolidating the Industrial Research Promotion Fund, the Austrian Space Agency, the Bureau for International Research and Technology Cooperation and the Technology Impulse Society. FFG is the gateway to the international aerospace industry for Austria's industry and science sectors and aims to strengthen its international standing in these key technologies. It acts as the central contact point for the co-ordination of all aeronautical and space-related activities in Austria. The agency supports the participation of Austrian researchers in international and bilateral aerospace collaborations. It implements Austrian aeronautical space policy and represents Austria in international aeronautical and space organizations. The focus of the agency is on managing the contributions of Austria to the programmes of the **European Space Agency** (ESA).

The tasks of the FFG are: to manage and finance research projects in the business and science sectors; to manage co-operative programmes and projects with European and international partners; to represent Austria's interests at relevant European and international institutions; to provide consultation and support to Austria's involvement in European space programmes; to provide support and strategy development for Austrian decision-makers; and to improve public awareness of the importance of R&D.

Avio (ASI), Italy, www.aviogroup.com

Avio is an Italian aerospace engine manufacturer, founded in 1908. It is 30% owned by the **Agenzia Spaziale Italiana** (ASI). Avio developed a role as a subsystems and component maker, participating in aeronautical and space programmes from R&D to manufacture and assembly. In the space sector, Avio develops propulsive systems with solid and liquid propellant. Avio designed, qualified and produced more than 180 solid propellant boosters, more than 3,500 engines for Ariane 3 and 4 and contributed to the Ariane 5.

Under the **European Space Agency** (ESA) Avio, in co-operation with ASI, developed the Vega launcher for launch of small and medium satellites to low Earth orbit (LEO). Avio is also a contractor for the P-80 Solid Propellant Engine Demonstrator. Further, Avio produced, under US licence, solid propellant rocket engines for a European multiple launcher rockets system (MLRS). Avio is also committed to the Italian–French programme for the development and production of ground-to-air missiles, ASTER 15 and ASTER 30.

Azerbaijan National Aerospace Agency (ANASA), Azerbaijan, www.mdi.gov.az

Azerbaijan National Aerospace Agency was established in 1975 as the governmental body to co-ordinate space research programmes with scientific and

Azerbaijan National Aerospace Agency (ANASA)

commercial goals. ANASA's space programme is carried out through international co-operation. The programme has included a sequence of satellite missions through international co-operation with other states. ANASA plans to make proposals for assembly and production of Very Small Aperture Satellite Terminals (VSAT)—a small Earth station for satellite transmission.

Azerbaijan is preparing, for 2011, a satellite mission called AzerSat 1, which will provide military monitoring, mobile telephone communications and television, radio and internet broadcast capabilities. The system will be accessible in Europe and central and south-east Asia. There is a preliminary agreement with companies in several states, including Georgia and Moldova, to use AzerSat 1 in the future.

B

Boeing Company, USA, www.boeing.com

The Boeing Company, founded in Seattle, Washington in 1916, is an aerospace company and manufacturer of commercial jetliners and military aircraft. Boeing's space business has expanded over the years, acquiring Rockwell International in 1996 and merging with McDonnell Douglas in 1997. Boeing corporate headquarters are in Chicago, Illinois, since 2001.

In the space arena, Boeing designs and manufactures electronic and defence systems, satellites, launch vehicles, and advanced information and communications systems. As a service provider to the **National Aeronautics and Space Administration** (NASA), Boeing operates the Space Shuttle and the International Space Station (ISS). Boeing has customers in more than 90 states and employs more than 158,000 people across the USA and in 70 countries. In 2009, Boeing's revenue was US $68,000m.

Programmes and projects in space and intelligence systems, include Global Positioning System IIF (GPS IIF) and SkyTerra satellites. In regard to space exploration, Boeing Launch Systems works with commercial customers and the Delta family of launch vehicles and Boeing provides payload processing for the Space Shuttle, ISS, expendable launch vehicles and other payload programmes at Kennedy Space Center, Florida. Boeing is a subcontractor to NASA's space flight operations contractor for the Space Shuttle, **United Space Alliance**. As the original developer and manufacturer of the Space Shuttle orbiters, Boeing is responsible for orbiter engineering, major modification design, engineering support to operations, including launch and overall Space Shuttle systems and payload integration services (note that US policy calls for termination of the Space Shuttle programme by 2010–11). Boeing is responsible for design, development, construction and integration of the ISS and assisting NASA in operations as the prime contractor. The company is also responsible for integrating the systems, procedures and components of all the participating countries in the ISS co-operative enterprise (note that in April 2010 the Barack Obama Administration committed the USA to ISS operations until 2020). NASA's Constellation programme calls for a new fleet of spacecraft and rockets, within which Boeing is responsible for producing and delivering avionics systems and upper stage for the Ares I rocket. However, in April 2010 the Obama Administration called for the termination of the Ares I project.

Bolivarian Agency for Space Activities—see Agencia Bolivariana para Actividades Espaciales (ABAE), Bolivia.

Brazasat Commercial Space Services, Brazil, www.brazsat.com.br

Brazasat Commercial Space Services was founded in 1991 by a group of private investors for the commercialization of space activities in Brazil. The company supplied equipment and services in support of the Brazilian Complete Space Mission, which includes the development of the Centro de Lançamento de Alcântara (Alcântara Launch Centre, at www.cla.aer.mil.br), launch vehicles and satellites. Brazasat objectives are to promote development and partnerships in satellite-based telecommunications and remote sensing, and information technology projects of direct benefit to the Brazilian space programme and Brazilian society. Services are primarily focused on broadband transmission of data, telephony, networking, broadcasting and internet applications via satellite.

Brazasat has established partnerships with **Astrotech Corporation**, Boeing Space Systems, Goodrich Corporation, **MacDonald, Dettwiler and Associates**, **European Aeronautic Defence and Space Company Astrium**, Raytheon, **Space Systems Loral**, Mitsubishi Corporation, Shin Satellite (now Thaicom Public Co. Ltd, at www.thaicom.net), Fokker Space (now **Dutch Space**), Lockheed Martin, and **GeoEye**. Brazasat also partners with the **National Aeronautics and Space Administration** (NASA) and played a role in the development of Brazilian microgravity experiments, which flew on board various NASA Space Shuttle missions, and Brazilian hardware for the International Space Station.

Brazilian Space Agency—*see* **Agência Espacial Brasileira (AEB), Brazil.**

British Interplanetary Society (BIS), United Kingdom, www.bis-spaceflight.com

The British Interplanetary Society was formed in 1933 and its programmes entail projects designed to promote, encourage and support international space activities. BIS has promoted ideas on space exploration at technical, educational and popular levels. It serves people involved with space in a specialized and general capacity, undertaking educational activities on space topics, encouraging technical and scientific space studies and fostering fundamental space research, technology and applications. BIS publications include *Spaceflight*, *Journal of the British Interplanetary Society* (JBIS) and *Space Chronicle*.

Bulgarian Aerospace Agency (BAS), Bulgaria, www.space.bas.bg

The Bulgarian Aerospace Agency began with the participation of Bulgarian scientists in space research, which started in 1969 when the Scientific Group of Space Physics of the Bulgarian Academy of Sciences was created. In 1975, the Scientific Group established the Central Laboratory for Space Research, which allowed Bulgarian scientists to gain significant experience through participation in Russian-led Interkosmos, Salyut and Mir programmes. A Bulgarian astronaut flew in space in 1979 aboard the Russian Soyuz spacecraft. The Space Research Institute at the Bulgarian Academy of Sciences succeeded the Central Laboratory for Space Research in 1987, which eventually became BAS.

BAS established three scientific research laboratories: the Laboratory of Aerospace Optics, the Laboratory of Aviation Equipment and the Laboratory of Navigation and Communication. It also has a mechanical workshop and the COSMOS Scientific-Production Enterprise. BAS is focused on fundamental and applied investigations in space physics, astrophysics, image processing, remote sensing, life sciences, scientific equipment and spacecraft development; space-related experiments; investigation on control systems; and education.

In the areas of space physics and astrophysics, for example, the Solar-Terrestrial Influences Institute (STIL-BAS), an independent scientific organization founded in 1990 by the management council of BAS, studies the area of solar-terrestrial influences as possible reasons for global change and also to meet the social needs for multidisciplinary study of the relationship between events on the Sun and their effects on space. STIL-BAS operates four space experiments: three on the International Space Station, and the Radiation Dose Monitor (RADOM) experiment on the Indian Chandrayaan-1 space probe.

C

Canadian Aeronautics and Space Institute (CASI), Canada, www.casi.ca

The Canadian Aeronautics and Space Institute is a non-profit, professional, scientific and technical organization devoted to the art, science and engineering of aeronautics, astronautics and associated technologies in Canada. It focuses on communications and networking among the space community in Canada and assists members in developing skills, exchanging information and sharing talents in their areas of interest.

The origin of CASI dates to 1954 when the Montréal-based Institute of Aircraft Technicians, the Ottawa Aeronautical Society and the Canadian section of the US Institute of Aeronautical Sciences merged to form the Canadian Aeronautical Institute. In 1962, the Canadian Astronautical Society in Toronto and the Montréal-based Astronautical Society of Canada merged with the Aeronautical Institute to become CASI. CASI serves 1,600 members as of 2010. Member sections cater to special interest groups, such as aerodynamics, aircraft design and development; astronautics; flight mechanics and operations; navigation; propulsion; remote sensing; and structures and materials. CASI holds annual conferences, workshops and symposia.

Canadian Space Agency/ Agence Spatiale Canadienne (CSA/ASC), Canada, www.asc-csa.gc.ca/eng/default.asp

The Canadian Space Agency was established in 1990 by the *Canadian Space Agency Act* and operates as a government department. The president of CSA/ASC is the equivalent of a deputy minister and reports to the Minister of Industry. Functions of the CSA/ASC include space programmes, space technologies, space science, the Canadian Astronaut Office and space operations. The mandate of the CSA/ASC is to promote the peaceful use and development of space, to advance the knowledge of space through science and to ensure that space science and technology provide social and economic benefits for Canadians. As of 2009, CSA/ASC employed 600 people with an annual budget of US $340m.

Space projects involving CSA/ASC include: Earth observations (RADARSAT, SCISAT, ODIN, CLOUDSAT, ENVISAT, TERRA, UARS and VIKING); satellite communications (MSAT, NIMIK and ANIK series); science (ALOUETTE, CASSIOPE, MOST, HERSCHEL, INTERBALL-2, JWST,

FUSE, NOZOMI, PHOENIX, PLANCK, AKEBONO and ISIS); and the International Space Station (ISS) Mobile Servicing System. The main task of the Canadian Astronaut Office is to develop, support, train and fly space missions. Canadian astronauts have flown on US Space Shuttle and ISS missions.

Center for Defense Information (CDI), USA, www.cdi.org

The Center for Defense Information provides analysis on components of US national security, international security and defence policy. CDI promotes discussion and debate on security issues, such as nuclear weapons, space security, missile defence, small arms and military transformation. Since 1972, CDI has produced research and published numerous books and monographs. To provide objective analysis, CDI does not hold policy positions and accepts no **US Government** or defence industry funding.

Since 1972 the *Defense Monitor*, published by CDI, has provided information about military programmes and key international security issues. The *Defense Monitor* is distributed to 25,000 individuals, including members of the US Congress, US State Department officials, students and faculty at US military academies and war colleges, members of national organizations and representatives of the media. CDI also addresses strategic, political, technical and economic questions surrounding the weaponization of space through analyses, news and data for policy-makers, media and others interested in this international security issue.

Center for Strategic and International Studies (CSIS), USA, csis.org

The Center for Strategic and International Studies is a public policy institution and non-profit organization founded in 1964. It was affiliated with Georgetown University until 1987. The foreign policy think tank focuses on national security issues and conducts policy studies and strategic analyses with attention to technology, public policy, international trade and finance and energy. The space initiatives of CSIS deal with understanding the role of civil space development in national and international security and foreign policy. Areas of study include space technology development and public policy; global space development; the Human Space Exploration Initiative, which examines questions relating to governance, finance and public support; and smart power through space, which explores the role that space can have in strengthening national and international security through diplomatic and co-operative means.

CSIS employs more than 200 people and has an annual operating budget of US $30m., with the majority of the money funded from corporate, foundation and individual contributions. Of note, is the Defense-Industrial Initiatives Group of CSIS, which provides research into the defence industry for government and corporate customers. Since the 1970s CSIS has published the

Washington Quarterly, a journal on global changes and the impact on public policy, and the *Freeman Report*, a foreign policy periodical that focuses on economics and international security in Asia and south-east China.

Central R&D Institute of Robotics and Technical Cybernetics (RTC), Russia, www.rtc.ru/index-en.shtml

The Central R&D Institute for Robotics and Technical Cybernetics was established in 1968 as the Special Design Bureau of Technical Cybernetics at the then Leningrad Polytechnical Institute, later the Saint Petersburg State Polytechnical University. RTC activity concerns control systems of objects, robotics, photon engineering, special instrument making, laser technologies, and intellectual real time control technologies with use of telecommunications systems (telenetics), including altimeters for soft landing systems and life-support systems for spacecraft and mobile robotics. RTC space robotic projects include an on-board manipulator system (OBMS), an adaptive walking robot TSIRKUL, a functional model of a space manipulator and electromechanical modules for space applications.

Central Scientific Research Institute of Machine Building (TsNIIMash), Russia, www.tsniimash.ru/index.htm

The Central Scientific Research Institute of Machine Building is a Russian company formed originally as the Scientific Research Institute (NII)-88 in 1946, changing its name to TsNIIMash in 1967. It is involved in the development of rocket and space technology, the production of long-range ballistic missiles and air defence missiles and propulsion units. TsNIIMash is the chief agency of space works and is under the jurisdiction of the **Russian Federal Space Agency** (Roscosmos).

Centre for Space Science and Applied Research (CSSAR), China, english.cssar.cas.cn

The Centre for Space Science and Applied Research of the **Chinese Academy of Sciences** (CAS) was founded in 1987 by merging the former Institute of Space Physics, which was the Institute of Applied Geophysics founded in 1958, and Centre for Space Science and Technology founded in 1978. In 1995 CSSAR merged with the General Establishment of Space Science and Application of CAS. CSSAR is engaged in space science research, space system design and assembly and science and technology development of flight instruments for satellites.

There are 500 employees at CSSAR, including academics, researchers and engineers. Doctoral and Master's degrees can be awarded in space physics, and Master's degrees can be awarded in spacecraft design and computer application through CSSAR. CSSAR also focuses on human space flight and is involved with the Geospace Double Star Exploration Program (DSP) and the Moon Exploration Programme of China.

Centre for the Development of Industrial Technology—*see* **Centro para el Desarrollo Technológico Industrial (CDTI), Spain.**

Centre National d'Etudes Spatiales (CNES), France, www.cnes.fr

The Centre National d'Etudes Spatiales (National Centre for Space Studies) is the national space agency of France. CNES was established in 1961 under the French presidency of Charles de Gaulle, who saw in the conduct of space programmes a key to strategic independence and to larger influence in international relations for France. The power derived from space was linked to the development of ballistic missiles and to the maintenance of a strong aerospace industry with skilled engineers and technological know-how. Other purposes of the national space effort involve scientific research and knowledge-oriented programmes to be conducted in French and European frameworks.

CNES is a public agency overseen by the Ministry of Research and the Ministry of Defence of the French Government. CNES conducts military programmes alongside civilian programmes. Although CNES is national, its operational level of action is European. From the beginning of the space effort in the early 1960s, the French Government recognized that it could never conduct significant space programmes with the budgets that were appropriated at the national level. For the French, European partners had to be convinced that space was important, and that they must act together.

CNES remains a leading actor in European space by retaining its R&D capacity and continuing to propose new ideas and programmes. The propositions made by CNES go through either the different **European Space Agency** (ESA) boards for programme development in navigation, observation and telecommunications or through the French national authorities for proposing programmes at the **European Union** (EU) level. French President Nicolas Sarkozy discussed space policy in 2008, at the time when France held the rotating EU presidency. Sarkozy's efforts helped to facilitate a commitment to space policy among smaller European countries.

CNES has several facilities in France, with headquarters in Paris and large-scale technical facilities in Toulouse, which is a large cluster for space activities with companies, administrative entities and universities. An additional facility operated by CNES is the European spaceport in French Guiana (known as the Guiana Space Center). The budget of CNES in 2010 was US $2,300m. Of this, about US $930m. of the CNES budget was given to ESA. With the remainder, France conducts programmes in co-operation with non-European countries. With the exception of a few sensitive military demonstrator programmes, there are practically no programmes at CNES conducted in a strictly national framework.

Centre National de la Cartographie et de la Télédétection (CNTC), Tunisia, www.cnt.nat.tn

Created in 1988, the Centre National de la Cartographie et de la Télédétection (National Centre for Cartography and Remote Sensing) is a non-administrative public establishment under the trusteeship of the Tunisian Ministry for National Defence. CNCT exploits remote sensing to contribute to national projects concerning agriculture, natural resources and environmental monitoring, urban management and sustainable development. CNCT undertakes studies for pilot projects using remote sensing techniques and employs a multidisciplinary team of engineers and technicians specializing in topography, geodesy, cartography and the techniques of data processing. The Centre disseminates knowledge to partners, laboratories and research centres of the European Mediterranean states and to other technical programmes.

Centre of Applied Space Technology and Microgravity—see Zentrum für Angewandte Raumfahrttechnologie und Mikrogravitation (ZARM), Germany.

Centre Royal de Télédétection Spatiale (CRTS), Morocco, www.crts.gov.ma

The Centre Royal de Télédétection Spatiale (Royal Centre for Remote Sensing), created in 1989, is the Moroccan national institution responsible for the promotion, use and development of remote sensing applications in Morocco. CRTS co-ordinates and carries out the national programme of remote sensing in collaboration with ministerial departments, private operators and universities. CRTS uses operational systems to collect, produce and analyse data from Earth observation satellites and other sources. It also operates national archiving facilities.

CRTS develops applications and methods in space technologies and related disciplines—remote sensing, mapping, navigation and telecommunications. It provides training and education opportunities in space technologies and maintains partnerships for research with universities and institutions. Since its inception, CRTS has implemented several projects involving remote sensing for natural resource monitoring, and for environmental protection and planning. The applications of these projects provide strategic support to decision-makers.

Centro de Investigación y Difusión Aeronautico Espacial (CIDA-E), Uruguay, www.globalsecurity.org/space/world/uruguay/agency.htm

The Centro de Investigación y Difusión Aeronautico Espacial (Aeronautics and Space Research and Dissemination Centre) was created in 1975 as a government agency to study and promote the study of aviation and space issues, to disseminate results of the research and studies and to help create public awareness. CIDA-E advises and works with the Air Force, directorate-general of civil aviation and other public and private entities related to aerospace.

The tasks of CIDA-E are to examine aviation and space issues; to organize courses and seminars; to maintain communication with similar centres abroad; and to prepare publications on legal, technical and scientific activities related to space and aviation. CIDA-E is a member of the **United Nations** Committee on the Peaceful Use of Outer Space (COPUOS). CIDA-E publishes articles and an annual journal of international circulation, organizes academic activities and participates in international forums.

Centro Italiano Ricerche Aerospaziali (CIRA), Italy, www.cira.it

In 1984, the Centro Italiano Ricerche Aerospaziali (Italian Aerospace Research Centre) was incorporated. CIRA's task is to define and implement the Italian national programme for Aerospace Research (PRORA) through participation in European programmes and international conferences. CIRA's objectives are to promote training and knowledge in aerospace; to develop research projects in collaboration with the scientific community; and to develop the capacity for modelling, simulation and testing for aerospace and space systems.

Centro para el Desarrollo Technológico Industrial (CDTI), Spain, www.cdti.es

The objective of the Centro para el Desarrollo Technológico Industrial (Centre for the Development of Industrial Technology) is to help Spanish companies with the generation and development of technology. CDTI is the representative of Spain to the European Union (EU) Space Council, where Europe's space policy is defined and ratified. Management of funds dedicated to space in Spain through the **European Space Agency** (ESA) and the EU is the responsibility of CDTI.

CDTI, with about 250 employees, manages Spain's civil space budget (US $60m. in 2009), to carry out space projects in co-operation with other space agencies. CDTI has signed agreements to carry out bilateral programmes with the **National Aeronautics and Space Administration**, the **Russian Federal Space Agency** (Roscosmos), the **Centre National d'Etudes Spatiales**, and the **Canadian Space Agency**.

Chilean Space Agency—*see* Agencia Chilena del Espacio, Chile.

China Aerospace Corporation (CASC), China, www.nti.org/db/china/casc.htm

The China Aerospace Corporation was established in 1993 and constitutes two primary organizations: the China Aerospace Machinery and Electronics Corporation (CAMET) and the China Aerospace Science and Technology

Corporation. CASC is headquartered in Beijing and has more than 100,000 employees, 30% of whom are technicians, engineers and researchers. CASC was formerly known as the No. 5 Research Academy of the Ministry of National Defence, established in 1956, and was subsequently called the Seventh Ministry of Machinery Building Industry. Following this, it was known as the Ministry of Aerospace Industry (MASI) until the establishment CASC.

In 1998, CASC underwent organizational change following policy guidelines adopted at the National People's Congress of China. The administrative and regulatory functions of CASC are under the control of the **Commission for Science, Technology and Industry for National Defence** (COSTIND). A new organization within COSTIND, called the State Aerospace Bureau, oversees the administrative functions of China's aerospace industry. The goal of this reorganization was to transform CASC into an industrial enterprise group exclusively engaged in R&D and production for China's aerospace industry.

China Great Wall Industry Corporation (CGWIC), China, www.cgwic.com

Established in 1980, China Great Wall Industry Corporation is an organization authorized by the Chinese Government to provide satellite in-orbit delivery services, commercial launch services and aerospace technology applications. CGWIC is engaged in the global business development of China's aerospace industry. From 1985 to 2009 CGWIC conducted 29 commercial launch missions for 35 satellites.

Since 2007, CGWIC provides satellite deployment services to international customers. CGWIC is devoted to development of aerospace technology applications in the fields of satellite applications, electronics and information technology, advanced material, green energy and petroleum chemical engineering. CGWIC products include the Dongfanghong (DFH) communications satellite series, NigcomSat-1 and VeneSat-1 communications satellite programmes, PakSat-1R, NigComSat-1R and CAST2000.

China National Space Administration (CNSA), www.cnsa.gov.cn

The China National Space Administration was established as a government institution to develop and fulfil China's international obligations, with the approval of the Eighth National People's Congress of China. The Ninth Congress assigned CNSA as an internal structure of the **Commission for Science, Technology and Industry for National Defence**.

CNSA responsibilities entail the signing of governmental agreements in the space area on behalf of Chinese organizations; intergovernmental scientific and technical exchanges; enforcement of space policies; and management of space science, technology and industry. China has signed governmental space co-operation agreements with Brazil, Chile, France, Germany, India, Italy, Pakistan, Russia, Ukraine, the United Kingdom and the USA. To illustrate, a Sino-Brazilian co-operation programme for satellite-based Earth observations, the CBERS-1 and CBERS-2 satellites, provides an example of co-operation between China and developing countries.

There are four departments under CNSA: the Department of General Planning, Department of System Engineering; Department of Science, Technology and Quality Control; and Department of Foreign Affairs. Services and satellites of CNSA include the Dongfanghong (DFH)-3 (third-generation telecommunications satellite) and the ZiYuan (ZY)-1 Earth resource satellites. It is estimated that in 2009 China allocated US $1,800m. for all civil, commercial and military space programmes.

China Satellite Launch and Control General (CLTC), China, www.nti.org/db/China/cltc.htm

The China Satellite Launch and Control General is located in Beijing and was established in 1966. CLTC, an arm of the **Commission for Science, Technology and Industry for National Defence**, oversees China's three launch sites and all launch and tracking activities. CLTC is also known as the China Satellite Launch and Tracking, Telemetry and Control General; China Satellite Launch and Telemetry, Tracking and Control General Co.; and the national space and missile tracking command centre. With a team of engineers and technical personnel, CLTC has launched more than 40 satellites for China and other countries. CLTC started providing launch services for commercial satellites in 1986 and has more than 10 launch service contracts signed with foreign companies. The CLTC tracking network was created in 1966 by Department 701 with the participation of the Southwest Institute of Electronics Technology (SWIET) of the Ministry of the Electronics Industry in China.

Chinese Academy of Launch Vehicle Technology (CALT), China, www.nti.org/db/China/calt.htm

The Chinese Academy of Launch Vehicle Technology was established in 1957 as one of nine design academies under the **China Aerospace Corporation**. It employs 27,000 people in 13 research institutes and six factories. CALT is also known as the First Academy of the Chinese Aerospace Corporation, Chinese Academy of Launch Vehicle Technology, China Carrier Rocket Technology Research Institute, China Launching Vehicle Technology Research Institute and Aerospace 1st Academy. CALT is China's organization for the research, development and production of space launch vehicles, among which is the Long March (CZ) family of rockets.

Chinese Academy of Sciences (CAS), China, english.cas.cn/ACAS

The Chinese Academy of Sciences, established in 1949, is an academic institution and R&D centre located in Beijing, China. The mission of CAS is to conduct research in basic and technological sciences; to undertake nationwide integrated surveys on natural resources and the environment; to provide the country with scientific data and advice for governmental decision-making; to undertake government-assigned projects for social and economic development; to initiate personnel training; and to promote China's high-technology activities. The **Centre for Space Science and Applied Research** operates under CAS.

Chinese Society of Astronautics, China, www.csaspace.org.cn

The Chinese Society of Astronautics was created in 1979 to promote and improve the quality of space science and technology development and to contribute to the national development of China's space programmes and projects. The main tasks of the society are to provide academic exchanges; to host academic conferences, lectures and exhibitions; to promote the exchange and dissemination of science, technology and programme management experience; and to conduct youth activities in space science and technology education. Local space institutes of the society exist throughout China and are responsible for carrying out these tasks. The society publishes the *Aerospace Journal* and the *Space Exploration* magazine.

Coalition for Space Exploration, USA, spacecoalition.com

The mission of the Coalition for Space Exploration is to inform the public about the value and benefits of space exploration to build support and funding for the **National Aeronautics and Space Administration** and to ensure that the USA remains a leader in space. The coalition works to increase support for space exploration. Membership in the coalition is open to companies and organizations that support the exploration of space—human, science and robotics—and are willing to commit funding or resources toward advocacy.

Colombian Space Commission—*see* Comisión Colombiana del Espacio (CCE), Colombia.

Comisión Colombiana del Espacio (CCE), Colombia, www.cce.gov.co

The Comisión Colombiana del Espacio (Colombian Space Commission) was created by a presidential decree in 2006 with the mission to consult, co-ordinate, direct and plan space activities and space policies in Colombia. The primary objective of CCE is satellite development for research and applications in Earth observations to establish economic and social progress for Colombia.

Comisión Nacional de Actividades Espaciales (CONAE), Argentina, www.conae.gov.ar

The Comisión Nacional de Actividades Espaciales (National Space Activities Commission) was created in 2001 as a specialized agency of Argentina to understand, design, execute, control, manage and administer space projects throughout the country. CONAE plans, executes and evaluates the uses of space science and technology with activities in the areas of ground infrastructure, satellite missions, information systems, access to space and institutional development. CONAE is in charge of tracking, telemetry and control (TT&C) of Argentinian satellites. It is also in charge of the signal reception and processing images captured by Argentinian and international satellites.

In 1994, and revised in 1997, CONAE implemented activities to be carried out with its objectives. Each review included new knowledge and technological developments for space applications and local environmental and social requirements. The national space programme addresses sensing, generation, transmission, processing, archiving, distribution and use of space information. Argentinian satellites, include a scientific spacecraft, to study solar physics and astrophysics, launched in 1996; a technological model of the SAC-C mission, launched in 1998; and the Argentine Earth Observing Satellite, launched in 2000. The 2009 budget for CONAE was US $70m.

Comisión Nacional de Investigación y Desarrollo Aeroespacial (CONIDA), Peru, www.conida.gob.pe

The Comisión Nacional de Investigación y Desarrollo Aeroespacial (National Aerospace Research and Development Commission) is the governing body of space activities in Peru. The commission, also known as the Space Agency of Peru, is focused on the national development of aerospace technologies. CONIDA's mission is to promote research, develop science and space technology for national interests, and create services to drive national development. Missions of CONIDA include geomatics, radar interferometry (INSAR), development of scientific instrumentation and astrophysics dedicated to R&D and scientific programmes. CONIDA works to promote and develop peaceful research and work toward the country's progress in space; to advance studies, research and theoretical and practical work; to conclude co-operation agreements with similar institutions and agencies; to encourage the exchange of technology; and to propose national laws applicable to space.

Commission for Science, Technology and Industry for National Defence (COSTIND), China, www.costind.gov.cn

The Commission for Science, Technology and Industry for National Defence is a civilian ministry within the State Council of the People's Republic of China responsible for defence procurement. COSTIND was formed in 1982 to centralize Chinese defence procurement and technology, which had been distributed among several agencies. In 2008, COSTIND merged into a new bureaucracy called the Ministry of Industry and Information Technology (MIIT), renamed the State Administration for Science, Technology and Industry for National Defence (SASTIND).

Committee on Space Research (COSPAR)

In the late 1990s a reorganization of the Chinese defence industry took place. The main focus of this reorganization was to separate the purchase of weaponry, which became the responsibility of the General Armaments Division of the People's Liberation Army; the production and development of weaponry, which became the responsibility of several different enterprises; and the development of policy for these industries, which became the responsibility of COSTIND. Areas of space focus for COSTIND include: the State Aerospace Bureau, the **China National Space Administration** and the **China Satellite Launch and Control General**. See also **China Aerospace Corporation**.

Committee on Earth Observation Satellites (CEOS), www.ceos.org

The Committee on Earth Observation Satellites was established in 1984 in response to a recommendation from a panel on Remote Sensing from Space, under the aegis of the G7 Economic Summit of Industrialised Nations Working Group on Growth, Technology and Employment. This group recognized the multidisciplinary nature of satellite Earth observation and the value of co-ordination across all proposed national missions. Funding and resources required by activities are contributed in-kind by participating CEOS agencies. The work of CEOS spans the scope of activities required for international co-ordination of Earth observation programmes and the maximum utilization of Earth observation data, ranging from the development of detailed technical standards for data product exchange to the establishment of high-level interagency agreements on common data principles.

CEOS co-ordinates civil space-borne observations of Earth. Participating national agencies strive to address critical scientific questions and to plan satellite missions with the intent of avoiding unnecessary overlap. CEOS has three primary objectives in pursuing these goals: to optimize benefits of space-borne Earth observations through co-operation of its members in mission planning and in development of compatible data products, formats, services, applications and policies; to serve as a focal point for international co-ordination of space-related Earth observation activities; and to exchange policy and technical information and compatibility of observation and data exchange systems.

Committee on Space Research (COSPAR), France, cosparhq.cnes.fr

The International Council of Scientific Unions, later the International Council for Science (www.icsu.org), established the Committee on Space Research during an international meeting in the United Kingdom in 1958. COSPAR's initial Space Science Symposium was organized in Italy in 1960. COSPAR is an interdisciplinary scientific organization with a focus on the progress of research carried out with the use of space. The organization also plays an important role in space science international co-operation.

COSPAR objectives are to promote scientific research in space, with emphasis on the exchange of results and information; and to provide a global forum for the discussion of problems that affect space science research. These objectives are primarily achieved through the organization of scientific assemblies and publications, which include *Advances in Space Research* and *Space Research Today*. COSPAR plays a role in the development of new space disciplines, such as space life sciences and space physics, and in the development of scientific guidelines for space exploration, most notably the *COSPAR Planetary Protection Policy*, which was first formulated in 1964 and last amended in 2005.

Croatian Space Agency—*see* **Hrvatska Svemirska Agencija (HSA), Croatia.**

CSIRO Marine and Atmospheric Research (CMAR), Australia, www.cmar.csiro.au

CSIRO (Commonwealth Scientific and Industrial Research Organization) Marine and Atmospheric Research is a government research agency that works with satellite reception antennas. CMAR was formed in 2005 as a result of a merger between the former CSIRO Division of Marine Research and the CSIRO Division of Atmospheric Research. CMAR's research covers the areas of climate, weather and ocean prediction; the living atmosphere; marine ecosystems and resources; and integrated coastal and oceans management.

CMAR hosts the Australian National Fish Collection and the CSIRO Collection of Living Microalgae and manages the Research Vessel Southern Surveyor as the Marine National Facility. In 2006 CMAR employed 550 staff at various locations in Australia. To help implement its tasks, CSIRO Marine and Atmospheric Research has access to data from satellite reception antennas. The Tasmanian Earth Resources Satellite Station receives data from a number of environmental satellite instruments including MODIS, AVHRR and SeaWifs. The data is available to the research community and public via a number of web-based services.

Czech Space Office (CSO), Czech Republic, www.czechspace.cz

The Czech Space Office is a non-profit association created in 2003 and is the central contact point for the co-ordination of all space-related activities in the Czech Republic. It supports the participation of Czech researchers in international space collaborations, manages the relationship with the **European Space Agency** (ESA) and contributions of the Czech Republic to ESA programmes, provides information in space-related project specification, and provides training opportunities. CSO organizes seminars and workshops for professionals from various fields of space activities and organizes educational and public events devoted to space-related topics. The aim of Czech space activities is to use high-technology space projects to improve the quality of life of its citizens. Expenditure for space activities amount to about US $10m., including contributions to ESA and the **European Organisation for the Exploitation of Meteorological Satellites** (EUMETSAT).

CSO's work encompasses gathering and archiving information about Czech space projects and information on foreign space programmes having importance for development of the Czech space activities. It covers the management of databases of the Czech institutions, both academic and industrial, and data on space projects being undertaken in the Czech Republic. CSO is involved in the following space projects, fully or partly funded by the **European Union** (EU): European Research Area (ERA) project on Space Technologies Applications and Research for the regions and medium-sized countries; CASTLE Project on Clusters in Aerospace and Satellite Navigation Technology Applications linked to Entrepreneurial Innovation; and the Centre for Innovation and Transfer of Technology Prague Project of the Academy of Space Technology.

D

Danish Astronautical Society—*see* **Dansk Selskab for Rumfartsforskning, Denmark.**

Danish Space Consortium—*see* **Dansk Rumkonsortium, Denmark.**

Dansk Rumkonsortium, Denmark, www.space.dtu.dk/Myndighedsbetjening/Dansk_Rumkonsortium.aspx

The Dansk Rumkonsortium (Danish Space Consortium) is an association of major Danish organizations in space, including private companies and public institutions. Dansk Rumkonsortium aims, through co-operation, to maximize the industrial and research benefits of Danish space activities. The consortium works as a consulting party in research and innovation for the Ministry of Science, Technology and Development in relation to the formulation of Denmark's space policy and priorities for Danish investment in the **European Space Agency** (ESA). The National Space Institute at the Technical University of Denmark plays a central role in the consortium.

The Consortium is organized into a number of professional forums: astronomy, solar system, Earth observation, International Space Station (ISS), education and dissemination, and space industry and technology. The consortium, which in 2010 comprised 40 companies along with the Confederation of Danish Industries, forms the framework for exchange of new knowledge and technology and acts as an incubator for collaborative projects among public and private organizations.

Dansk Selskab for Rumfartsforskning, Denmark, www.rumfart.dk

The Dansk Selskab for Rumfartsforskning (Danish Astronautical Society) was founded in Copenhagen in 1949. It is devoted to the peaceful uses of space and constitutes the Danish section of the **International Astronautical Federation**. The society arranges public meetings, lectures and company visits. The society has established six working groups, which focus on specific disciplines of space research, such as satellite communications and navigation, human space flight and microgravity research, planetary exploration and space-based astronomy and Earth observation and space-based meteorology.

DEIMOS Space

The society publishes a magazine, *Dansk Rumfart* (Danish Space Flight), which covers space news of Danish relevance. Further, it has hosted public lectures on all aspects of space flight, participated in numerous exhibitions and events, and hosts events for the Danish space industry.

Dassault Aviation, France, www.dassault-aviation.com/fr/espace.html

Since the 1960s, Dassault Aviation has worked with pyrotechnics for military aircraft and space by designing, studying and qualifying their use in aeronautics and space activities for the French Ministry of Defence, **Centre National d'Etudes Spatiales**, **European Space Agency** (ESA) and other organizations. Dassault Aviation has developed components and equipment for Europe's Ariane 5 and Vega launch vehicles, and also for the European Automated Transfer Vehicle, Jules Verne, to the International Space Station.

In 2009 Dassault Aviation signed a contract with **European Aeronautic Defence and Space Company Astrium** to provide for pyrotechnics equipment for the Ariane 5. The equipment facilitates motor ignition, stages and payloads separation, and launcher neutralization and destruction. The pyrotechnic components developed for launchers have also been customized for satellites, such as European Remote Sensing, Helios, Huygens, Italsat and Spot. These devices perform a number of functions, including deployment of solar panels and antennas, apogee motor ignition and unlocking of mechanisms.

Defense Advanced Research Projects Agency (DARPA), USA, www.darpa.mil

The Defense Advanced Research Projects Agency is an agency of the US Department of Defense (DoD) responsible for the development of new technology for use by the military. DARPA was established during 1958 as the Advanced Research Projects Agency (ARPA) in response to the 1957 launch of Sputnik. The mission of DARPA is to advance US military technology. DARPA reports directly to senior DoD officials. The agency had an annual budget of US $3,000m. as of 2009. Space projects are conducted through DARPA's Strategic Technology Office. Of particular interest to DoD are innovative concepts that support responsive access to space, and the development of space control sensors and systems.

DEIMOS Space, Spain, www.deimos-space.com

DEIMOS Space was founded in 2001 by a group of European engineers to develop and provide high-technology systems and engineering solutions. The company transfers relevant technology to the space sector, and works with space businesses in the European market, such as **European Aeronautic Defence and Space Company (EADS) Astrium** and **Thales Alenia Space**, in addition to **Eutelsat Communications** and **European Organisation for the**

Exploitation of Meteorological Satellites. DEIMOS Space participated in projects for the **European Space Agency**, including the Precision Agile Control System, Interferometer Constellation Control, and Highly Integrated Control and Data Systems. The company also develops software for the **European Union**'s Galileo programme.

Departamento de Ciência e Tecnologia Aeroespacial (DCTA), Brazil, www.cta.br

The Departamento de Ciência e Tecnologia Aeroespacial (Department of Science and Aerospace Technology) is a sector of the Brazilian Air Force that co-ordinates aerospace efforts in technology and education. DCTA was established in 1969 as the Department of Research and Development. In 2006 the name was changed to the Command-Generality de Tecnologia Aeroespacial. The current name has been in effect since 2009.

DCTA co-ordinates and implements research projects that contribute to aeronautics and space activities in Brazil. DCTA is responsible for the Institute of Aeronautics and Space (www.iae.cta.br) and the Centro de Lançamento de Alcântara (Alcântara Launch Centre, www.cla.aer.mil.br). The Institute of Aeronautics and Space designed and produced successful suborbital launch vehicles for numerous scientific and technological experiments. In 1990 the Alcântara Launch Centre became operational for sounding rocket launches and planned satellite launching. The geographical location of the Centre near the equator provides for significant advantage in launching geosynchronous satellites, an attribute shared by the Guiana Space Centre of the **European Space Agency**.

In development since 1980, under the management of DCTA, is the Brazilian Satellite Launcher Vehicle (Veículo Lançador de Satélite—VLS). The test launches of VLS failed in 1997 and 1999, and the third complete test planned in 2003 exploded on the launch pad before the test and destroyed part of the Centre's infrastructure. Since then, development and flight testing has resumed with Russian and Ukrainian assistance. The next test flight of the complete VLS rocket is planned for 2012.

Department of Science and Aerospace Technology—*see* **Departamento de Ciência e Tecnologia Aeroespacial (DCTA), Brazil.**

Design Bureau of Transport Machinery (KBTM), Russia, www.kbtm.ru/english/start.htm

The Design Bureau of Transport Machinery was formed in 1946. Space activities of KBTM included developing ground equipment for space launch systems—more than 70 space complexes and 10 converted military complexes. The complexes have provided preparation and launch of more than 1,600 launch vehicles for various purposes. KBTM is in charge of the Plesetsk

Deutsches Zentrum für Luft und Raumfahrt (DLR)

and Svobodny Cosmodromes, the launch and technical complexes of Cosmos, Cyclone, Zenit, Angara and Rockot launch vehicles.

Deutsches Gesellschaft für Luft-und Raumfahrt-Lilienthal-Oberth (DGLR), Germany, www.dglr.de

The Deutsches Gesellschaft für Luft-und Raumfahrt-Lilienthal-Oberth (German Society for Aeronautics and Astronautics Lilienthal-Oberth) is a non-profit organization for the research, science and technology of aeronautics and space. With more than 3,000 members, DGLR represents the technical and functional areas of aviation and aerospace, acts as a link among the disciplines and promotes exchange among industrial firms, government agencies, research institutes and universities.

DGLR goals include: promoting technical and scientific work in the aerospace disciplines; promoting the training of professional talent; informing the public about the technical, economic and cultural importance of aerospace; and developing perspectives for aviation and aerospace. The attainment of these goals is accomplished through annual meetings, symposia and lectures; formation of committees to address specific topics; scientific and technical publications; co-operation with organizations and associations with similar objectives; and documenting the history of aerospace.

Deutsches Zentrum für Luft und Raumfahrt (DLR), Germany, www.dlr.de

The Deutsches Zentrum für Luft und Raumfahrt (German Aerospace Centre) is Germany's national research centre for aeronautics and space. Its R&D work in aeronautics, space, transportation and energy is integrated into national and international ventures. The German Government has given DLR responsibility for planning and implementing German space programmes and projects, and DLR represents Germany's interests in space.

Approximately 6,200 people work for DLR in 29 institutes and facilities at 13 locations in Germany. The DLR national budget was US $720m. in 2009. The purpose of DLR is exploration of Earth and the solar system; research aimed at protecting the environment; and development of environmentally friendly technologies to promote mobility, communications and security.

DLR is involved in, among others, the following products and space projects. The Cassini-Huygens mission of the **European Space Agency** and the **National Aeronautics and Space Administration**. The mission comprises the Cassini space probe with planned operations until 2017, and the Huygens lander, which completed its mission in 2005. In Germany, DLR, the Max-Planck Society (MPG), several universities and the German aerospace industry are involved with this mission. In 2008 the Columbus space laboratory was launched and subsequently attached to the International Space Station (ISS). Germany and DLR played a key role in the construction of Columbus and DLR is involved in operations of Columbus. European Mars Express, launched in 2003, is providing new data on the geology, mineralogy and atmosphere of Mars. The High Resolution Stereo Camera (HRSC), developed in DLR's Institute of Planetary Research, is mapping Mars in three dimensions. Robotic Components Verification on the ISS technological experiment developed by DLR was fitted on the outer platform of the ISS in 2005. The Rosetta mission, being undertaken by ESA, aims to research the history of how the solar system was formed by investigating a comet. DLR played a major role in building the lander and runs the lander control centre. The Scanning Imaging Absorption Spectrometer for Atmospheric Cartography atmospheric sensor, on board the ESA satellite ENVISAT, measures the relevant concentrations of trace gases for air quality, the greenhouse effect and ozone chemistry, and was contributed by DLR and the Netherlands Agency for Aerospace Programmes (NIVR). The launch of the German radar satellite TerraSAR-X to map Earth was a joint venture of DLR and the **European Aeronautic Defence and Space Company Astrium**. The European Venus Express space mission intended to learn more about Venus's atmosphere, structure, dynamics and composition. The DLR Institute of Planetary Research and DLR's Visual Information Systems equipment developed the camera and spectrometers for the mission.

DigitalGlobe, USA, www.digitalglobe.com

DigitalGlobe provides commercial, high-resolution (less than 1 m panchromatic resolution) imagery products and services from its satellite constellation. The imagery supports a wide variety of uses from mapping and analysis to navigation technology.

DigitalGlobe collects imagery for a variety of purposes, including identifying natural resources; understanding Earth's environmental condition; protecting homelands and borders; responding to emergencies and natural disasters; planning investments in multimillion-dollar infrastructure development; and monitoring oil and gas pipelines and facilities. Products include Basic Satellite Imagery, Standard Satellite Imagery, Orthorectified Satellite Imagery, Orthorectified Aerial Imagery, CitySphere and Basic Stereo Pair Imagery. DigitalGlobe remote sensing satellites include QuickBird and WorldView-1.

The entire commercial Earth observation sector generated US $2,000m. in sales of data and value-added services in 2009. The sector is led by DigitalGlobe, **GeoEye** of the USA and **Spot Image** of France. In 2009, DigitalGlobe's revenue was US $280m., and in 2010 DigitalGlobe forecasted revenue of US $330m. The principal customer for DigitalGlobe is the **National Geospatial-Intelligence Agency** of the USA.

DMC International Imaging, United Kingdom, www.dmcii.com

DMC International Imaging provides for the Disaster Monitoring Constellation (DMC). This constellation is unique in that each satellite is independently owned and controlled by a separate state. States involved in DMC include: Algeria, the People's Republic of China, Nigeria, Spain, Turkey and the United Kingdom (UK). DMC satellites are designed and built by **Surrey Satellite Technology Ltd** of the UK with the support of the **United Kingdom Space Agency**. The main function of DMC International Imaging is to provide imaging capability to the partner states. The DMC consortium participates in the *International Charter: Space and Major Disasters*. In addition, DMC members encourage the use of DMC data for scientific and commercial applications.

Dutch Space, the Netherlands, www.dutchspace.nl

The roots of Dutch Space are in the Fokker Aircraft company, established in 1912. Fokker Aircraft started with satellite space activities in the 1960s. In the 1970s–80s, Fokker Space initiated the development of solar panel technology, becoming Dutch Space in 2002. Dutch Space is the largest space company in the Netherlands and is a supplier for the international space industry. Major activities of Dutch Space are control systems and engineering, robotics, instruments, recovery systems, software and simulation, testing and integration, structures and mechanisms, such as with solar panel technology, and a number of other advanced products.

Dutch Space has participated in a number of specific space missions and products, including: the Rosetta space probe; the **European Space Agency** mission with satellites Herschel Space Observatory and Planck; the **National Aeronautics and Space Administration** Dawn probe to investigate two asteroids, Ceres and Vesta; the International Space Station (ISS); the Huygens probe; the Mars Sample Return mission; the X-ray Multi-Mirror Mission-Newton Observatory; the Sloshsat Facility for Liquid Experimentation and Verification in Orbit; Ulysses; Giotto; the European Hipparcos satellite; the Infrared Space Observatory; and the Graphical Astronomy and Image Analysis Tool.

E

Ecuadorian Civilian Space Agency—*see* **Agencia Espacial Civil Ecuatoriana (EXA), Ecuador.**

Eisenhower Center for Space and Defense Studies, USA, www.usafa.edu/df/dfps/csds/index.cfm web.mac.com/rharrison5/Eisenhower_Center_for_Space_and_Defense_Studies

The Eisenhower Center for Space and Defense Studies was established in 2005. Recognizing the need for coherent space policy, the Eisenhower Center was formed with four main goals: to develop an intellectual foundation for space policy; to define a curriculum of space policy studies for higher education; to establish space policy as an integral element of national security policy and national security strategy; and to develop a generation of young people with a propensity for space activities.

The Eisenhower Center functions as a space policy think-tank at the US Air Force Academy (USAFA, www.usafa.af.mil), bringing together space and policy experts to discuss a range of topics at various forums. From 2005–2010 the Eisenhower Center developed and organized Asia, space and strategy workshops; the Future of Space Commerce workshop; the Space Solar Power workshop and concept development; space situational awareness workshops; national space forums focused on space policy challenges; space studies seminars; transatlantic space co-operation workshops; and a space deterrence workshop and study. In these efforts, the Eisenhower Center collaborated with a variety of governmental and non-governmental organizations, including the **Center for Strategic and International Studies**, the **Center for Defense Information**, **Futron Corporation**, **George C. Marshall Institute**, **National Aeronautics and Space Administration**, **National Security Space Office**, Office of the Secretary of Defense, **Space Policy Institute**, **Secure World Foundation**, and **US Air Force Space Command**. In addition, the Eisenhower Center edits and publishes the *Space and Defense* journal, compiled and edited a *Space Defense Policy* textbook and curriculum and sponsors policy research.

Emirates Institute for Advanced Science and Technology (EIAST), United Arab Emirates, www.eiast.ae

The Emirates Institute for Advanced Science and Technology pioneers space systems and services, and applications based on those space systems.

Euroconsult

The EIAST space programme includes three main themes. First, is Earth Observation—EIAST's programme with DubaiSat-1 designed to support infrastructural development. The second is Satellite Navigation—EIAST is working with international organizations in the field of satellite navigation. It is a member of the International Committee for Global Navigation Satellite Systems, and through its membership, EIAST is working with other satellite navigation providers and vendors to find interoperability among systems. Third, is Satellite Communication—EIAST is exploring this satellite technology.

Eurisy, France, www.eurisy.org

Eurisy was created in 1989 as an independent, non-governmental organization, fostering collective action for bridging space and society. Eurisy membership includes 40 governmental space offices and space agencies, international organizations, research institutions, institutions of higher education and private businesses involved in activities related to space. It raises awareness among decision-makers, civil society and other users of the strategic importance of space for sustainable economic, environmental and social development policies. Eurisy also helps to advance European space education goals.

Eurockot Launch Services, Germany, www.eurockot.com

Eurockot Launch Services provides commercial launch services with the Rockot launch system to operators of low Earth orbit satellites. Formed in 1995, the company is located in Germany and is owned by **European Aeronautic Defence and Space Company Astrium**. Eurockot serves an international market and operates from launch facilities in Plesetsk Cosmodrome in Russia.

Eurockot provides services to act as the contractual partner of the customer; provides an end-to-end mission management service, including interface control and detailed and documented mission analysis and mission design on-site management of prelaunch operations at the launch site; and a comprehensive range of services varying from public relations support to launch risk coverage arrangements.

In 2000 the Rockot launch system was commissioned as an operational launch vehicle. Eurockot has launched a number of spacecraft: the **European Space Agency** (ESA) Salty Mission in 2009; SERVIS-1 in 2003 and SERVIS-2 in 2009; ESA environmental satellite Gravity Field and Steady-State Ocean Circulation Explorer launch in 2009; the Korean KOMPSAT-2 Earth observation satellite in 2006; and the **National Aeronautics and Space Administration** Gravity Recovery and Climate Experiment satellite in 2002.

Euroconsult, France, www.euroconsult-ec.com

Euroconsult provides consulting, research, market analyses and forecasts to support business planning and strategic decision-making for more than 560 clients in 50 countries. Clients include satellite operators and service providers, satellite manufacturers and launch service providers, equipment providers and integrators, government agencies, media and broadcasting companies, and banks and investors.

The purpose of Euroconsult is to respond to client needs through the development of research reports and other decision-making tools. To meet these needs, Euroconsult relies on data research capabilities, methods of analysis, analytical proprietary databases and regular contact with industry decision-makers. Euroconsult provides government policy reviews; programme performance assessments with independent reviews of programme efficiency; return-on-investment prospects; socio-economic benefit analysis and cost-benefit analysis; due diligence capability to serve the needs of lenders, investors and merger and acquisition targets; strategic planning and market analysis; and comprehensive reports and strategic planning tools for business leaders, analysts, government agencies and banks and investors.

European Aeronautic Defence and Space Company (EADS) Astrium, www.astrium.eads.net

European Aeronautic Defence and Space Company has a number of divisions, including EADS North America, Airbus, Eurocopter, EADS Defense and Security and EADS Astrium. There are three main areas of activity for EADS Astrium: EADS Astrium Satellites for spacecraft segments and ground segments; EADS Astrium Space Transportation for launchers, propulsion systems and orbital infrastructure; and EADS Astrium Services for the development and delivery of satellite services. These activities cover civil and military satellite systems for telecommunications, Earth observation, science and navigation and defence systems and space industrial services. With an employee base of 15,000 and organizations in France, Germany, the United Kingdom and Spain, EADS Astrium is a partner to national, international and commercial customers.

EADS was formed in 2000 through a series of aerospace mergers in Europe (Aérospatiale Matra from France, DaimlerChrysler from Germany and Construcciones Aeronáuticas from Spain), to provide for the economy of scale to develop large-scale systems successfully. In 2003 EADS acquired Astrium, a wholly owned subsidiary of EADS. Know as EADS Astrium since 2003, the company reported revenue in 2009 of more than US $6,400m., an increase of 21% over the previous two years.

European Conference for AeroSpace Sciences (EUCASS), Belgium, www.eucass.eu

The European Conference for AeroSpace Sciences, a non-profit Belgian association, was created in 2006 by European scientists and engineers for scientific communication, technical activities and exchanges among researchers and industry end users world-wide. The goal of EUCASS is to provide a forum for aerospace researchers through conferences. EUCASS aims to attract young scientists to the profession and to develop workforce mobility and co-operation in the European countries. EUCASS also provides a venue for exchanges among decision-makers in governments and industry.

EUCASS addresses the sciences of aerospace. These disciplines have been assembled into five groups, each placed under the supervision of a committee. The five groups are surveyed in EUCASS conferences. They include flight dynamics and guidance, navigation and control; system integration dealing with space launch activities; structures and materials; flight physics; and propulsion physics.

European Organisation for the Exploitation of Meteorological Satellites (EUMETSAT), www.eumetsat.int

The European Organisation for the Exploitation of Meteorological Satellites is a meteorological and environmental satellite intergovernmental organization serving the interests of Europe's National Meteorological Services, the citizens of Europe and world-wide customers. The **European Space Agency** supported the original formation of EUMETSAT in 1984. The relationship between the two organizations has evolved and they are now partners.

EUMETSAT operates meteorological satellites that monitor the atmosphere and ocean and land surfaces, which deliver satellite data and images. This information is supplied to users world-wide and assists meteorologists to identify and monitor weather and climate. EUMETSAT consists of 26 member states in Europe that are parties to the *EUMETSAT Convention*, which entered into force in 1986.

The purposes of EUMETSAT are to maintain the existing range of meteorological and climate monitoring data, products and services; to develop new satellite systems, including the ground infrastructure, product processing, archiving and dissemination systems; and to work with the partners of the European meteorological infrastructure to maximize the impact of satellite data for weather prediction. EUMETSAT systems include Meteosat satellites, the global overview of which are complemented by the observations provided by the polar orbiting Metop satellite and the marine observer, Jason-2, a joint project of space agencies in Europe and the USA.

European Space Agency (ESA), www.esa.int

The European Launcher Development Organisation (ELDO) and the European Space Research Organisation (ESRO) were the first attempts at European co-operation in space. Created in 1962, ELDO worked to build the Europa space launch vehicle, and ESRO focused on satellites. ESRO was successful, but ELDO failed as the different stages of the Europa rocket were built by different European countries without proper integration mechanisms. Following the failures of ELDO, policy-makers at the highest level met in a series of conferences and decided to renew their efforts. As a result, ESA was created in 1975 as a multinational international organization to shape the development of Europe's space capability by co-ordinating the financial and intellectual resources of its members. ESA is the interlocutor of the many national space agencies in Europe, and in 2010 there were 18 member states: Austria, Belgium, Czech Republic, Denmark, Finland, France, Germany, Greece, Ireland, Italy, Luxembourg, the Netherlands, Norway, Portugal, Spain, Sweden, Switzerland and the United Kingdom. Canada also sits on the council of ESA and takes part in some projects under a co-operation agreement. Hungary, Romania and Poland are participating in the Plan for European Co-operating States, while other countries are in negotiation with ESA about joining this initiative. Not all member countries of the **European Union** (EU) are members of ESA and not all ESA member states are members of the EU. ESA is an independent multinational organization, although it maintains close ties with the EU through an ESA–EU Framework Agreement. The two organizations share a joint European Strategy for Space and have together developed European Space Policy.

ESA's purpose is to provide for, and promote, co-operation among European states in space research and technology, and their space applications, with a view to their being used for scientific purposes and for operational space applications systems. ESA programmes cover Earth observations, human space flight, launchers, navigation, space science, space engineering, spacecraft operation, technology development and innovation and telecommunications. About 2,000 people work for ESA. Financial contributions from the member states, calculated in accordance with each country's gross national product, fund ESA. In 2009, ESA appropriated funding of US $5,200m., while its budget for 2010 stood at US $5,000m.

ESA operates a number of technical centres, which encompass European Space Research and Technology Centre in the Netherlands, for the technical preparation and management of ESA space projects; European Space Operations Centre in Germany, for tracking and control of satellites; an ESA centre in Italy, to co-ordinate satellite operations through more than 20 ground stations and facilities throughout Europe; European Astronaut Centre in Germany, for the training of European astronauts; and European Space Astronomy Centre in Spain, for services to astronomical research projects. ESA also owns the launch facilities at the Guiana Space Centre, Europe's spaceport. The centre launches the Ariane 5, Soyuz and Vega space launch vehicles. See also **Arianespace**.

European Space Policy Institute (ESPI), Austria, www.espi.or.at

The European Space Policy Institute is a think tank that provides an independent view and analysis on mid-term to space policy issues relevant to the use of space in Europe. ESPI supports a network of experts and centres of excellence working with ESPI in-house analysts, including the European Space Policy Research and Academic Network, to contribute to decision-making; increase awareness of space technologies and applications with user communities; and support students and researchers in their space-related work. ESPI produces studies commissioned by its members and by external institutions.

Founding members of ESPI include the **European Space Agency** and the **Austrian Research Promotion Agency**. Other members include **Arianespace**, Belgian High Representation for Space Policy, **Centro para el Desarrollo Technológico Industrial**, **European Aeronautic Defence and Space Company Astrium**, **European Organisation for the Exploitation of Meteorological Satellites**, Eutelsat, **Centre National d'Etudes Spatiales**, **Deutsches Zentrum für Luft und Raumfahrt**, **Agenzia Spaziale Italiana**, **Norwegian Space Centre**, European Satellite Operators Association (www.esoa.net), **Telespazio**, and **Thales Alenia Space**.

European Union (EU) ec.europa.eu

A more recent evolution in the European space landscape, during the past decade, is the emergence of the European Union as a space actor in Europe. The EU is the leading authority on two key programmes: Galileo, for positioning, navigation and timing services; and the Global Monitoring for Environment and Security (GMES) programme concerning Earth observations from space. A process of reorganization of European space was underway in 2010, when the **European Space Agency** (ESA) could become the EU executive agency for space, with the EU Commission formulating space policy and providing for programmatic decisions. The EU Commission decisions are dictated by the governments of the 27 state members of the EU.

The entry into force, in December 2009, of the Treaty on the Functioning of the European Union, known as the Lisbon Treaty, gave a mandate to the EU Commission to exercise its right to implement space policy for the EU. In this regard, the Lisbon Treaty states that the EU is to implement a European space policy through joint initiatives, research and technological development and co-ordination of the efforts needed for the exploration and exploitation of space; to contribute to attaining the objectives referred to in the implementation of a European space policy through the formation of European space programmes; and to establish relations with ESA to implement European space programmes.

The EU and ESA had no official link until 2003, when a framework agreement was signed that arranged for ministerial-level space councils, the EU–ESA joint secretariat and the high-level Space Policy Group. These joint bodies developed the 2007 European Space Policy, which also provided the legal basis for programmatic co-operation between the EU and ESA, including ESA's role as implementing agency for the EU in developing Galileo and GMES. EU funding for space is expected to grow. In 2009 the EU allocated

US $1,600m. to space. This was in addition to the US $5,200m. for ESA space programmes and projects in 2009 and US $2,500m. for national space efforts in Europe in 2009, primarily by France, Germany, Italy, Spain and the United Kingdom.

Eutelsat Communications, France, www.eutelsat.com

The history of Eutelsat Communications started in 1977 with the creation of an intergovernmental organization, the European Telecommunications Satellite Organisation (EUTELSAT), to develop and operate a satellite-based telecommunications infrastructure for Europe. The *EUTELSAT Convention* entered into force in 1985 with 17 European states as members. By 2001, 49 European states were parties to the Convention.

The mission of EUTELSAT was to ensure that the EUTELSAT satellite fleet provides coverage for all member states; to enable member states to be sure that their operators, service providers and broadcasters have equitable access to EUTELSAT's services; to ensure the continuity of the collectively owned rights and obligations of the organization under international law, in particular the rights to use radio frequencies and orbital locations, which were assigned collectively to the member states and the organization by the **International Telecommunication Union**; and to monitor relevant developments in regulations and agreements, and ensure that EUTELSAT is in a position to comply with their provisions, while adhering to the interest of its member states.

EUTELSAT started operations with the launch of its first satellite in 1983. The organization started by providing space segment capacity for basic telecommunications and audiovisual services to Europe. From these activities, EUTELSAT expanded to the provision of all types of analogue and digital television and radio broadcasting services, business telecommunications services, multimedia communications, messaging and positioning services and access to broadband internet. Initially established to address satellite communications demand in Western Europe, EUTELSAT developed its infrastructure to expand coverage to additional markets, such as Central and Eastern Europe, the Middle East, the African continent, and large parts of Asia and the Americas from the 1990s.

In the context of the privatization of the telecommunications sector in Europe, the intergovernmental operations and activities of EUTELSAT were transferred to a private company in 2001 called Eutelsat, SA. In 2005, the principal shareholders of Eutelsat, SA grouped their investment in a new entity, Eutelsat Communications, which in 2010 was the holding company of the group owning 95% of Eutelsat, SA.

As of 2009 Eutelsat Communications had a fleet of 26 satellites, expansion into mobile satellite services, a workforce of 650 and revenue of more than US $1,250m. Eutelsat is based in France and it operates subsidiaries in Belgium, Brazil, People's Republic of China, Germany, Italy, Portugal, Poland, the United Kingdom and the USA.

Federal Aviation Administration Office of Commercial Space Transportation (FAA-AST), USA, www.faa.gov

The Office of Commercial Space Transportation (AST) was established in 1984 in the Office of the Secretary of Transportation within the Department of Transportation. AST was transferred to the FAA in 1995. AST regulates the commercial space transportation industry, to ensure compliance with international obligations of the USA, and to protect the public health and safety, safety of property and national security and foreign policy interests of the USA; encourages, facilitates and promotes commercial space launches and re-entries by the private sector; recommends appropriate changes in US federal statutes, treaties, regulations, policies, plans and procedures; and facilitates the strengthening and expansion of the US space transportation infrastructure.

AST issues FAA licences for commercial launches of orbital rockets and suborbital rockets. The first US-licensed launch was a suborbital launch of a Starfire vehicle in 1989. Since then, AST has licensed almost 200 launches. Active space launch licences as of 2010, include Boeing Launch Services, Orbital Sciences Corporation, Sea Launch, Lockheed Martin and Space Exploration Technologies (SpaceX).

AST also issues licences for the operations of non-federal government launch sites, known as spaceports. The first launch from a licensed, non-federal facility was that of the National Aeronautic and Space Administration's Lunar Prospector on-board a Lockheed Martin Athena 2 in 1998, from Spaceport Florida. Since 1996, AST has issued site operator licences for the eight spaceports: California Spaceport at Vandenberg Air Force Base (www.californiaspaceauthority.org); Spaceport Florida at Cape Canaveral Air Force Station (www.spaceflorida.gov/index.php); Mid-Atlantic Regional Spaceport at Wallops Flight Facility in Virginia (www.marsspaceport.com); Mojave Airport and Spaceport in California (www.mojaveairport.com); Kodiak Launch Complex on Kodiak Island, Alaska (www.akaerospace.com); Oklahoma Spaceport (www.okspaceport.state.ok.us/spaceauthority.html); Spaceport America in New Mexico (www.spaceportamerica.com)—there are current plans for suborbital space tourism with **Virgin Galactic** at Spaceport America; and Cecil Field Spaceport in Jacksonville, Florida (www.cecilfield spaceport.com).

In 2004, the *Commercial Space Launch Amendments Act* of 2004 was enacted by the US Congress. To comply with the act, AST published updated FAA regulations that implement requirements for licences. These regulations include *Final Rule on the Requirements for Amateur Rocket Activities* in 2008; *Final Rule on Human Space Flight Requirements for Crew and Space Flight Participants* in 2007; *Final Rule on Experimental Permits for Reusable Suborbital Rockets* in 2007; and *Safety Approval Final Rule* in 2006. Of note, is the evolution in regulations to deal with 'new space' entrants in commercial space transportation, which include plans for suborbital human space flight.

Federation of American Scientists (FAS), USA, www.fas.org

The Federation of American Scientists, a non-profit organization, was founded in 1945 by scientists who had worked on the US Manhattan Project to develop atomic bombs and recognized that science had become central to public policy questions. They believed that scientists have a responsibility to warn the public and policy leaders of dangers from scientific and technical advances and to show how policy can increase the benefits of new scientific knowledge. FAS projects are organized under three main programme areas: strategic security, Earth systems and educational technologies.

In the area of space, the FAS Space Policy Project promotes national security and international stability by providing the public and decision-makers with information and analysis on civil and military space issues, policies and programmes. The project is dedicated to increasing international co-operation in space as a means of improving global co-operation to solve problems on Earth. The project also focuses on specific policy questions related to advanced technology weapons, such as ballistic missile proliferation, commercial space development and military space systems.

Finmeccanica, Italy, www.finmeccanica.com

Finmeccanica is the main Italian industrial group operating globally in the aerospace, defence and security sectors. It was established in 1948, and as of 2009 had a workforce of more than 73,000 people and revenue of US $24,000m. In the space arena it provides satellite and space services through the Space Alliance. In 2007 Finmeccanica and Thales established the Space Alliance, which comprised **Telespazio** (67% Finmeccanica, 33% Thales) and **Thales Alenia Space** (67% Thales, 33% Finmeccanica). The focus for the Space Alliance is on the design, development and manufacture of telecommunications, Earth observation and remote sensing satellites for civil and military use.

French Aerospace Industries Association—*see* Groupement des Industries Françaises Aéronautiques et Spatiales (GIFAS), France.

French Institute of Space History—*see* **Institut Français d'Histoire de l'Espace (IFHE), France.**

Futron Corporation, USA, www.futron.com

Futron Corporation, founded in 1986, is a technology consulting firm. Its primary competencies include engineering analysis, risk management and space industry analysis. Futron's clients include many sectors of the space industry, such as the **Federal Aviation Administration Office of Commercial Space Transportation**, the **National Aeronautics and Space Administration**, and the US Department of Defense. Futron works with solutions for aerospace, telecommunications and technology enterprises, and combines business, technical and management expertise to provide solutions for improved decisions, performance and results.

G

General Organization of Remote Sensing (GORS), Syria, www.gors-sy.net

The General Organization of Remote Sensing was established in 1986, replacing the **National Remote Sensing Centre**, which was established in 1980. GORS is responsible for land surveying using remote sensing techniques. Its purpose is to offer training for specialists of different scientific disciplines related to remote sensing techniques; to establish centres and institutions for training and applications of remote sensing; to perform scientific studies, research and experiments that are related to remote sensing; to provide remote sensing data; to propose national legislation for remote sensing activities and to follow their execution; and to address international activities related to remote sensing. GORS represents Syria in international meetings, conferences and symposia that are related to remote sensing and the organization co-operates with Syrian universities and other Arab states. GORS conducts projects concerning geo-engineering and integrated planning; agriculture and environment; geological and hydrological research; and archaeology and tourism.

GeoEye, USA, www.geoeye.com

GeoEye was formed in 2006 when ORBIMAGE purchased the assets of Space Imaging. ORBIMAGE, which was a commercial spin-off of **Orbital Sciences Corporation**, began performing geospatial services for the **US Government** in 1987. Space Imaging was established in 1995 as a commercial remote sensing provider through investments made by Lockheed Martin. Space Imaging operated a high-resolution, commercial remote sensing satellite system, IKONOS. In 2007 GeoEye acquired MJ Harden, enabling GeoEye to include aerial imagery and specialized geospatial processing services.

GeoEye assets include remote sensing satellites GeoEye, IKONOS and OrbView; aerial mapping aircraft; and a network of partners and satellite receiving ground stations. GeoEye's geospatial information data products support national intelligence, defence and security, air and marine transportation, oil and gas, environmental monitoring, online mapping, insurance and risk management, urban planning and emergency preparedness. GeoEye, along with **DigitalGlobe** of the USA, dominates the US commercial remote sensing market, which generated US $2,000m. in sales world-wide in 2009. Total revenue for GeoEye in 2009 was US $270m., an 85% increase on 2008.

The principal customer for GeoEye is the **National Geospatial-Intelligence Agency** of the USA.

Geo-Informatics and Space Technology Development Agency (GISTDA), Thailand, new.gistda.or.th

Thailand has been involved in satellite remote sensing since the launch of the **National Aeronautics and Space Administration**'s Landsat programme in 1971 (landsat.gsfc.nasa.gov). Involvement took place through the Thailand Remote Sensing Programme (TRSP) under the National Research Council of Thailand. In 1979 TRSP became part of the Thailand Remote Sensing Centre. TRSC serviced and promoted utilization of remote sensing data for management of natural resources and the environment in Thailand.

In order to enhance the utilization in remote sensing and Geographic Information System, GISTDA was established in 2000 as a public organization, which assumed responsibility for space technology and geo-informatics applications. One key programme is the Thailand Earth Observation Satellite (THEOS), which was developed by GISTDA working with the European Aeronautic Defence and Space Company Astrium. THEOS was launched by Russia in 2008.

George C. Marshall Institute, USA, www.marshall.org

The George C. Marshall Institute, named after the Second World War military leader George C. Marshall, was established in 1984 in Washington, DC, as a non-profit think-tank to conduct technical assessments of scientific issues with an impact on public policy. The Marshall Institute provides policymakers with technical analyses on a range of public policy issues. In the 1980s and 1990s, the Marshall Institute was engaged primarily in lobbying in support of the Strategic Defense Initiative.

The Marshall Institute addresses space security and national defence issues. US military operations in Iraq and Afghanistan have confirmed the importance of space-based assets to US military power. Only a few times in history has a technical advance had such an impact on national security and the balance of power. The advent of sea power was one such development, air power was another and space power was a third. In response to these developments, the Marshall Institute examines issues affecting US military space policy, including budgets and programmes, policy and arms control and threats to space assets.

German Aerospace Centre—see Deutsches Zentrum für Luft und Raumfahrt (DLR), Germany.

German Society for Aeronautics and Astronautics Lilienthal-Oberth — see Deutsches Gesellschaft für Luft-und Raumfahrt-Lilienthal-Oberth (DGLR), Germany.

Global Network against Weapons and Nuclear Power in Space (GN), USA, www.space4peace.org

The Global Network against Weapons and Nuclear Power in Space was founded in 1992 by the joint efforts of the Florida Coalition for Peace and Justice, the Citizens for Peace in Space of Colorado Springs, Colorado, and the New York-based journalism professor, Karl Grossman. The inaugural meeting of GN was held in Washington, DC. Since its founding, GN has met annually to bring together activists who are interested in space issues. About 150 organizations are affiliated with GN.

GN's mission is to apply space technology to social and environmental needs on Earth; explore alternative technology paths for space power and propulsion; solve problems on Earth instead of creating imbalances and conflicts in space; prevent confrontation and enhance international co-operation in space; ban space weapons and space military installations by national and international laws; avoid oversized, costly and risky space projects; ban the use of nuclear power in space; encourage and foster global democratic debate about space exploration and colonization; and strengthen existing international space laws that call for collective use of celestial bodies.

Group on Earth Observations (GEO), www.earthobservations.org

The Group on Earth Observations was established in response to calls for action by the 2002 World Summit on Sustainable Development and by the Group of Eight (G8) leading industrialized countries. These high-level meetings recognized that international collaboration is essential for exploiting the growing potential of Earth observations to support decision-making in an increasingly complex and environmentally stressed world.

GEO is dedicated to the Global Earth Observation System of Systems (GEOSS). GEOSS, along with the Integrated Global Observing Strategy, facilitates the sharing and applied usage of global, regional and local data from satellites, ocean buoys, weather stations, and other surface and airborne Earth observing instruments. GEOSS supports policy-makers, resource managers, science researchers and many other experts and decision-makers.

GEO is a voluntary partnership of 80 governments and the European Commission, and 56 intergovernmental, international and regional organizations with a mandate in Earth observation or related issues. GEO is governed by a plenary composed of all members and participating organizations. It meets at least once a year at the level of senior officials and periodically at the ministerial level. An executive committee oversees GEO's activities. GEO committees and working groups implement 10-year plan.

GEO's 10-year implementation plan, adopted in 2005 for the period 2005–2015, outlines its goals for building GEOSS in respect to nine Societal Benefit Areas: disasters, health, energy, climate, water, weather, ecosystems, agriculture and biodiversity.

Groupement des Industries Françaises Aéronautiques et Spatiales (GIFAS), France, www.gifas.asso.fr

The origins of the Groupement des Industries Françaises Aéronautiques et Spatiales (French Aerospace Industries Association) date to 1908, when a group of aviation pioneers in France sought to bring an industrial and commercial dimension to aviation. This led to the creation of the Chambre Syndicale des Industries Aéronautiques. In 1975, the chamber acquired the name Groupement des Industries Françaises Aéronautiques et Spatiales. In 2010, GIFAS had more than 250 members. GIFAS analyses and defends trade interests by examining the economic, social, financial, environmental and technical regulations of the aerospace industrial sector and promotes the French aerospace industry and its products by participating in trade shows and exhibitions. Active members of GIFAS include prime contractors and large system suppliers—Airbus, Avions de Transport Regional, **Arianespace**, **European Aeronautic Defence and Space Company Astrium**, **Dassault Aviation**, Eurocopter, Matra BAE Dynamics Alenia, **Safran Group** and **Thales Alenia Space**, as well as small and medium-sized aerospace companies.

Groupement Luxembourgeois de l'aéronautique et de l'espace (GLAE), Luxembourg, www.glae.lu

The Groupement Luxembourgeois de l'aéronautique et de l'espace (Luxembourg Aeronautic and Space Industry Association) was founded as a non-profit organization in 2005. GLAE offers Luxembourg companies, representing 450 employees in direct or indirect relation with the field of space activities, the opportunity to create synergies at the national and European level for both the private and public sectors. GLAE, constituted within the Luxembourg Federation of Employers (FEDIL), has the objective of providing a permanent link among its members, advising them and defending their shared professional, economic and social interests.

H

Henry L. Stimson Center, USA, www.stimson.org

Founded in 1989, the Henry L. Stimson Center, based in Washington, DC, is a non-profit, non-partisan institution devoted to enhancing international peace and security through analysis and outreach. The center's work is focused on global security: strengthening institutions for international peace and security; building regional security; and reducing weapons of mass destruction and transnational threats. The center's approach is geared toward providing policy alternatives, solving problems and overcoming obstacles to a secure world. Through research and analysis, it seeks to understand complex issues and engage policy-makers, policy implementers, non-governmental institutions and other experts.

The center deals with space issues through its Space Security Program. The goals of the programme are to increase public awareness about the dangerous consequences of flight testing and deploying space weapons; provide policy-makers, legislators, negotiators and non-governmental organizations with information to construct wise space security choices; and offer a pragmatic alternative to space weapons.

There are three major areas of activity to realize these goals. First, is a code of conduct. The USA and other nations have endorsed and practised codes of appropriate conduct at sea, on the ground and in the air. A code of conduct for responsible spacefaring nations also makes sense at a time when military, commercial and exploratory activities in space are growing significantly. Of note is that the center released a working draft of a *Code of Conduct for Responsible Spacefaring Nations* in 2007. The **European Union** has also issued a draft Code of Conduct for outer space activities (*European Council draft Code of Conduct for outer space activities*, 17 December 2008). Second, space assurance: to ensure the continued preservation and growth of space capabilities, nations would be well served by adopting a comprehensive strategy to ensure the safety and security of life-saving satellites. The center's research focuses on ways a strategy of space assurance might be conceived and implemented. Finally, space diplomacy: diplomacy can establish norms that are in the national security interests of the USA because they clarify responsible behaviour and facilitate responses to irresponsible behaviour. The center explores various diplomatic options for enhancing space security, including a code of conduct and a verifiable treaty banning destructive methods against space objects.

Hrvatska Svemirska Agencija (HSA), Croatia, www.csa.hr

The Hrvatska Svemirska Agencija (Croatian Space Agency) was established in 2002 to co-ordinate Croatian space research programmes with scientific and commercial goals. HSA co-operates with the **European Space Agency**. In 2010 HSA started the development of its space programme with plans to launch a research satellite in 2013.

Hungarian Astronautical Society—*see* **Magyar Asztronautikai Társaság (MANT), Hungary.**

I

IHI Corporation, Japan, www.ihi.co.jp

IHI Corporation, previously known as Ishikawajima-Harima Heavy Industries until 2007, is a Japanese manufacturer of rockets for scientific observation, launch vehicles for satellites and defence-related systems and other R&D in space-related equipment. IHI employs 7,600 people, with net sales of US $14,800m. in 2009. Space projects of IHI, among others, focus on sounding rockets, development of Japanese space launch vehicles and the Japanese Experiment Module for the International Space Station.

ImageSat International, Israel, www.imagesatintl.com

ImageSat International is a commercial provider of high-resolution, satellite Earth imagery focused on government markets and their defence forces for national security and intelligence applications. The company has offices in Cyprus and Israel. The offices in Israel supervise the construction for ImageSat of the Earth Remote Observation Satellites (EROS) and the operation of ImageSat's ground control station. Among the diverse applications for EROS imagery are air, naval and ground forces; homeland security and border control; infrastructure; mapping; and environment and disaster control.

In 2000, ImageSat launched EROS A, on board a Russian Start-1 launch vehicle and in 2006 ImageSat launched EROS B. ImageSat International offers various types of panchromatic images characterized either by Imaging Technique—the manner in which the satellite payload is used to acquire the imagery data—or Processing Level—the level of post-processing performed on the ground after acquisition of data.

Indian Space Research Organisation (ISRO), India, www.isro.org

India's national space programme is primarily civilian, with a focus on the application of space technology as a tool for socio-economic development of the country. The basic aim of India's space programme is to utilize space technologies for development, such as with communications, meteorology and natural resource management. Since the 1990s, Indian space efforts have also included commercial development, navigation, astronomy and planetary missions.

The Indian space programme started in 1963 with the launching of sounding rockets. In 1975 India's first satellite was launched by the Soviet Union. India became a spacefaring power in 1980 when it launched an indigenous satellite on the Indian Satellite Launch Vehicle from its own launch site. Initially, India's space programme was under the aegis of the Department of Atomic Energy in 1962, with the creation of the Indian National Committee for Space Research.

ISRO was formed under the Department of Atomic Energy in 1969, and was brought under the Department of Space in 1972. A space commission was also established in the same year, which reported directly to the prime minister. The Department of Space along with ISRO operates four independent projects: the Indian National Satellite Space Segment Project; the National Natural Resource Management System; the National Remote Sensing Agency; and the Physical Research Laboratory. The Department of Space also sponsors research in various academic and research institutions. This allows ISRO to interact with various educational institutes and outsource research.

In 2010 ISRO had various operating divisions throughout India. These divisions deal with space systems, propulsion, communications, telemetry and tracking, research, launches and other facets of the space programme. The major achievements of the space programme have been in the area of the domestic design, production and launching of remote sensing and communications satellites. Throughout the years, ISRO established a strong infrastructure for remote sensing and communications satellite systems, with launcher autonomy for both polar and geostationary satellite launch. In 1992 ISRO established a commercial company called **Antrix Corporation Ltd**. This company markets space and telecommunications products of ISRO.

In 2008 India launched a lunar robotic programme called Chandrayaan. Chandrayaan-1, which was launched in 2008, successfully completed its mission, and Chandrayaan-2 is planned for launch in 2012. India may also attempt to send a human mission to the Moon by 2020 and has plans for human space flight by 2014–2016. This added dimension of undertaking human space missions needs to be viewed not as a policy shift, but as a natural progression of capability with ISRO and the national space programme of India. India has plans to send robotic spacecraft to study Mars and is in the process of developing a satellite-based navigation system. The national space programme of India was funded at US $1,100m. in 2009.

Indonesian National Institute of Aeronautics and Space (LAPAN), www.lapan.go.id

The Indonesian National Institute of Aeronautics and Space was established in 1962 for R&D of space technology. LAPAN utilizes satellites for economic development and disaster mitigation and for climate change, communications and satellite-based positioning and navigation. Activities of LAPAN include R&D, design and engineering sciences; development of national policy and law (LAPAN performs the duties of the secretariat to National Aeronautics and Space of Indonesia (DEPANRI), which is a national body co-ordinating programmes between various institutions and directing policies relating to space issues); information dissemination; and development of science and technology resources.

Inmarsat, United Kingdom, www.inmarsat.com

Inmarsat was originally the International Maritime Satellite Organization. From 1979 to 1999, Inmarsat functioned as a self-financed international organization of states on the basis of the *Convention on the International Mobile Satellite Organization* (*Inmarsat Convention*), which entered into force in 1983.

Inmarsat was established in 1979 by the International Maritime Organization (www.imo.org), a **United Nations** specialized agency that was itself established in 1948, and entered into force in 1959, to implement the *International Convention for the Safety of Life at Sea*. The purpose of Inmarsat was to establish a satellite communications network for the maritime community to facilitate maritime security and the efficiency of shipping.

In 1976 Comsat—the Communications Satellite Corporation that was established as a government charted company in the USA to develop commercial satellite communications, and was, for a time, a majority owner and manager of **Intelsat SA** (Comsat was acquired by Lockheed Martin in 2000)—launched the satellite, MARISAT, to provide mobile services to the US Navy and other maritime customers. Inmarsat initially leased MARISAT satellite transponders beginning in 1982. By 1990, Inmarsat had developed the first generation satellite systems, INMARSAT satellites, for maritime, aeronautical and land mobile communications services. In 2007, Inmarsat developed a satellite-based telephone system.

Inmarsat's assembly of member governments agreed to privatize Inmarsat in 1998. From 1999 to 2003 Inmarsat was completely privatized, and an associated intergovernmental body, the International Mobile Satellite Organization (www.imso.org), was established to ensure that the privatized Inmarsat continued to meet public service obligations, including obligations relating to the Global Maritime Distress Safety System (GMDSS). The privatization of Inmarsat was primarily financed by various private equity firms. In 2009, Inmarsat revenues exceeded US $1,000m.

Institut for Rumforskning og-teknologi Danmarks Tekniske Universitet (DTU Space), Denmark, www.space.dtu.dk

The Danish Space Research Institute, a government research institute established in 1968, merged with the research units of the Danish National Survey and Cadastre to create the Danish National Space Centre (DNSC) in 2005. In 2007 parts of the Danmarks Tekniske Universitet (Technical University of Denmark, DTU) and the Department of Informatics and Mathematical Modelling at DTU merged with DNSC to form the Institut for Rumforskning og-teknologi Danmarks Tekniske Universitet (National Space Institute, also known as DTU Space).

DTU Space functions in four key areas: space-related graduate education; space science and environmental science research; industrial sector collaboration; and public sector consultancy. DTU Space offers graduate degree programmes through the School in Space Science and Technology of DTU, which includes geodesy and geomatics. The research divisions of DTU Space include astrophysics, geodesy and geodynamics, measurement and instrumentation, microwaves and remote sensing, solar system physics and Sun-climate studies. In these areas DTU Space is involved with numerous space projects in Europe and the USA.

DTU space also helps with the commercial development of space science and technology. There are four areas of focus: development of space science as a new business area; deliveries of products and services to specific space projects; development of new products or services exploiting satellite navigation and Earth observation satellites; and commercial benefit of spin-off technologies originally developed for space science and space research.

DTU Space consults the government sector on the benefits of satellite-based systems. The Ministry of Science, Technology and Innovation, the Land Consolidation Agency under the Danish Ministry of the Environment, the Greenland Home Rule Authorities and the US Department of Defense make use of DTU Space consulting. DTU Space gives advice on monitoring and control work and carries out operational assignments within environment, security and transportation.

Institut Français d'Histoire de l'Espace (IFHE), France, ifhe.free.fr

The Institut Français d'Histoire de l'Espace (French Institute of Space History) was created in 1999 to promote the identification and conservation of heritage and documentary material in connection with space activities, including oral histories, heritage protection, publication of papers and conferences. Partners of IFHE include the **Centre National d'Etudes Spatiales**, the **European Space Agency** and the French Defence Historical Service, which is the record centre for the French Ministry of Defence.

Institute for Space Applications and Remote Sensing (ISARS), Greece, www.space.noa.gr

The Institute for Space Applications and Remote Sensing is one of the five research institutes of the National Observatory of Athens. ISARS was founded in 1955 under the name of Ionospheric Institute. In 1990 the institute was renamed Institute of Ionospheric and Space Research and in 1999 took its current title to reflect its expanded activities, which cover a variety of aspects of space research and applications. ISARS is supervised and supported by the

General Secretariat of Research and Technology of the Greek Ministry of Development.

Instituto Nacional de Pesquisas Espaciais (INPE), Brazil, www.inpe.br

The Instituto Nacional de Pesquisas Espaciais (National Institute for Space Research) was created in 1971 to develop satellite data collection and remote sensing capabilities and to help develop the Centro de Lançamento de Alcântara (Alcântara Launch Centre, www.cla.aer.mil.br). The mission of INPE is to develop science and technology in space and the terrestrial environment, and offer products and services for Brazilian space efforts. INPE develops satellites for imaging and monitoring of the Amazon region, for meteorological studies as part of the Global Precipitation Measurement Program and for astrophysical and microgravity scientific research. Space missions conducted by INPE include: China–Brazil Earth Resources Satellite programme; the Amazonia-1 satellite; EQUARS and MIRAX scientific satellite programmes; the SCD satellites as part of an environmental data collection programme in Brazil; the PRODES project to monitor the Brazilian Amazon forests; and the DEGRAD system to map Amazon forest degradation.

Instituto Nacional de Técnica Aeroespacial (INTA), Spain, www.inta.es

The Instituto Nacional de Técnica Aeroespacial (National Institute for Aerospace Technology) was created in 1942. It is a public research organization specializing in aerospace research and technology development for Spanish aerospace activities. The mission of INTA is to focus on technologies that can be applied to the aerospace field; to aerospace applications; to provide technical assessment and services; and to act as a technological centre. INTA activities, include studies in aerodynamics and propulsion, development of rocket and missile motors and R&D of renewable energy sources. INTA develops technological activities within the areas of space sciences, payloads and aerospace applications. The institute is also involved with Earth observations and remote sensing. Space missions and projects of note include: Spasolab, an official laboratory of the **European Space Agency**; and small satellite development (Minisat 1, launched in 1997, and NANOSAT 1, launched in 2005).

Intelsat SA, Luxembourg, www.intelsat.com

Of the many benefits gained from space enterprises, telecommunications was the first commercially viable industry to develop. The creation of a satellite communications framework was not a foregone conclusion because of technical problems associated with reaching geostationary orbit and the time delay of signals to and from that orbit. Simultaneously, the ownership and management of what was envisioned to be a global communications network

Intelsat SA

was similarly contested, with regard to the question of government ownership, regulated monopoly or private competition.

With the development of satellite and rocket technology by the early 1960s, communications satellites became a foreign policy tool in the technological and political competition of the cold war. In 1961 US President John Kennedy challenged the USA to establish communications satellites to operate on a global basis. What resulted was the Communications Satellite Act of 1962, which created a government chartered corporation called the Communications Satellite Corporation (Comsat). Half of the Comsat shares were owned by the telecommunications companies, with the remaining 50% available to the public. In 2000, Comsat was acquired by **Lockheed Martin Space Systems Company** as the result of privatization in the communications satellite sector (see below, regarding the *Open-Market Reorganization for the Betterment of International Telecommunications Act*).

Comsat, representing the interests of US telecommunications companies, set out to create an international consortium of countries for participation in a non-governmental, non-profit satellite telecommunications international body. The Soviet Union rebuffed the idea at the outset, while the Europeans joined. This led to Intelsat, which was created in 1964 on an interim basis by its initial member states. Intelsat was formally established as an intergovernmental organization in 1973 on entry into force of an intergovernmental agreement, *Agreement Relating to the International Telecommunications Satellite Organization* (*INTELSAT Agreement*).

Intelsat priced service independent of route, essentially using heavy traffic routes to subsidize lighter traffic ones, usually to developing states. In this manner, Comsat and Intelsat balanced commercial and geopolitical interests to bring into existence a model of global telecommunications (Inmarsat also played a key role in mobile satellite communications services). By 2000, (Intelsat was privatized in 2001), Intelsat had 144 member states, and more than 200 states came to rely on Intelsat satellite communications.

By the late 1980s, a world-wide trend emerged toward deregulation of telecommunications. In 1988, the first private communications satellite was launched, yet services could not be provided to the public telephone companies of states that were signatories to the Intelsat consortium. The deregulatory trend and the emergence of private communications satellites led the US Congress to enact in 2000 the *Open-Market Reorganization for the Betterment of International Telecommunications Act*. This act called for Intelsat to transfer its satellites and financial assets to private investors and become largely independent of former signatories, which were typically government-owned companies. Intelsat became a private company in 2001, with unanimous consent of its 144 member states, by transferring substantially all of its assets and liabilities to a new company established for this purpose, Intelsat Ltd, and its subsidiaries. The same member states established the **International Telecommunications Satellite Organization** in 2001, to oversee the public service obligations of Intelsat.

Until 2008 Intelsat was owned by more than 200 shareholders representing telecommunications network operators from more than 140 states. Intelsat acquired PanAmSat in 2006. In 2008 Intelsat Ltd announced the successful closing of its majority acquisition by BC Partners, a private equity firm based in the United Kingdom. In 2009 Intelsat Ltd changed its domicile to Luxembourg and, as a result of this migration, the company became known as Intelsat SA. Intelsat supplies video, data and voice connectivity in approximately 200 countries and territories with a satellite fleet of more than 50 that covers more than 99% of the world's population. Intelsat SA revenue as of 2009 was US $2,500m.

Inter-Agency Consultative Group for Space Sciences (IACG)

The Inter-Agency Consultative Group for Space Sciences was established in 1981 as a multilateral organization involving European, Japanese, Soviet and US scientists, and their respective national space institutes and national space agency representatives. Its initial mission goals, which were successfully realized, were to co-ordinate nationally approved missions to rendezvous with and scientifically study Halley's Comet. Participating scientists within IACG conceptualized mission co-ordination that allowed for the optimization of mission objectives for each national spacecraft that encountered Halley's Comet. The IACG focused on co-ordination of science to the exclusion of augmented or integrated hardware development that would necessitate some degree of technology transfer.

These aspects were reflected in IACG terms of reference including no technology transfer, no exchange of funds and an advisory role to member space agencies. On this basis, the IACG was able to function as a viable multilateral organization that reinforced historical patterns of collaboration in data exchange and analysis prevalent in space science missions since the beginning of the space age and the International Geophysical Year. Annual meetings of the IACG were held from the group's formation in 1981 to mission implementation with Halley's Comet in 1986. These meetings were informal with delegations composed of senior scientists and managers from the **European Space Agency** (ESA), the former Japanese Institute for Space and Astronomical Science, the Soviet Institute for Space Research, and the **National Aeronautics and Space Administration** (NASA), and also professional astronomers from the International Halley Watch.

The success of IACG was viewed as a compelling reason for the organization to extend its duration. This took place through co-ordination of solar terrestrial science programmes underway in ESA, Japan, Russia and NASA. However, this second phase of co-operation proved to be problematic. The expansion of the IACG, because of the size and complexity of co-ordinating numerous satellites and co-ordination for solar terrestrial science, caused the group to become less an informal community of scientists and more a formalized bureaucracy. Yet, the IACG was deficient in the requisite organizational mechanisms, like a formalized organizational charter specifying decision-making procedures and distributions of financial resources. In 1996 the IACG adopted solar system exploration as a subject of international co-operation; however, national interest for IACG co-ordination declined in Europe and the USA and by 2002 the IACG ceased to function.

Inter-Agency Space Debris Coordination Committee (IADC), www.iadc-online.org

The Inter-Agency Space Debris Coordination Committee is an international governmental forum for the co-ordination of activities related to the issues of human-made and natural debris in space. The primary purposes of the IADC are to exchange information on space debris research activities among member space agencies, to facilitate opportunities for co-operation in space debris research, and to identify debris mitigation options. In the area of debris mitigation, the IADC has put forward technical guidelines for member organizations to follow. These guidelines have been adopted on the agenda of the **United Nations** Committee on the Peaceful Uses of Outer Space (COPUOS) for consideration as legal principles for the international community.

IADC comprises a steering group and four specified working groups covering measurements, environment and database, protection and mitigation. IADC member agencies include: **China National Space Administration, Deutsches Zentrum für Luft und Raumfahrt, European Space Agency, Centre National d'Etudes Spatiales, Agenzia Spaziale Italiana, Indian Space Research Organisation, Japan Aerospace Exploration Agency, National Aeronautics and Space Administration, National Space Agency of Ukraine, Russian Federal Space Agency** (Roscosmos), and **United Kingdom Space Agency**.

International Academy of Astronautics (IAA), Sweden, iaaweb.org

The International Academy of Astronautics is a non-governmental organization established in Sweden in 1960. IAA works to promote the development of astronautics for peaceful purposes, to recognize individuals who have distinguished themselves in a related branch of science or technology and to provide a programme through which members may contribute to endeavours for aerospace science.

IAA encourages international scientific co-operation through scientific symposia and meetings and the work of specialized commissions: Space Physical Sciences; Space Life Sciences; Space Technology and System Development; Space Systems Operations and Utilization; Space Policy, Law and Economy; and Space and Society, Culture and Education. IAA publishes the journal *Acta Astronautica* and co-sponsors, along with the **International Astronautical Federation** and the **International Institute of Space Law**, the annual International Astronautical Congress, of which the 61st Congress was held in 2010.

International Association for the Advancement of Space Safety (IAASS), the Netherlands, joinspace.org/iaass

The International Association for the Advancement of Space Safety, established in 2004 in the Netherlands, is a non-profit organization dedicated to furthering co-operation and scientific advancement in the field of space systems safety. IAASS membership is open to any professional who has an interest in space safety. The mission of IAASS is to advance the science and application of space safety; to improve communication, dissemination of knowledge and co-operation among interested groups and individuals in this field and related fields; to improve understanding and awareness of the space safety discipline; to promote and improve the development of space safety professionals and standards; and to advocate the establishment of safety laws, rules and regulatory bodies at national and international levels for civil use of space.

International Astronautical Federation (IAF), France, www.iafastro.com

The International Astronautical Federation, based in Paris, France, was founded in 1951. It is an international non-governmental, non-profit organization consisting of 198 members from 57 countries (as of 2010) and governed by a constitution. Members of IAF include space agencies, space companies, societies, associations, universities, institutes and non-profit organizations.

IAF encourages the advancement of knowledge about space and the development and application of space assets for the benefit of humankind. The purposes of IAF are to promote public awareness and appreciation of space activities, the exchange of information on space programme developments and plans, workforce development, the recognition of achievements in space programme co-operation and the use of space by developing countries for human development. To meet these purposes, IAF organizes the annual International Astronautical Congress, which held its 61st annual meeting in 2010, in addition to specialized meetings and symposia throughout the year.

International Charter Space and Major Disasters, www.disasterscharter.org

Following the UNISPACE III conference (see **United Nations**), held in Austria in 1999, the European and French national space agencies initiated the *International Charter Space and Major Disasters* with the **Canadian Space Agency** (CSA). The charter aims to provide a system of space data acquisition and delivery to those affected by natural or human-made disasters, through authorized users. Each member agency has committed resources to support the provisions of the charter and is helping to mitigate the effects of disasters.

Members include: **Agence Spatiale Algérienne**, Argentine Space Agency, Canadian Space Agency, **China National Space Administration**, **European Space Agency**, **Centre National d'Etudes Spatiales**, **Indian Space Research Organisation**, **Japan Aerospace Exploration Agency**, **National Oceanic and Atmospheric Administration**, **National Space Organization** of Taiwan,

International Institute of Space Law (IISL)

National Space Research and Development Agency of Nigeria, **TÜBİTAK Space Technologies Research Institute, United Kingdom Space Agency** and the US Geological Survey. There are also commercial partners, which include **Spot Image** of France, **DigitalGlobe** and **GeoEye** of the USA, and the Disaster Monitoring Constellation International Imaging of the UK.

International Institute for the Unification of Private Law (UNIDROIT), Italy, www.unidroit.org

Established in 1926 as an auxiliary organ of the League of Nations, the International Institute for the Unification of Private Law was, following the demise of the League, re-established in 1940 on the basis of a multilateral agreement, the Unidroit Statute. As of 2010 there were 63 member states. UNIDROIT is an independent, intergovernmental organization with headquarters in Rome, Italy. The purpose of UNIDROIT is to study needs and methods for modernizing, harmonizing and co-ordinating private and commercial law among states and groups of states.

In the area of space, UNIDROIT established a committee of governmental experts for the preparation of a draft *Protocol on Matters Specific to Space Assets*. This protocol addresses the difficult task of applying the benefits of the UNIDROIT Convention on *International Interests in Mobile Equipment* to space assets, which are increasingly being financed by private-sector investors, rather than central governments, which are subject to a myriad of existing regulations under international treaties, and which are often physically located beyond terrestrial jurisdictions. The *Protocol on Matters Specific to Space Assets* is under consideration by an intergovernmental negotiation process, which includes representation by private-sector financiers and the space industry.

International Institute of Air and Space Law, the Netherlands, law.leiden.edu/organisation/publiclaw/iiasl

The International Institute of Air and Space Law, founded in 1985, is an international academic research and teaching institute, specializing in legal and policy issues regarding aviation and space activities. The objective of the institute is to contribute to the development of aviation and space law and related policy by conducting and promoting research and teaching at the postgraduate level. It forms an integral part of the Leiden Law School of Leiden University.

International Institute of Space Law (IISL), the Netherlands, www.iislweb.org

The International Institute of Space Law was founded in 1960 by the **International Astronautical Federation** (IAF). It replaced the Permanent Committee on Space Law, which the IAF had created in 1958, and since then the IISL has held annual colloquia on space law in conjunction with the International Astronautical Congress, the proceedings of which are published by the **American Institute of Aeronautics and Astronautics**. In 2007 the IISL was formally established as an independent association under national law of the Netherlands, and its structure has been improved to better fulfil its role.

Space law is an area of the law that encompasses national and international law governing activities in space. The objectives of IISL include co-operation with organizations and institutions in the field of space law; the holding of meetings, colloquia and competitions on juridical and social science aspects of space activities; the preparation and commissioning of studies and reports; and the publication of books and proceedings. IISL issues statements on legal issues and addresses topics that are of interest to the space community. Also, IISL presents reports on its activities to the legal subcommittee and contributes to the **United Nations** *Highlights in Space* report. In co-operation with the European Centre for Space Law (www.esa.int/SPECIALS/ECSL) of the **European Space Agency**, IISL organizes an annual space law symposium on topical space law issues for the delegates and staff attending the annual session of the COPUOS legal subcommittee.

International Launch Services (ILS), USA and Russia, www.ilslaunch.com

International Launch Services traces its roots to 1993, when what was then Lockheed Corporation of the USA established a venture to market Russian Proton launch vehicles. This international partnership involved Lockheed and two Russian companies, **Khrunichev State Research and Production Space Centre**, and **RSC Energia**. The joint venture, called Lockheed Khrunichev Energia International (LKEI), signed its first launch customer in 1993.

Lockheed merged with Martin Marietta in 1995, creating Lockheed Martin, and Martin Marietta's Commercial Launch Services organization, which manufactured and launched the Atlas rockets, was then consolidated with LKEI, creating a new entity called ILS. As of 2008, ILS headquarters are located in Reston, Virginia.

ILS captures 40%–50% of available missions annually. In 2010 the company provided commercial launch services only on Proton, with exclusive rights to market Khrunichev's Angara next-generation vehicle when it becomes available. In May 2008, Khrunichev acquired Space Transport's shares in ILS, making Khrunichev State Research and Production Space Centre of Moscow the majority shareholder.

In 2010, ILS had a staff of about 60 people and had signed contracts for more than 100 launches valued at US $8,000m. The Proton vehicle launches commercial ILS missions and Russian Government payloads from the Baikonur Cosmodrome, which is operated by the Russian Space Agency (Roscosmos), under lease from the Republic of Kazakhstan. Proton launches all Russian geostationary and interplanetary missions under Khrunichev. ILS is responsible for marketing and managing the Proton system, which has been used in more than 340 launches and has a reliability record of 94%.

International Promotional Association for Spaceflight—*see* **Internationaler Förderkreis für Raumfahrt (IFR), Austria.**

International Space University (ISU), France, www.isunet.edu

The International Space University provides graduate-level training to the space community at its main campus in Strasbourg, France, and at locations around the world. ISU offers a two-month space studies summer programme and a one-year Master's programme covering disciplines related to space programmes and enterprises—space science, space engineering, systems engineering, space policy and law, business and management and space and society. Since its founding in 1987, ISU has graduated more than 2,700 students from 100 countries.

ISU develops leaders of the world space community by providing inter-disciplinary educational programmes to students and space professionals in an international, intercultural environment. ISU also serves as an international forum for the exchange of knowledge and ideas on issues related to space and space applications. ISU programmes impart skills for space initiatives in the public and private sectors, promote understanding and co-operation, foster a network of students, teachers and alumni, and encourage the development of space for peaceful purposes.

International Telecommunication Union (ITU), Switzerland, www.itu.int

The International Telecommunication Union is an international organization, part of the **United Nations** system of organizations for information and communications technology issues, and a focal point for governments and the private sector in developing networks and services. ITU co-ordinates the shared global use of the radio spectrum; promotes co-operation in assigning geostationary Earth orbits (GEO) for telecommunications satellites; works to improve telecommunications infrastructure in the developing world; and establishes world-wide standards that foster interconnection of communications systems.

ITU is located in Geneva, Switzerland, and the membership base includes 191 states and more than 700 other members and associates, primarily regional and international telecommunication satellite organizations and companies. ITU also assists in mobilizing technical, financial and human resources, and strengthens emergency communications for disaster mitigation.

Organized by the ITU, World Radiocommunication Conferences (WRC) are held every two to four years. WRC is charged with reviewing and, if necessary, revising the *Radio Regulations*. *Radio Regulations* is the international treaty governing the use of the radio frequency spectrum, GEO satellite use of that spectrum and GEO orbital slot allocations. Radio frequency spectrum is a limited natural resource that is shared among states on a regional and global basis. The WRC is charged with managing the international use of radio frequency spectrum in a rational and equitable manner. Further, the decisions of the WRC have a significant impact on a national basis. In the US case, for example, the Federal Communications Commission established an advisory committee before each WRC to voice US interests during the process. ITU has a specific Space Services Department to manage all procedures and data regarding the *Radio Regulations*.

International Telecommunications Satellite Organization (ITSO), USA, www.itso.int

The International Telecommunications Satellite Organization is an intergovernmental organization charged with overseeing the public service obligations of **Intelsat SA**. Its mission is to ensure that Intelsat provides public telecommunications services, including voice, data and video, on a global and non-discriminatory basis. Headquartered in Washington, DC, ITSO has 150 member countries. The creation of ITSO resulted from the efforts of nations to join the USA in 1964 to establish a global communications satellite system. ITSO transferred its global satellite system, including geostationary-orbital locations, landing rights and the brand-name of Intelsat, to the privatized Intelsat SA in 2001. ITSO's mission is to act as the supervisory authority of Intelsat, and promote international telecommunications services.

Internationaler Förderkreis für Raumfahrt (IFR), Austria, www.ifr-raumfahrt-gesellschaft.de

The Internationaler Förderkreis für Raumfahrt (International Promotional Association for Spaceflight) was founded as a non-profit organization in 1969 in Salzburg, Austria, by Hermann Oberth and Wernher von Braun among others. The mission of IFR is to promote the ideas and visions of these astronautical pioneers and to promote the dissemination and public acceptance of space flight.

IFR awards the Hermann Oberth medal, the Wernher von Braun medal, the Hermann Oberth honorary ring and the Wernher von Braun honorary ring for outstanding merits in the field of space flight. IFR is a member of the International Astronautical Federation and also organizes an annual astronautical conference designed for the general public, presenting in all fields of astronautics, including economy, legal issues, medicine and environmental aspects, science, technology, politics and management.

Iran Space Council (ISC)

The Iran Space Council was created in 2005 in the pursuit of peaceful space activities for the development of culture, technology and science. ISC is formally responsible for Iran's space policy and the **Iranian Space Agency** operates under its guidance. The council is chaired by the president of Iran. Iran has pursued space capabilities since 1977, when it began to develop communications satellites to help the country modernize.

Official aims of the programme focus on using space and related technologies to develop Iran's economic and technological potential. More specifically, accessing space allows Iran to expand its telecommunications industries, monitor natural disasters and extend the influence of Iranian culture by widening the broadcast area of Iranian television and radio programmes.

Iranian Space Agency, isa.ir

The Iranian Space Agency was established in 2003. ISA is officially mandated by the Iranian Government to conduct research, design and implementation in the field of space technology to develop communications networks and remote sensing capabilities. The president of ISA is one of the deputies of the Ministry of Communication and Information Technology. ISA works in co-operation with the **Iran Space Council** (ISC).

Iran declared in 2005 that the Government would allocate US $500m. during the following five years, to 2010, to the space programme. In 2010 ISA implemented the launch of an Iranian Kavoshgar-3 rocket with living organisms on-board. The launch was successful and the organisms returned to Earth unharmed. As of 2010 Iran, through ISA and ISC, had launched three satellites: one in 2005 on a Russian Kosmos rocket, a second in 2008 on a Chinese rocket, and a third in 2009 on a domestically built Safir rocket.

Ireland Space Programme

Ireland does not have a separate national space programme, choosing to participate in European Space programmes. In addition, there exists bilateral co-operation among Irish scientific research teams and teams in other countries, such as Russia and the USA, on space science missions.

Israel Aerospace Industries (IAI), www.iai.co.il

Israel Aerospace Industries was established in 1953 as Bedek Aviation Company. IAI develops both military and commercial aerospace technology for the Israeli Ministry of Defence and other global customers. IAI objectives are to conduct R&D for the global defence industry in advanced technologies, systems and services for aeronautics, space and electronics; to strive towards a long-term balance among military and commercial endeavours; and to contribute to the security and economy of Israel. IAI revenue in 2009 was US $4,000m.

IAI developed its capabilities in the modification, upgrade and improvement of fighter and commercial aircraft and helicopters, engines and electronics systems and development of both military and commercial aerospace technology.

In the space sector, IAI develops satellites, ground stations and space launchers. IAI designs, manufactures and operates the AMOS communications satellites and lightweight and high-resolution remote sensing satellites, TecSAR (making use of Synthetic Aperture Radar) and OPTSAT. Of note, IAI offers orbit insertion services using the SHAVIT family of satellite launchers developed by IAI. Israel's first space launch took place in 1988.

Israel Space Agency (ISA), www.most.gov.il

The history of Israel in space started in 1988 with the launch of an Israeli reconnaissance satellite, Ofeq 1, by the indigenously developed SHAVIT space launch vehicle. The **Israeli Space Programme** focuses on satellite and launch vehicle development. ISA, established in 1983, is a governmental body within the Israeli Ministry of Science and Technology, which co-ordinates all space programmes with civil, scientific and commercial goals. ISA is involved in different stages of research, development, construction, launching and operations in a series of space projects.

One leading project of ISA includes the Tel-Aviv University Ultra Violet Experiment. This is an ultraviolet telescope for astronomical observations, developed in the 1990s in collaboration with India. A second project of note entails the design and construction of small satellites for scientific purposes. This is joint project by ISA and the **Centre National d'Etudes Spatiales**. In addition to co-operation with France and India, ISA has also reached co-operative agreements with Canada, Germany, the Netherlands, Russia, Ukraine, the USA and the **European Space Agency**.

Israeli Society of Aeronautics and Astronautics, www.aerospace.org.il

The Israeli Society of Aeronautics and Astronautics was established in 1951. The society directs its activities towards three goals: advancing the professional activities in Israel in the scientific and technological aspects of aeronautics and astronautics; distributing information and knowledge in Israel in these fields; and leading co-operative relations with international organizations in these fields. The society co-operates locally with the Israeli Technion and other universities, and with other scientific associations to realize its goals.

The society participates in organizing an annual conference on aerospace sciences. To promote aerospace, the society supports the publication of *Israel Aerospace Magazine*, published quarterly since 1972. The society represents Israel in the International Council of the Aeronautical Sciences (see www.icas. org) and in the **International Astronautical Federation**.

Israeli Space Programme

The Israeli Space Programme was established in 1981 because of national security needs and the desire to be self-sufficient in high-technology areas. The

Italian Space Agency

failure of Israel's strategic posture in the early 1970s, which led to the 1973 Yom Kippur war, convinced Israeli officials that an indigenous means of acquiring intelligence data through reconnaissance satellites was essential for national security. This conclusion was reinforced by the lack of satellite intelligence available to Israel from the USA during the Yom Kippur war. Since 1981 Israel has developed a number of indigenous capabilities, which include reconnaissance and remote sensing satellites (Ofeq, TecSAR, OPTSAT and EROS), satellite-based telecommunications (AMOS), and a space launch vehicle (SHAVIT).

As of 2010, Israel was one of nine states that had indigenously developed and launched satellites. Other states with both confirmed space launches and indigenous satellite development include Russia (since 1957), the USA (since 1958), France (since 1965), Japan (since 1970), China (since 1970), the United Kingdom (since 1971), India (since 1980) and Iran (since 2009). The **European Space Agency** (ESA) has conducted space launches since 1979 with the Ariane programme (both France and the UK contribute to this effort and do not maintain national efforts).

Israel's space capabilities have reflected its general emphasis on advanced technology in the economy and on the importance of maintaining a qualitative edge in the conflict with states hostile to its interests. Israel has come to realize that its economic growth, national security and international credibility are inextricably linked with space. The Israeli approach to space is pragmatic and focused on niche technologies that satisfy national security needs, such as remote sensing and Earth observation, communications, small satellites and operationally responsive technologies. Israel has developed a space industrial base characterized by a skilled workforce, technological capacity and advanced manufacturing capabilities.

The key players in Israel's space programme include the Israel Space Agency, which serves as a co-ordinating organization for co-operation and scientific programmes; Israeli military industries with involvement in hardware development for civil, commercial and military space; and the Israeli Ministry of Defence and Israeli Defence Forces, which are in charge of military space operations.

Italian Aerospace Research Centre—*see* **Centro Italiano Ricerche Aerospaziali (CIRA), Italy.**

Italian Association for Aerospace—*see* **Associazione Italiana per l'Aerospazio (AIAD), Italy.**

Italian Association of Aeronautics and Astronautics—*see* **Associazione Italiana di Aeronautica e Astronautica (AIDAA), Italy.**

Italian Space Agency—*see* **Agenzia Spaziale Italiana (ASI), Italy.**

J

Japan Aerospace Exploration Agency (JAXA), Japan, www.jaxa.jp

In 2003, the Institute of Space and Astronautical Science (ISAS), the National Aerospace Laboratory of Japan (NAL) and the National Space Development Agency of Japan (NASDA) merged into one independent administrative institution to form the Japan Aerospace Exploration Agency. JAXA administers space exploration activities for Japan, from basic R&D to applications. The space objectives of JAXA are to establish systems for natural disaster management and Earth observations; to develop space science robotic projects related to lunar, planetary and asteroid exploration; to establish technologies for future lunar utilization; and to develop space transportation systems, including launch vehicles and orbital transfer vehicles.

The objectives of JAXA are accomplished through a number of mission areas, which are realized through the development of a number of indigenous satellites and spacecraft. These include communications and weather observation, and astronomical observation and space development. Japanese satellites in orbit in 2010 are performing missions in a wide range of areas. For example, they have been playing a role in assessing and analysing abnormal weather patterns. For the purpose of planetary exploration, plans are underway to send probes to the Moon and Mars.

JAXA mission areas include: space transportation systems—HII space launch vehicle, and HII orbital transfer vehicle for cargo resupply to the International Space Station (ISS); human space activities—assembly of ISS with the Japanese Experiment Module, known as KIBO, and training of ISS Japanese astronauts; utilization with Satellites—Earth observation satellites, telecommunications satellites and navigation satellites; and space science research—robotic probes for lunar and planetary research, including missions to the Moon, asteroids, Venus, Mercury and Mars, and astronomical observation satellites.

Japanese space policy is framed by new approaches and by the reform of space organizations. In addition to the creation of JAXA in 2003, the new *Japanese Basic Law for Space Activities* was established in 2008. In accordance with the Basic Law, a new minister and a new strategic headquarters were established for space activities. In 2009, the strategic headquarters announced the new *Japanese Basic Plan for Space Activities*. This plan states the use of space for diplomacy and for national security purposes (new policy developments for Japan), and programme direction to meet the challenges of space, such as those posited by the USA in civil space, Europe in regard to space and national security and the People's Republic of China in relation to its lunar exploration plans.

Japan Society for Aeronautics and Space Sciences (JSASS), Japan, www.jsass.or.jp

The mission of the Japan Society for Aeronautics and Space Sciences is to provide the opportunity for fundamental and applied research in aeronautics and space science and to contribute to the further advancement of academia in Japan in aeronautics and in space science. JSASS co-operates with 31 aerospace-related societies from 14 different countries. JSASS activities include publishing a journal, conducting conferences, granting awards and recommending candidates for science and technology grants for academic training. In the space area, JSASS focuses on astrodynamics, space systems and space utilization.

Johns Hopkins University Applied Physics Laboratory (APL), USA, www.jhuapl.edu

The Applied Physics Laboratory is a non-profit centre for engineering. APL is a division of the Johns Hopkins University and R&D. APL was organized to develop critical Second World War technology in 1942. It works on more than 600 programmes with annual funding of about US $1,000m.

APL solves research, engineering and analytical problems that present critical challenges in the civilian and national security space areas. APL's Space Department produces space flight hardware and software systems, and conducts space science and engineering for both civilian and military customers. There are two associated business areas within the Space Department: Civilian Space Business Area and National Security Space Business Area.

The Civilian Space Business Area makes critical contributions to the missions of its major sponsor, the National Aeronautics and Space Administration, to meet the challenges of space science. APL conducts research and space exploration, and develops and applies space science, engineering and technology, including the production of spacecraft, instruments and subsystems. The Civilian Space Business Area focuses primarily on the science discipline of space physics and planetary science. APL has designed, developed and launched 64 spacecraft and more than 150 space instruments.

Programmes in APL's National Security Space Business Area focus on space solutions to critical military problems of the US Department of Defense. APL develops and conducts experimental missions, builds space instruments, and produces applications for warfighter needs. Programme areas include space weather, small satellite development, operationally responsive space, modelling and assessment of environmental phenomena, space threat awareness and characterization and space environmental monitoring.

K

Kawasaki Heavy Industries (KHI), Japan, www.khi.co.jp

The origin of Kawasaki Heavy Industries dates back to 1878. In 1969 Kawasaki Dockyard, Kawasaki Rolling Stock Manufacturing and Kawasaki Aircraft merged to become Kawasaki Heavy Industries Ltd. KHI participation in space development began with work for the National Space Development Agency of Japan, later the **Japan Aerospace Exploration Agency**, on rocket launch complex development, an acoustic test facility and an experimental geodetic satellite.

Since the 1990s, KHI has been responsible for the development and production of payload fairings and payload attach fittings for the Japanese H-II rocket, the construction of the launch complex for the H-II and other operations services for the H-II. KHI participated in the development of reusable launch vehicles for spacecraft and robotics projects, such as the Japanese Experiment Module for the International Space Station, the HOPE-X experimental orbiting plane, the docking mechanism for the Engineering Test Satellite and the Next Generation of Unmanned Space Experiment Recovery System Re-Entry Module. Kawasaki is also involved in the development of human space technology, including the training of astronauts. KHI employs more than 32,000 people, and expected revenue for 2010 was US $12,500m. inclusive of all areas of business.

Khrunichev State Research and Production Space Centre, Russia, www.khrunichev.ru

Khrunichev State Research and Production Space Centre was established during the Second World War to produce heavy bomber aircraft. In 1951, a space-related design bureau was added, and in the late 1970s, Khrunichev was transferred to the Salyut Design Bureau. In the 1980s, Khrunichev became a part of **RSC Energia**, and then an independent entity in 1988. Khrunichev is part of **International Launch Services** that provides sales and management of satellite launches on the Proton. Khrunichev developed Salyut, Almaz and Mir space stations for Russia.

Khrunichev's mission is R&D aimed at creating, upgrading and operating launchers, launch vehicle upper stages, space station modules and spacecraft, including the Proton, Rockot, Angara, Baikal and the Kosmos launch vehicles, the Breeze upper stage, the Zarya, Zvezda and multipurpose lab modules for the International Space Station, and the Yakhta, KazSat-1 and 2, Monitor E, Sterkh, Express MD 1 and 2 and Mozhaets-5 small spacecraft.

Korea Aerospace Research Institute (KARI), Republic of Korea, www.kari.re.kr

The Korea Aerospace Research Institute, founded in 1981, is the aeronautics and space agency of the Republic of Korea (South Korea). The main functions are R&D on satellites, space launch and sounding rockets and aircraft; technical support for Korean aerospace industries; and assistance to the government's policy in the aerospace field. KARI projects include the Korea Space Launch Vehicle (KSLV), the Arirang-1 satellite and the Korea Multi-Purpose Satellite (KOMPSAT) programme.

South Korea and Russia co-operated in 2004 to develop the KSLV. Russia is also helping to build a spaceport: the Korea Space Centre. In 2009, a test flight of the KSLV-I failed. Through the Korean astronaut programme, Russia has trained South Korean astronauts and one was sent to the International Space Station in 2008. The KOMPSAT programme is a series of multipurpose satellites, for which KARI is the programme manager, and it will perform the assembly, integration and test of the flight hardware. KOMPSAT 1 is an Earth resources sensing and scientific experiment satellite. Supporting the KOMPSAT 1 programme is **Northrop Grumman Aerospace Systems** of the USA, which will assist in integration and testing.

Korea Astronomy and Space Science Institute (KASI), Republic of Korea, www.kasi.re.kr

The Korea Astronomy and Space Science Institute is the national astronomy research institute of the Republic of Korea (South Korea). KASI was established in 1974 as the Korean National Astronomical Society; the name changed to KASI in 2005. KASI established the Sobaeksan Optical Astronomy Observatory in 1978, the Bohyunsan Optical Astronomy Observatory in 1996 and the Korean VLBI Network (KVN) in 2001. KASI performs research in optical, radio, theoretical and observational astronomy and conducts astronomy in space. The mission of KASI is to advance understanding of the universe through telescopic observations and theoretical analysis; to discover new astronomical phenomena and progress research areas in space science; to promote international research activities by collaborating with foreign astronomers; and to disseminate astronomical information for scientific purposes.

Korean Committee of Space Technology (KCST), Democratic People's Republic of Korea, www.kari.re.kr

The Korean Committee of Space Technology is the state-controlled space agency of the Democratic People's Republic of Korea (North Korea). It was founded in the 1980s and is connected to the Artillery Guidance Bureau of the Korean People's Army. KCST is responsible for all operations concerning space exploration and the construction of satellites. In 2009, North Korea acceded to the *Outer Space Treaty* and the *Registration Convention* after a previous declaration of preparations for a new satellite launch (acceded or accession refers to the act of joining a treaty by a party that did not take part in its negotiations.

KCST operates the Musudan-ri and Pongdong-ri rocket launching sites, Baekdusan-1 and Unha (Baekdusan-2) launchers and Kwangmyŏngsŏng satellites. In 2009 North Korea announced more ambitious future space projects, including its own human space flights and development of a crewed, partially reusable launch vehicle. KCST also implements research in the field of space sciences, remote sensing and defence research projects and promotes education in the field of space sciences.

Also of note is that North Korea developed an indigenous space launch vehicle based on Chinese and Iranian technology, called Unha-2. In 2009, North Korean attempted to launch a North Korean satellite. While North Korea claimed mission success, according to the US military, which tracks all missile launch and space objects, the mission failed. The USA, Japan, South Korea and Europe view North Korea's space launch programme as means for North Korea to develop a long-range ballistic missile; however, North Korea portrays its space programme as peaceful and within international norms, and notes its accession to the *Outer Space Treaty* prior to the space launch.

L

Lavochkin Research and Production Association (NPO Lavochkin), Russia, www.laspace.ru

Lavochkin Research and Production Association, also called NPO Lavochkin, was founded in 1937 as part of the Lavochkin Aircraft Design Bureau. It is a Russian aerospace company that develops and manufactures space launch vehicle upper stages and satellites and interplanetary probes, such as Phobos Grunt, a sample return mission to Phobos (moon of Mars) scheduled for launch by 2012. NPO Lavochkin is also a contractor for military space programmes, including early warning and missile defence satellites.

Lockheed Martin Space Systems Company, USA, www.lockheedmartin.com/ssc

Lockheed Martin Space Systems Company is a major operating unit of Lockheed Martin Corporation (LMC). LMC was formed in 1995 with the merger of Lockheed Corporation and Martin Marietta Corporation. In 1996 Lockheed Martin acquired the defence electronics and systems integration businesses of Loral Corporation (see **Loral Space & Communications**). Lockheed Martin traces its roots to the early days of aviation. In 1909 aviation pioneer Glenn L. Martin organized a company around the airplane construction business and built it into an airframe supplier to US military and commercial customers. Martin Marietta was established in 1961 when the Glenn L. Martin Company merged with American-Marietta Corporation, a leading supplier of building and road construction materials. In 1913 Allan and Malcolm Loughead, name later changed to Lockheed, flew the first Lockheed plane over San Francisco Bay. The modern Lockheed Corporation was formed in 1932 after the fledgling airplane company was reorganized.

As of 2010 LMC employs 140,000, including 65,000 scientists and engineers and 23,000 professionals in information technology careers. Lockheed Martin is the world's second largest defence contractor by revenue as of 2010. As of 2009 85% of Lockheed Martin's revenues came from the US Department of Defense and other US Federal Government agencies, such as the **National Aeronautics and Space Administration**. In addition to LMC, the top 10 defence contractors by revenue include: BAE Systems of the United Kingdom,

Boeing Company of the USA, **Northrop Grumman Aerospace Systems** of the USA, General Dynamics of the USA, Raytheon of the USA, **European Aeronautic Defence and Space Company Astrium** of Europe, **Finmeccanica** of Italy, L-3 Communications of the USA, and Thales Group. Total revenue for LMC in 2009 was US $45,200m.

Lockheed Martin Space Systems Company comprises two primary business units: Space Systems and Michoud Operations; and joint ventures, United Space Alliance and United Launch Alliance. LMC Space Systems comprises several core lines of business, which include: human space flight, global communications systems, commercial space, sensing and exploration systems, missile defence systems, strategic missiles, commercial launch systems, surveillance and navigation systems and special programmes.

LMC Space Systems Company designs, develops, tests, manufactures and operates advanced-technology systems for national security, civil and commercial customers. Chief products include human space flight systems; remote sensing, navigation, meteorological and communications satellites and instruments; space observatories and interplanetary spacecraft; laser radar; fleet ballistic missiles; and missile defence systems. Michoud Operations builds the external fuel tanks for the Space Shuttle launch system. The organization is involved in the R&D of systems designed to upgrade the Space Shuttle and future space launch systems.

The space products of LMC Space Systems Company include: 2001 Mars Odyssey; Advanced Extremely High Frequency; Athena; Cassini; Defense Meteorological Satellite Program; Defense Satellite Communications System; Genesis; Geostationary Operational Environmental Satellite R-Series; Global Positioning System; Gravity Probe B Relativity Mission; Hubble Space Telescope; International Space Station (ISS); Lunar Prospector; Mars Global Surveyor; Mars Reconnaissance Orbiter; Mars Science Laboratory Aeroshell; Milstar; Mobile User Objective System; Orion; Phoenix Mars Lander; Polar-Orbiting Operational Environmental Satellites; Rosetta Spacecraft; Solar Array Flight Experiment; Space-Based Infrared System High; Spitzer Space Telescope; Stardust Stardust-NExT, and Viking.

Loral Space & Communications, USA, www.loral.com

Loral Space & Communications was formed from the remnants of Loral Corporation. Loral Corporation started as Loral Electronics in 1948 and became involved with communications satellites in 1961. From 1987 to 1994, the corporation expanded in the aerospace sector with the acquisitions of Goodyear Aerospace Corporation and Ford Aerospace. Loral Corporation initiated the Globalstar project (www.globalstar.com) in 1991 for mobile satellite services (see **Satellite Communication Service Companies**). In 1996 Loral Corporation divested its defence electronics and system integration businesses to **Lockheed Martin Space Systems Company**, and in the process Loral Space & Communications was established.

Loral Space & Communications is a satellite communications company with activities in two primary areas: satellite services and satellite manufacturing. The satellite manufacturing business is conducted by **Space Systems Loral**. Loral Space & Communications owns 56% of XTAR (www.xtarllc.com), a joint venture between Loral and HISDESAT (www.hisdesat.es), a consortium comprising leading Spanish telecommunications companies, including Hispasat (www.hispasat.com) and agencies of the Spanish Government. Through its XTAR-EUR satellite, XTAR provides X-band services to government users in the USA, Spain and other allied countries. As of 2007 Loral also owns 64% of **Telesat** of Canada, which operates a fleet of satellites for telecommunications services. Revenue for Loral Space & Communications was at the US $1,000m. level.

Luxembourg Aeronautic and Space Industry Association—*see* **Groupement Luxembourgeois de l'aéronautique et de l'espace (GLAE), Luxembourg.**

M

MacDonald, Dettwiler and Associates Ltd (MDA), Canada, www.mdacorporation.com

MacDonald, Dettwiler and Associates, an aerospace and information technology company, was incorporated in 1969. As of 2009, MDA employed more than 3,200 people and had revenue of US $1,000m. The Space Missions unit of MDA supports human space flight through the development of advanced robotics. MDA developed the Space Shuttle's Canadarm and the Mobile Servicing System for the International Space Station. MDA also developed an extension to the Canadarm for on-orbit inspection of the Space Shuttle.

MDA's satellite solutions provide responsive access to space through the development of small satellite missions and turn-key systems for Earth observation, surveillance of space, information delivery and space science. MDA is a supplier of commercial satellite payloads, systems and subsystems for communications and remote sensing satellites. A significant commercial satellite developed by MDA is the Canadian RADARSAT programme. RADARSAT is a commercial Earth observation synthetic aperture radar platform. MDA served as the prime contractor for RADARSAT-2 and holds exclusive distribution data rights to RADARSAT-1 and RADARSAT-2. There are also plans for MDA to develop systems for on-orbit servicing and planetary missions to explore the Moon and Mars.

Magyar Asztronautikai Társaság (MANT), Hungary, www.mant.hu

The Magyar Asztronautikai Társaság (Hungarian Astronautical Society) is a civil organization founded in the 1980s, with more than 400 members. It appeals to Hungarian space researchers and others who are interested in space technology and uses of space, recruiting people for space research and engineering, and using space research for human applications. MANT makes use of space research and applications to educate about physics, astronomy, medicine, biology, space law, flight technology, meteorology and material sciences.

Malaysian National Space Agency—see Agensi Angkasa Negara (ANGKASA), Malaysia.

Mars Society, USA, www.marssociety.org

The Mars Society is an international, non-profit space advocacy organization based in the USA. It was founded by Robert Zubrin in 1998. Since that time, the Mars Society has grown to more than 4,000 members and 6,000 associate supporters across more than 50 countries. The purpose of the Mars Society is to further the goal of the exploration and settlement of Mars. Objectives include public outreach to instil the vision of pioneering Mars, support of government funded Mars exploration programmes, and conducting Mars exploration missions on a private basis.

Mexican Space Agency—*see* **Agencia Espacial Mexicana (AEXA), Mexico.**

Mitsubishi Electric Corporation, Japan, global.mitsubishielectric.com

The company was founded in 1921, when Mitsubishi Shipbuilding, now **Mitsubishi Heavy Industries** spun-off a factory that made electric motors for ocean-going vessels into a new company called Mitsubishi Electric Corporation. Mitsubishi Electric had revenue of US $37,400m. in 2009 and more than 100,000 employees. The company has been involved in the development of space technologies since the 1960s.

In the area of space technology, Mitsubishi Electric manufactures satellite-related technologies that include solar panels, antennas, amplification, tracking, control and ground station systems. Mitsubishi Electric developed the DS2000 standard satellite platform, a high-speed communications platform for the commercial communications satellite market. Since 1968 Mitsubishi Electric has been involved in the supply of international communications satellite hardware, contributing to **Intelsat SA** satellites. In addition, Mitsubishi Electric is participating in the development of the GLOBALSTAR satellite project (globalstar.com) for mobile, satellite-based communications.

Mitsubishi Heavy Industries (MHI), Japan, www.mhi.co.jp

Mitsubishi Heavy Industries was created in 1964. As of 2010, revenue exceeded US $34,000m., and the company had more than 63,000 employees. MHI is engaged in the development and production of aerospace products for the defence and civil sectors. In the civil space sector, MHI is the producer of the H-IIA and H-IIB space launch vehicles of Japan and provides space launch services. MHI is also involved in the International Space Station (ISS) programme with the crewed space facility, Japanese Experiment Module (JEM) or KIBO, which was launched by the US Space Shuttle. KIBO consists of the pressurized module, the exposed facility, the remote manipulator and the experiment logistics module. MHI is also involved in the Japanese H-II Transfer Vehicle for the ISS to deliver daily goods, such as water, food, clothing and experimental equipment, to the JEM/KIBO.

N

National Aeronautics and Space Administration (NASA), USA, www.nasa.gov

The National Aeronautics and Space Administration is an independent agency of the US Federal Government, i.e., existing outside of federal executive departments. As an independent agency, NASA is directly accountable to the Office of the President, primarily the Office of Science and Technology Policy and the National Security Council, which develops and co-ordinates civil space policy for the president, and to the US Congress through annual authorizations and appropriations and special congressional hearings that address issues relevant to NASA.

The *National Aeronautics and Space Act*, legislated by Congress and signed into law in 1958 by the president, created NASA and established the objectives toward which NASA must work. This act provided NASA with powers for decision-making. For example, NASA can independently enter into international agreements, with the co-ordination of the US Department of State, with foreign national space agencies and governments and into separate agreements—Space Act Agreements—with international and national organizations, such as commercial space companies, to license or develop specific technologies or concepts of interest to NASA. NASA is also vested with federal authority to establish its own internal rules for governance, management and human resources. These rules are established by policy directives and procedural requirements.

The *National Aeronautics and Space Act* established NASA and NASA centres from a number of distinct organizations that existed at the time, varying from federally funded research and development efforts for aviation technology and military facilities engaged in missile, rocket and satellite development, and R&D expertise, to academic R&D centres.

The NASA workforce in 2009 was 18,600 civil service employees, located at headquarters and the NASA centres. Of the 18,600 NASA civil service employees, 60% were scientists and engineers, the majority in the latter category, 30% were managers and 10% were technicians. The private aerospace sector supports 90% of all NASA work. About 43,500 contractors directly support NASA work at headquarters and the NASA centres. In addition to these contractors, the aerospace industries in the USA employed a workforce of 650,000 as of 2009, of which 12% supported space missions across civil, commercial and military sectors.

Civil space spending grew dramatically from 1962 to a peak in NASA funding in 1966 because of the Apollo programme, which was given the highest national priority, and was linked at that time to the national security and prestige of the USA in the Cold War. From 1967 to 1972 the budget declined to less than 50% of the spending levels for Apollo. This decline elevated cost considerations as the principal factor in programmes and projects. The fixed nature of NASA's budget since 1982, the year that the Space Shuttle was declared operational, to 2010 has intensified the conflict that exists within NASA between human space flight and science. The continued demand for large-scale human space flight programmes forces NASA to commit more than 50% of its budget to this area at the expense of science programmes and projects.

In regard to future budgets for civil space, there is a trend towards more money for NASA. The annual NASA budget for 2009 was US $17,600m., while for 2010 funding was supported at US $18,700m. and for 2011 funding is expected to rise to US $19,000m. Under the current budget proposals, as of 2010, NASA was to receive additional increases in subsequent years, reaching US $21,000m. in 2015. This money is to be directed to better fund aeronautics research, exploration systems and science and to seed the development of commercial space transportation for humans and cargo to low Earth orbit (LEO).

NASA is organized through four enterprises: aeronautics research focused on fundamental research on air and space vehicles; exploration systems with a focus on the development of new human space flight systems and robotic missions that serve as precursors to human space exploration; science with a focus on Earth, heliophysics, planetary science and astrophysics missions; and space operations with a focus on the operations of existing human space flight systems (as of 2010, the Space Shuttle and International Space Station). Each enterprise is focused on a set of NASA programmes and projects. These programmes and projects are administered by NASA headquarters and implemented by the NASA centres.

NASA-led human space flight missions during the course of its history, since 1958, include Mercury, Gemini and Apollo programmes of the 1950s to 1970, to the Space Shuttle and International Space Station (ISS) programmes from the 1980s to the present (US national space policy calls for termination of the Space Shuttle by 2011 and commitments to support ISS operations until 2020). Successful planetary science missions until 2009, encompass Pioneer (ten missions) to the Moon, Jupiter, Saturn and Venus (1958–78); Mariner Missions (eight missions) to Mars, Mercury and Venus (1962–73); Lunar Orbiter (five missions) to the Moon (1966–67); Surveyor (five missions) to the Moon and preparation for Apollo (1966–68); Ranger (three missions) to the Moon (1964–65); Viking (two missions) to Mars surface (1975); Magellan to Venus (1989); Mars Global Surveyor (1996); Mars Pathfinder (1996); NEAR-Shoemaker to asteroid surface (1996); Cassini to Saturn and Titan (1997); Deep Space Technology demonstrator (1998); Lunar Prospector (1998); Stardust to collect cometary material (1999); Galileo to Jupiter (2001); Genesis to collect solar wind (2001); Mars Odyssey (2001); Comet Nucleus Tour

(2002); Mars Exploration Rovers (2003); MESSENGER to Mercury (2004); Phoenix to Mars surface (2004); Rosetta to study comets (2004); Deep Impact to study comets (2005); Mars Reconnaissance Orbiter (2005); Pluto-Kuiper Belt Mission (2006); Dawn to asteroids Vesta and Ceres (2007); and Lunar Reconnaissance Orbiter (2009). In addition, NASA developed and operated the great space-based observatories, which are a series of space-borne observatories designed to conduct astronomical studies. Observatories include the Hubble Space Telescope (operational), Compton Gamma Ray Observatory (mission ended in 2000), Chandra X-Ray Observatory (operational), and Spitzer Space Telescope (operational). The next generation space-based observatory is the James Webb Space Telescope, which is under development and planned for launch in 2014 (www.jwst.nasa.gov). Finally, there are the large-scale Earth observation platforms that are part of NASA, Earth Observing System and other Earth science programmes.

NASA also functions to develop space commercial capabilities as directed by national space policy. A programme of note is the Commercial Crew and Cargo Program that manages Commercial Orbital Transportation Services (COTS) partnership agreements with US commercial industry. The primary COTS partners, as of 2010, are **Orbital Sciences Corporation** and Space Exploration Technologies (SpaceX). Other partners within the Commercial Crew and Cargo Program are involved with the Commercial Crew Development (CCDev) activity. While COTS concerns commercial cargo transportation services to LEO, CCDev deals with developing crewed commercial services to LEO. CCDev is funded through Space Act Agreements between NASA and commercial partners.

National Aerospace Laboratory (NLR), the Netherlands, www.nlr.nl

The National Aerospace Laboratory was founded in 1961. NLR activities in the field of space exploration contribute to technological innovations for space and technology development. NLR also supports policy-makers in the Netherlands. NLR participates in a number of space projects, including satellite navigation for Europe, Earth observation with the **European Union**'s Global Monitoring for Environment and Security programme and testing and simulation for other European space systems.

National Aerospace Research and Development Commission—*see* **Comisión Nacional de Investigación y Desarrollo Aeroespacial (CONIDA), Peru.**

National Authority for Remote Sensing and Space Sciences (NARSS), Egypt, www.narss.sci.eg

The National Authority for Remote Sensing and Space Sciences is an outgrowth of the Remote Sensing Centre, established in 1971 as a US–Egyptian joint project that was affiliated to the Egyptian Academy of Scientific Research and Technology. In 1994, NARSS was established as an organization under the State Ministry of Scientific Research to promote the use of space technology for the development of the country and to introduce high-technology capabilities in regional planning and other applications. The missions of NARSS are to provide technology in the fields of remote sensing and peaceful application of space sciences, and to build the capability to utilize these technologies to support national development. NARSS is working to develop the Egyptian space programme, including the EgyptSat 1 remote sensing satellite, which was launched by Ukraine in 2007.

National Center for Remote Sensing, Air and Space Law, USA, www.spacelaw.olemiss.edu

The National Center for Remote Sensing, Air and Space Law serves the remote sensing and space industry communities by addressing and conducting research and education related to the legal aspects of applying remote sensing, air and space technologies to human activities. The areas of focus of the centre, include remote sensing law, national space law, international space law and aviation law. Research activities address the need for space law in areas that include deployment and operations of satellites; commercial infrastructure, data policies and intellectual property; privacy and liability; use of imagery as legal evidence; environmental issues; and licensing. In addition, the space exploration law initiative of the centre addresses legal issues raised by domestic and international space exploration. Issues include the establishment of a permanent lunar base; improving policy and legal stability for commercial space investments; reducing legal and regulatory barriers to international space co-operation; and assessing European Commission regulatory actions affecting space activities.

National Centre for Cartography and Remote Sensing—*see* Centre National de la Cartographie et de la Télédétection (CNTC), Tunisia.

National Centre for Space Studies—*see* Centre National d'Etudes Spatiales (CNES), France.

National Geospatial-Intelligence Agency (NGA), USA, www1.nga.mil

The origins of the National Geospatial-Intelligence Agency date to the digital geospatial work of the Defense Mapping Agency and the US Army Topographic Engineer Center. These efforts, and other efforts carried out by various governmental organizations in the geospatial area, merged to form the National Imagery and Mapping Agency (NIMA) in 1996. In 2003, NIMA became the NGA.

NGA is a US Department of Defense combat support agency and a member of the US Intelligence Community (IC). NGA develops imagery and map-based intelligence solutions for US national defence, homeland security and safety of navigation. It provides global support to IC mission partners through NGA representatives stationed around the world and geospatial intelligence support for global world events, natural disasters and military actions.

NGA provides geospatial intelligence (GEOINT) for decision-makers, warfighters and natural disaster responders. GEOINT involves the exploitation and analysis of imagery and geospatial information to describe, assess and visually depict physical features and geographically referenced activities on Earth. NGA's Office of GEOINT Sciences provides analysis of global positional information, including imagery and mapping for navigation, safety, intelligence, positioning and targeting in support of national security objectives.

National Institute for Aerospace Technology—*see* **Instituto Nacional de Técnica Aeroespacial (INTA), Spain.**

National Institute for Space Research—*see* **Instituto Nacional de Pesquisas Espaciais (INPE), Brazil.**

National Oceanic and Atmospheric Administration (NOAA), USA, www.noaa.gov

The National Oceanic and Atmospheric Administration was created in 1970 as part of the US Department of Commerce. Primarily related to weather issues, NOAA interacts with the space sector through the satellites that it operates. NOAA operates two types of satellite systems for the USA—geostationary satellites and polar-orbiting satellites. These satellites provide for imaging Earth for the National Weather Service (www.weather.gov) and monitoring conditions in space, including observing solar flare activity from the Sun (Space Weather Prediction Center at www.swpc.noaa.gov/index.html).

Geostationary satellites monitor the Western Hemisphere and polar-orbiting satellites circle Earth and provide global information. NOAA satellites enable long-term observations, 24 hours a day, seven days a week. Data from satellites is used to measure the temperature of the ocean, which is a key indicator of climate change. Satellite information is also used to monitor coral reefs, harmful algae blooms, fires and volcanic ash.

NOAA promotes and enables civil and commercial uses of Earth observations and remote sensing satellites through the NOAA Satellite and Information Service, known also as the National Environmental Satellite, Data and Information Service (NESDIS, www.nesdis.noaa.gov). In the civil area, NESDIS acquires and manages operational environmental satellites, operates the NOAA National Data Centers, provides data and information services including

National Reconnaissance Office (NRO)

Earth system monitoring, performs official assessments of the environment and conducts related research.

For commercial uses, NESDIS issues licences and regulates the private remote sensing industry. Pursuant to the 1992 *Land Remote Sensing Policy Act* and administration policy on foreign access to remote sensing space capabilities of 2003, responsibilities have been delegated from the Secretary of Commerce to the assistant administrator for NOAA Satellite and Information Services (NOAA/NESDIS) for the licensing of companies to operate private space-based remote sensing systems. The licensing process is based on NOAA's *Final Rule on the Licensing of Private Land Remote-Sensing Space Systems* of 2006.

NOAA licensees as 2010 include: AstroVision—AVStar (www.astrovision.com); Ball Aerospace—SAR (Synthetic Aperture Radar); **DigitalGlobe**—EarlyBird-1, QuickBird-1, QuickBird-II follow-on, M-5 and WorldView; DISH Operating—EchoStar-11 (operations for the DISH Network, www.dishnetwork.com); **GeoEye**—OrbView-2, OrbView-3, IKONOS, IKONOS Block II, GeoEye-1, 2 and 3; Northrop Grumman—Continuum and Trinidad; Technica—EaglEye (company plans call for a remote sensing system to be launched in 2013, www.technicainc.com).

National Office for Aerospace Studies and Research—*see* **Office National d'Etudes et Recherches Aérospatiales (ONERA), France.**

National Reconnaissance Office (NRO), USA, www.nro.gov

A space bureaucracy of impressive size is tied to the US Intelligence Community (IC). The IC is centred on the activities of the National Reconnaissance Office. NRO has operated a succession of ever-more sophisticated remote sensing satellites, since 1959, capable of photographing objects on Earth, piercing cloud cover and camouflage with Synthetic Aperture Radar, and determining the material out of which potentially threatening objects are made (by studying the composition of objects, such as tanks and aircraft, intelligence officers can estimate their capabilities). Other reconnaissance satellites eavesdrop on voice and visual communications. An intricate web of intelligence agencies of the IC, including the Central Intelligence Agency (CIA) (see **US Government**) and the National Security Agency (NSA), utilize information produced by NRO operated reconnaissance satellites.

The NRO was established in 1961 after problems emerged with the US Air Force (USAF) satellite reconnaissance programme at the time. NRO coordinated the USAF and CIA, and later the Navy and NSA, reconnaissance activities. NRO is a unique intelligence agency as it is not a distinct organization in its own right, but composed of parts of other organizations, primarily the USAF and CIA. The NRO, unlike most other intelligence agencies, is primarily an intermediary organization; it neither establishes intelligence requirements nor processes the intelligence data that it collects. For much of its early existence, intelligence requirements were established by the US Intelligence Board and imagery was evaluated by the CIA's National Photographic Interpretation Center. Signals intelligence was evaluated by components of the military services, such as Strategic Air Command or NSA. In 2010 intelligence requirements were established by the National Foreign Intelligence Program, the **National Geospatial-Intelligence Agency** interpreted imagery and the NSA evaluated signals intelligence.

During the Cold War NRO's mission focused primarily on nuclear arms control verification based on National Technical Means of Verification (NTMV). Since the end of the Cold War, and the declassification of NRO in 1992, the office has adapted to a new mission focused on a collection of a wide and difficult array of intelligence targets, rather than a priority on the NTMV function.

Since 1992 the NRO has come under oversight and criticism. Of note, was the 2000 *Report of the National Commission for the Review of the National Reconnaissance Office*. The US Congress, which commissioned the report, criticized the increased openness of NRO and called for a return to the intense secrecy that marked its early years, which also shielded NRO from congressional oversight. Furthermore, there were calls for a return to the focus on serving the needs of the president, not the warfighter, thus reversing the post-Cold War trend toward expanding the customer base for satellite intelligence information. Despite the report and views, these recommendations were not adopted by NRO.

NRO's first photo-reconnaissance satellite programme was the Corona programme from 1959 to 1972, the existence of which was declassified in 1995. NRO missions, since 1972, have been classified, and portions of many earlier programmes remain unavailable to the public. Since 1972 NRO has developed and operated a series of reconnaissance satellites for imagery known as the Key Hole (KH) series (Corona was a code name for KH-1, 2, 3 and 4). The 2010 operational system is KH-12.

In 1999, NRO embarked on a project with Boeing called the Future Imagery Architecture to create a new generation of imaging satellites. Due to management and technical problems, the programme, valued at more than US $25,000m. in contract work to Boeing, was terminated in 2005. That same year, the contract shifted to Lockheed Martin, which was asked to restart production of the KH reconnaissance satellite systems with new upgrades beginning with KH-13 (KH-13, which was launched, is reported to have malfunctioned and re-entered Earth's atmosphere in 2007).

National Remote Sensing Centre (NRSC), Mongolia, www.icc.mn

The National Remote Sensing Centre was established in 1980 to represent the Government of Mongolia in space technology and applications, develop multilateral co-operation with other space organizations and contribute to the development of remote sensing applications in Mongolia. NRSC has the following operational activities: acquisition of satellite data from the **National Aeronautics and Space Administration**'s (NASA) Earth Observing System and **National Oceanic and Atmospheric Administration** (NOAA) weather satellites; providing data to users; and monitoring natural resources and hazards. NRSC also conducts natural resource mapping, land cover change analysis and related research activities.

National Security Space Institute (NSSI), USA, www2.peterson.af.mil/nssi

The National Security Space Institute is the US Department of Defense's focal point for space education and training, complementing existing space education programmes at Air University (www.au.af.mil/au), the Naval Postgraduate School (www.nps.edu) and the Air Force Institute of Technology (www.afit.edu). The mission of NSSI is to expand efforts to assess the value of space contributions emphasizing US dependence on space, ideas contributing to freedom of action in space and integration of space forces.

NSSI grew from two pioneer organizations: the Space Tactics School (STS) and the Space Operations School (SOPSC). STS, which existed from 1994 to 1996, was absorbed by the US Air Force Weapons School in 1996. SOPSC, which ran from 2001 to 2004, extended beyond STS, and addressed teaching broader space concepts and systems.

The *Report of the Commission to Assess United States National Security Space Management and Organization* of 2001 (space.au.af.mil/space_commission), which was commissioned by the US Congress, amplified the need for more space education and training, noting the shortfall in growing space professionals at senior leadership level. The report served as a catalyst to help transform the SOPSC into the NSSI, which officially activated in 2004.

The NSSI had two main schools, the Space Professional School, which was responsible for the Space Professional continuing education courses, and the Space Operations School, dedicated to teaching advanced space concepts, deployment training and instruction to non-space professionals. In 2007, **US Air Force Space Command** decided to reorganize the NSSI into two schools: the Space Professional School of the NSSI under Air University; and the Space Operations School, renamed the Advanced Space Operations School (ASOpS), within the US Air Force Space Innovation and Development Center. In 2009, ASOpS was activated.

National Security Space Office (NSSO), USA, www.acq.osd.mil/nsso

The National Security Space Office was created in 2004 by merging the responsibilities and resources of the National Security Space Architect (NSSA), the National Security Space Integration Directorate (NSSI) and the Transformational Communications Office (TCO). NSSO was established to build on the functions and processes of the NSSA, the NSSI and the TCO, and to add new capabilities, including national security space-wide enterprise engineering and functional area integration.

NSSO facilitates the integration and co-ordination of defence, intelligence, civil and commercial space activities. It focuses on cross-space enterprise issues and provides decision-making support on requirements, strategies, plans, programmes, budgets, acquisitions and professional development activities to the Air Force, Congress, Department of Defense (DoD) Executive Agent for Space (see **US Air Force Space Command**), Director of National Intelligence, Intelligence Community (IC), Joint Chiefs of Staff, **National Reconnaissance Office**, Office of the President, Office of the Secretary of Defense, and US Strategic Command.

NSSA was the oldest of the constituent organizations comprising the new NSSO, established in 1998. The mission of NSSA was to develop and integrate space architectures across DoD and IC space mission areas. The NSSI Directorate was established in 2002 to integrate black (classified) and white (civil, public) space activities. The task of the NSSI Directorate was to define and track unified space capabilities and systems, and to monitor space systems acquisition. The TCO was established in 2002 as a direct result of the Transformational Communications Architecture prepared by NSSA and subsequent Transformational Communications studies. The mission of TCO was to develop a process, strategy and architecture that enabled the implementation of an inter-operable satellite communications network.

National Space Activities Commission—*see* **Comisión Nacional de Actividades Espaciales (CONAE), Argentina.**

National Space Agency of the Republic of Kazakhstan (KazKosmos), www.kazcosmos.kz

The National Space Agency of the Republic of Kazakhstan is a central, nongovernmental executive body under state regulation in the field of space activities. It carries out its activities in accordance with the constitution and the laws of Kazakhstan. Funding for KazKosmos comes from the national budget. Of note is that KazKosmos co-ordinates activities to lease to Russia the use of the Baikonur Cosmodrome space launch complex.

When the Soviet Union collapsed, the Baikonur Cosmodrome ended up in territory of Kazakhstan. In 1991, a *Joint Activity in Space and Exploitation* agreement, signed at the creation of the confederation of states of the former Soviet Union, recognized the importance of Baikonur and the need to maintain it for the benefit of all member states. Through negotiations, the Russian Federation and Kazakhstan signed a lease agreement, whereby Kazakhstan receives an annual fee for Russian use of the space launch complex. Russia has plans to develop a new Russian spaceport to eventually replace the use of Baikonur.

Other KazKosmos activities include the formation of a unified state policy in the field of space activities; the implementation of government regulations and co-ordination of activities in space; the formation and development of space industry in Kazakhstan; the establishment of conditions for a market of space technologies and services; the creation of a legal framework for space activities in Kazakhstan; the implementation of state control in the field of space activities; and the implementation of international co-operation in space activities.

National Space Agency of Ukraine (NSAU), www.nkau.gov.ua/NSAU/nkau.nsf

The National Space Agency of Ukraine was created by the Ukrainian Government in 1993 to co-ordinate national space activities. It oversees launch programmes and satellites, develops state policy concepts in the sphere of research and peaceful uses of space, organizes and develops space activities, contributes to national security and defence capability and organizes and develops Ukraine's co-operation with other international space organizations.

The First Programme of NSAU (1993–97) was called on to continue the research and industrial space-related potential for the benefit of the national economy and national security and to access the international market of space services. The Second Programme (1998–2002) was aimed at creating the internal market of space services, accessing international space markets by presenting in-house products and services and integrating Ukraine into the global space community. From 2003–2007, and to 2010, NSAU outlined the main goals, assignments, priorities and methods of maintaining and further developing space activities in Ukraine. NSAU is also advancing co-operation with European space programmes.

Programme areas include: scientific space research; remote sensing of Earth; satellite telecommunications systems; development of the ground-based infrastructure for navigation and information systems; space activities in the interests of national security and defence; development of base elements and advanced space technologies; and development of the space industrial base. The realization of programmes is linked to develop a national system for Earth observations; to meet national demands in the socio-economic sphere and in security and defence purposes; to introduce satellite systems and communications facilities into the telecommunications infrastructure of the state; to obtain fundamental knowledge on near-Earth space, the solar system, deep space and biological and physical processes in microgravity; to create and develop techniques for space access to realize national and international projects and to enable the development of an indigenous rocket to be employed on the global market of space transportation services; and to ensure the innovative development of the space sector in terms of advancing research and development of the space industrial base.

National Space Grant College and Fellowship Program (Space Grant), USA, www.nasa.gov/offices/education/programs/national/spacegrant/home

The **National Aeronautics and Space Administration** (NASA) initiated the National Space Grant College and Fellowship Program, also known as Space Grant, in 1989. Space Grant is a national network of colleges and universities. These institutions are working to expand opportunities to understand and participate in NASA's aeronautics and space projects. The Space Grant national network includes more than 850 affiliates from universities, colleges, industry, museums, science centres and state and local agencies. These affiliates belong to the 52 consortia in all 50 states of the USA, the District of Columbia and the Commonwealth of Puerto Rico, which fund fellowships and scholarships for students pursuing careers in science, technology, engineering and mathematics, as well as curriculum enhancement and faculty development. Member colleges and universities administer projects in their states.

National Space Institute of the Technical University of Denmark—see Institut for Rumforskning og-teknologi Danmarks Tekniske Universitet (DTU Space), Denmark.

National Space Organization (NSPO), Taiwan, www.nspo.org.tw

The National Space Organization, formerly known as the National Space Programme Office, is the civilian space agency of Taiwan, under the auspices of the National Science Council. NSPO is involved in the development of space exploration; satellite construction and development of related technologies and infrastructure, including the FORMOSAT series of Earth observation satellites; research in aerospace engineering, remote sensing, astrophysics, atmospheric sciences and information sciences; and national security.

In the area of satellite development, NSPO co-operates with the Chungshan Institute of Science and Technology (the Institute is the primary R&D institution of Taiwan's Ministry of National Defense Armaments Bureau, cs.mnd.gov.tw). The focus is on the development of a sounding rocket for upper atmospheric studies, which involves an emphasis on technological integration and miniaturization capabilities required for the planned development of constellations of small satellites.

National Space Research and Development Agency (NASRDA), Nigeria, www.nasrda.net

The National Space Research and Development Agency was established in 1998 by the Nigerian Government with an initial budget of US $93m. The focus areas of the national space programme include science, technology, remote sensing, satellite meteorology, communications, information technology and defence and security. Space missions encompass the NigComSat-1 communications satellite, NigeriaSat-1 satellite of The Disaster Monitoring Constellation (DMC, see **DMC International Imaging**), NigeriaSat-2 (to replace NigeriaSat-1), and NigComSat-1R (to replace NigComSat-1).

NASDRA is also responsible for space policy. The realization of Nigerian satellites has led to the development of national space policy within Nigeria to comply with the obligations of the Outer Space Treaty Regime. In 1967, Nigeria acceded to the *Outer Space Treaty*, but only recently developed space policy to carry out those obligations. This development is reflected in the 2009 accession of Nigeria to the *Registration Convention* and to the *Liability Convention*.

National Space Society (NSS), USA, www.nss.org

The National Space Society is an educational, non-profit organization dedicated to the creation of a spacefaring civilization. The National Space Institute of 1974 and the L5 Society of 1975 merged to form NSS in 1987. The society has more than 12,000 members and more than 50 chapters worldwide. The society publishes *Ad Astra* magazine. The mission of NSS is to promote social, economic, technological and political change to expand civilization beyond Earth. Steps toward this goal include advocacy for human space flight, commercial space development, space exploration, space applications, space resource utilization, robotic precursors for human space missions, defence against asteroids and space education.

Netherlands Institute for Space Research—*see* Stichting Ruimte Onderzoek Nederland (SRON), the Netherlands.

Netherlands Space Office (NSO), the Netherlands, www.nivr.nl

The Nederlands Instituut voor Vliegtuigontwikkeling en Ruimtevaart (Netherlands Agency for Aerospace Programmes, NIVR) was the official governmental space agency until 2009. In 2009 NIVR became the Netherlands Space Office. NSO represents the Netherlands to other national and international space agencies. National space policy in the Netherlands is primarily focused on international co-operation at the European level with the **European Organisation for the Exploitation of Meteorological Satellites**, **European Space Agency** (ESA), **European Union**, **Centre National d'Etudes Spatiales** and **Deutsches Zentrum für Luft und Raumfahrt** (DLR). NSO and ESA work together on astrophysics, planetary research, Earth observations, research in weightlessness of space and the International Space Station. Of the total national space budget in the Netherlands, 70% is given to ESA.

Northrop Grumman Aerospace Systems, USA, www.as.northropgrumman.com

The origins of Northrop Grumman date back to Northrop Aircraft Incorporated and Grumman Aeronautical Engineering Company. Northrop Aircraft Incorporated was formed in 1939 and, in 1959, the name was changed to Northrop Corporation. The Grumman Aeronautical Engineering Company started in 1930 and was known in the space sector for developing the Apollo Lunar Module. In 1994 Grumman was acquired by Northrop to form Northrop Grumman. Since then, Northrop Grumman has acquired numerous companies. Of note for space, was the attempted merger with Lockheed Martin that was not approved by the **US Government** in 1998, the acquisition of the space business of TRW and ownership of Scaled Composites (developing suborbital space systems for tourism, www.scaled.com).

Northrop Grumman Corporation more than 120,000 employees and annual revenue of US $30,000m. In 2009 Northrop Grumman announced several structural actions, which included streamlining its organizational structure to five sectors: aerospace systems, electronic systems, information systems, shipbuilding and technical services.

Northrop Grumman Aerospace Systems employs 24,000 people, with US $10,000m. in revenue. Aerospace Systems provide crewed and robotic aircraft, human and robotic space systems, missile systems and advanced technologies. Space-related products focus on Earth observations and remote sensing platforms, propulsion systems, satellite communications and space-based surveillance. With Earth observations and remote sensing, Northrop Grumman Aerospace Systems supports the development of the **National Aeronautics and Space Administration**'s (NASA) Earth Observing System and hyperspectral systems; in the area of satellite communications, projects include Milstar and the Advanced EHF payloads; and with space-based surveillance, projects entail the Defense Support Program and the Space Tracking and Surveillance System.

Northrop Grumman Aerospace Systems is also the prime contractor for the James Webb Space Telescope (www.jwst.nasa.gov). The telescope, part of the NASA's great observatories programme, is planned for launch and operations beginning 2014. The programme involves co-operation with the **European Space Agency** and the **Canadian Space Agency**.

Norwegian Space Centre (NSC), Norway, www.spacecentre.no

The Norwegian Space Centre is the national space agency of Norway organized as a government agency under the Ministry of Trade and Industry. NSC was established in 1987, and it promotes development, co-ordination and evaluation of national space activities and supports Norwegian interests in the **European Space Agency** (ESA). The principal goals of NSC are to create growth in the space sector, to meet national user needs and to attain an international position in space research and space-related ground infrastructure. Norway is involved in space research, such as Aurora research, solar physics and Sun–Earth studies. NSC supports space research through ESA scientific programmes.

NSC manages the Norsk Romsenter Eiendom, AS (Norwegian Space Centre Properties) and the **Andøya Rocket Range** (AAR). ARR is a commercial operation that derives its income from national and international tasks. In 1997 ARR was privatized with NSC holding 90% of the shares and Kongsberg Defence & Aerospace (www.kongsberg.com) holding the remaining 10%. Norsk Romsenter Eiendom, AS holds 50% of the shares in Kongsberg Satellite Services, AS (KSAT), which in turn owns the Tromsø Station (TSS) and the Svalbard Station (SvalSat). TSS operates a Local User Terminal of the satellite-based search and rescue system COSPAS/SARSAT for the Ministry of Justice of Norway. TSS also acquires Earth observation data, and performs ancillary services for national and international users including ESA programmes. SvalSat downloads data from and controls polar orbiting satellites for customers.

O

Office National d'Etudes et Recherches Aérospatiales (ONERA), France, www.onera.fr

The Office National d'Etudes et Recherches Aérospatiales (National Office for Aerospace Studies and Research, also known as Aerospace Lab) is a public research organization focused on research and technology in aviation, space and defence systems. ONERA has major facilities in France and employs about 2,000 people, including 1,500 scientists, engineers and technicians.

ONERA was established in 1946 around six key tasks: to direct and conduct research in aerospace; to promote research to the national and European industry; to produce and implement the associated test facilities; to provide benefits to industry and high-level expertise; to undertake measures of expertise for the benefit of France; and to train researchers and engineers.

ONERA research teams developed new technologies for supersonic and hypersonic propulsion, applying these technologies to experimental vehicles. ONERA played a role in the development of the Concorde supersonic transport, Mirage fighters, Airbus, helicopters and the Ariane space launch vehicle. ONERA has been France's representative in international scientific projects, especially with **National Aeronautics and Space Administration**, the US Air Force and with Japan and the People's Republic of China. In addition to space launch, ONERA also deals with the following areas in space: formation flying in Earth orbit, autonomy of space systems, orbital transfer vehicles, re-entry systems, monitoring Earth from space, knowledge of the space environment and synthetic aperture telescopes.

OHB Technology AG, European Space and Technology Group, Germany, www.ohb-technology.de

OHB Technology AG, known as the OHB Technology Group, was formed in 1981; its activities in the space sector commenced in 1985. Annual revenue is US $300m. OHB Technology Group encompasses numerous commercial entities among five business units: space systems and security; payloads and science; international space; space transportation and aerospace structures; and telematics and satellite services.

The space systems and security comprises OHB-System AG Bremen, Germany (founded in 1958 and part of the OHB Technology Group since 2002);

Orbital Sciences Corporation

OHB France SAS; STS Systemtechnik Schwerin GmbH Schwerin, Germany (established in 1993); and RST Radar Systemtechnik GmbH. Of note for this unit is OHB-System AG Bremen. Space missions led by OHB-System for the OHB Technology Group include all European work with the International Space Station (ISS) (Columbus laboratory and the automated transfer vehicle Jules Verne); SAR-Lupe, Germany's satellite-based radar reconnaissance system; Orbcomm, a satellite-based communications network; the Concept Demonstration Satellite, commissioned by the US Coast Guard; Quick Launch, consisting of six Orbcomm satellites; ABRIXAS, a project for the **Deutsches Zentrum für Luft und Raumfahrt** (DLR); and Satellite for Information Relay, a satellite-based, bi-directional communications system for digital data transfer.

The payloads and science unit produces applications that range from terrestrial observation and satellite navigation to scientific payloads for exploration and the ISS. Payloads and science involves: Kayser-Threde GmbH München, Germany (acquired by OHB Technology Group in 2007); VRS Verkehr Raumfahrt Systemhaus GmbH Leipzig, Germany; and RapidEye AG Brandenburg, Germany.

The international space unit is focused on the development of small satellites, human and robotic space systems and scientific payloads. It combines the space activities of the OHB Technology Group outside Germany, and includes Carlo Gavazzi Space SpA of Italy (acquired by OHB Technology Group in 2009, www.cgspace.it); Antares SCARL of Italy; LUXSPACE Sàrl Luxembourg (established by OHB Technology in 2005); SMP SA Toulouse, France; and ELTA SA Toulouse, France.

The space transportation and aerospace structures unit supplies components for aeronautics and aviation, and is a key supplier for the European Ariane 5 space launch vehicle. The unit comprises: MT Aerospace AG Augsburg, Germany (established in 2005); MT Mechatronics GmbH Mainz, Germany; MT Aerospace Guyane SAS Kourou, French Guiana; MT Aerospace Satellite Products of the United Kingdom; MT Mecatronica Ltda of Chile; and Arianespace.

The telematics and satellite services business unit is the exclusive marketing entity in Europe for communications services provided by the global ORBCOMM satellite system (OHB invested in this system in 2002, www.orbcomm.com). Entities within this business unit include: OHB Teledata GmbH Bremen, Germany; megatel GmbH Bremen, Germany; Telematic Solutions SpA Milan, Italy; ORBCOMM Deutschland AG Bremen, Germany; Timtec Teldatrans GmbH Bremen, Germany; and ORBCOMM Incorporated of the USA.

Orbital Sciences Corporation, USA, www.orbital.com

Orbital Sciences Corporation, established in 1982, focuses on satellites and space launch systems. Satellites include: small and medium geosynchronous

Earth orbit (GEO) satellites for communications and broadcasting; low Earth orbit (LEO) spacecraft that perform remote sensing and scientific research; spacecraft used for national security missions; and planetary probes to explore deep space. The Pegasus space launch vehicle developed by Orbital Sciences Corporation is a light-class to medium-class launch vehicle. Orbital employs 3,700 people and has substantial operational experience with hundreds of satellites, launch vehicles and other space-related systems delivered or under contract since 1982. Revenue totalled US $1,200m. in 2009.

Orbital is entering human space flight by providing the Launch Abort System for the **National Aeronautics and Space Administration**'s (NASA) Orion Crew exploration vehicle. Orbital is also developing commercial cargo resupply services for the International Space Station through the NASA Commercial Crew and Cargo Program and the associated Commercial Orbital Transportation Services partnerships (www.nasa.gov/offices/c3po/home).

P

Pakistan Space and Upper Atmosphere Research Commission (SUPARCO), Pakistan, www.suparco.gov.pk

The Pakistan Space and Upper Atmosphere Research Commission, the national space agency for Pakistan, was established in 1961. SUPARCO is mandated to conduct R&D in space science and space technology. It works toward developing indigenous capabilities in space technology and promoting space applications for socioeconomic development.

SUPARCO undertakes research and conducts pilot studies based on the applications of satellite remote sensing data and Geographic Information System technology for natural resources surveying, mapping and environmental monitoring; undertakes research studies in space and atmospheric sciences, including satellite meteorology, satellite radiance, troposphere and stratosphere studies, atmospheric pollution, satellite geodesy and astronomy; and undertakes research studies relating to the ionosphere and associated radio wave propagation and geomagnetism.

SUPARCO launches sounding rockets for upper- and middle-atmospheric research. In addition to this, SUPARCO is involved with activities, including: communication satellites for voice, video, television and digital data transfers; Earth observation satellites for various scientific and technological applications; the acquisition of satellite remote sensing data for Earth resources surveying; the acquisition of data for atmospheric and meteorological studies; transmitting and receiving signals from communications satellites; reception of signals under the satellite search and rescue COSPAS-SARSAT programme.

Planetary Society, USA, www.planetary.org

The Planetary Society was formed in 1980 with the purpose to endorse robotic space exploration. Although, in the late 1980s, it began to support human missions to Mars. The Planetary Society does not publish its current membership numbers, but claims to have members in more than 125 states world-wide, and a membership base of 120,000. The Planetary Society is non-governmental and non-profit, and is funded by donations and membership support. It publishes the *Planetary Report*.

In addition to being an advocate for space science programmes, the Planetary Society is a participant in several projects with partners around the world, including investigations of Mars and near-Earth objects, development of solar sails, the search for extra-solar planets, and optical and radio searches for extraterrestrial intelligence. In 2005 the Planetary Society attempted to launch Cosmos 1 to test the feasibility of solar sailing, but the space launch vehicle operated by the **Russian Federal Space Agency** (Roscosmos) failed. The solar sailing project was funded by donations and members of the Planetary Society. LightSail 1, a new solar sailing mission supported by the Planetary Society, is under development as of 2010. The Planetary Society also supports the Search for Extraterrestrial Intelligence programme, which was a **National Aeronautics and Space Administration** programme until 1992.

Pratt & Whitney Rocketdyne (PWR), USA, www.pratt-whitney.com

Pratt & Whitney Rocketdyne is a rocket engine provider and a United Technologies Company (www.utc.com). In 2005, Pratt & Whitney Space Propulsion and Rocketdyne Propulsion & Power merged to form PWR. PWR supplies propulsion systems for Atlas and Delta expendable launch systems of the USA, the main engines for the Space Shuttle and specialized engines for defence space systems in the USA. PWR developed the Redstone engine that took US astronauts into space, followed by Atlas, which placed astronaut John Glenn in Earth orbit. Within the same decade, PWR engines powered astronauts to the Moon and back. In 2009, PWR revenue was more than US $12,000m., though this was both for space propulsion and aircraft engines.

R

Romanian Cosmonautics and Aeronautics Association—*see* **Asociația Română pentru Cosmonautică și Aeronautică (ARCA), Romania.**

Romanian Space Agency (ROSA)—*see* **Agenția Spațială Română (ASR), Romania.**

Royal Centre for Remote Sensing—*see* **Centre Royal de Télédétection Spatiale (CRTS), Morocco.**

RSC Energia, Russia, www.energia.ru

RSC Energia (formally, the S. P. Korolev Rocket and Space Corporation Energia) is a Russian company engaged in space flight systems. The focus of the company is on the development of space technologies, including robotic systems for various applications and rocket systems for spacecraft orbital insertion. In the past, RSC Energia has been known as Special Design Bureau-1 (OKB-1), Central Design Bureau of Experimental Machine Building (TsKBEM) and Scientific-Production Association (NPO) Energia. In 1994, the enterprise assumed its current name of RSC Energia.

In the 1940s and 1950s RSC Energia produced ballistic missiles—mobile land-use tactical missiles, submarine-launched ballistic missiles and strategic intercontinental ballistic missiles. Since the 1950s, RSC Energia has initiated astronautic activities. In addition to the development of the Sputnik satellites, other early projects were aimed to create space technology systems, such as Soyuz and Interkosmos.

More recent activities of RSC Energia, include: the development of the Russian segments of the International Space Station (ISS) (this includes the Functional Cargo Block Zarya, the service module Zvezda and the docking module Pirs); operations of the Progress transfer vehicle for cargo to low Earth orbit that supported Russian space stations Salyut and Mir, and currently supports cargo resupply to the ISS; operations of the Soyuz space launch vehicle since 2000 in support of the ISS programme; development of upper stages for the Proton and Zenit space launch vehicles; development and operations of **Sea Launch**, which is a commercial joint venture with **Boeing Company** of the USA, **Yuzhnoye State Design Office** of Ukraine and Aker ASA of Norway (www.akerasa.com).

RUAG Space, Switzerland, www.ruag.com

RUAG Space is a subsidiary of RUAG, a Swiss aerospace, technology and defence company founded in 1998 in Switzerland. RUAG employs about 6,000 people and has production sites in Switzerland, Austria, Germany, Hungary and Sweden. RUAG Space is a supplier of space products to the industry. In 2008 RUAG Space acquired Saab Space and Austrian Aerospace, and in 2009 it purchased Oerlikon Space. RUAG Space employs 1,100 people and had revenue of US $245m. in 2009. The company is involved in product areas, including launcher structures and separation systems, satellite structures, digital electronics for satellites and launchers, satellite communication equipment and satellite instruments.

Russian Federal Space Agency (Roscosmos), Russia, www.federalspace.ru

The Russian Federal Space Agency was established in 1992 to provide administration of Russian space assets and to manage international space co-operation in joint space projects and programmes, such as the International Space Station (ISS). Roscosmos is responsible for the co-ordination of the activities at the Baikonur Cosmodrome space launch centre. Baikonur, which is located in Kazakhstan, is leased by Russia from Kazakhstan.

In 2006 the federal space programme put forward a 10-year plan for 2006–15, which covers specific programmes, timelines, funding and implementation. The strategic intent of the national space programme is to increase the standard of living, foster economic growth and provide for national security. The 10-year space programme plan is modified as needed to meet national priorities and budgetary parameters. In Russia, space projects are prioritized in two key documents—space policy priorities and military priorities. Close to two-thirds of all Russian space projects have national security priorities and implications, hence the reason for dual funding of programmes and projects from the civil Russian Federal Space Agency, Roscosmos, and the Ministry of Defence. In addition to military space, space launch capabilities are prioritized in Russia (Russian space launchers captured 40% of the global total in 2009).

Other areas of concern for Roscosmos include upgrades and operations of the Russian Global Orbital Navigation Satellite System (GLONASS); Earth observation and meteorological satellite systems; support of telecommunications throughout Russia; support for commercial satellite services; ISS development and support; R&D in space technologies, spacecraft and space launch vehicles; maintenance of space infrastructure, such as Baikonur and ground systems; market research, including forecasting of space industry developments; implementation of state policy to the space sector; and support for space science projects and deep space robotic observation.

Roscosmos has intergovernmental space co-operation agreements with more than 19 states, including Argentina, Brazil, India, Japan, Sweden, the USA and the member states of the **European Space Agency**. Roscosmos contributes to the activities of the United Nations Committee on Peaceful Uses of Outer Space, the **Committee on Space Research**, the **Inter-Agency Space Debris Coordination Committee**, the **Committee on Earth Observation Satellites**, and the **International Astronautical Federation**. Commercial launch services represent one of the most competitive areas of Russian activities in the world space market. Commercial services of Russian space launchers are marketed by joint ventures between Russia and foreign partners: **International Launch Services** with **Lockheed Martin Space Systems Company** of the USA, **Sea Launch** and **STARSEM, The Soyuz Company**.

S

Safran Group, France, www.safran-group.com

The roots of the Safran Group date back to 1905–1912, when Gnome and Rhone was created to manufacture aircraft engines. In 1925 the Société d'Applications Générales de l'Electricité et de la Mécanique (Sagem) was established. Sagem expanded into telephony and communications in the 1940s, and later into avionics and electronics. In 1945 Gnome and Rhone was nationalized in France, and assumed the name Snecma. Since then, Snecma has diversified into missile and space launch development, and modern aircraft engines. In 2005 Sagem and Snecma merged to form the Safran Group. The group's aerospace business covers space launch propulsion, rocket motors, jet engines and aircraft equipment. As of 2009 the Safran Group employed 55,000 people, and had revenue of US $14,500m.

In the space sector, Snecma (www.snecma.com) of the Safran Group designs, develops, produces and markets engines for space launch vehicles and satellites. Snecma was involved in the development of the Diamant space launch vehicle. Diamant was a French expendable launch system and it was the first space launch vehicle not built by the USA or Russia. Out of 12 launch attempts between 1965 and 1975, nine were successful. Diamant launched Astérix, the first French satellite, into orbit in 1965. Despite the success, France terminated Diamant in favour of the European Ariane space launch vehicle in 1975.

Snecma is the propulsion prime contractor on Europe's Ariane space launch vehicle, and it is involved with the solid rocket motor of Europe's Vega launcher. Snecma is also involved with electric propulsion systems based on plasma thrusters for satellites and other spacecraft, mainly through international partnerships. In the area of space propulsion, Snecma primarily works with national space agencies, the **European Space Agency**, the **Centre National d'Etudes Spatiales** and the **National Aeronautics and Space Administration**. In addition, Snecma develops and manufactures solid rocket motors for the strategic missiles in France's nuclear deterrent force.

Samara Space Centre, Russia, www.samspace.ru

State Research and Production Space Rocket Centre 'TsSKB-Progress'—the Samara Space Centre—is a Russian enterprise focused on the development,

Satellite Communication Service Companies

manufacture and operation of space launch vehicles. The legacy of the Samara Space Centre dates to the beginnings of the space age with the development of the Vostok, Molniya and Soyuz space launch vehicles. In 1961, Vostok launched cosmonaut Yuri Gagarin into space. Vostok rockets remained in operations until 1991. The Molniya launch vehicle was first used in 1960, and a Molniya-M version was in production and operation as of 2010. The Soyuz series launch vehicle began operation in 1963 and is used to launch human crews and cargo to the International Space Station. In 1996 **STARSEM, The Soyuz Company**, was created to provide commercial launch services with the Soyuz family of space launch vehicles from the European Guiana Space Centre.

TsSKB-Progress was initiated under the guidance of Soviet General Designer of rocket-space systems, S. P. Korolev. TsSKB-Progress was established by combining the Samara Progress Plant and the Central Specialized Design Bureau TsSKB. The Vostok, Molniya and Soyuz space launch vehicles were derived from ballistic missiles, specifically the Russian R-7. The R-7 booster launched Sputnik in 1957. In 1996 TsSKB and the Samara Progress Plant created the State Research and Production Space Rocket Centre, known as the Samara Space Centre. The Samara Space Centre conducts its activity according to the **Russian Federal Space Agency** (Roscosmos). In addition to operations of the Molniya and the Soyuz, the Samara Space Centre develops and launches remote sensing and Earth observation satellites.

Satellite Communication Service Companies

Since 2001, satellite manufacturing, ground equipment and satellite value-added services, primarily to support satellite telecommunications, have been the only viable set of commercial space activities that can exist on market forces. Global revenues for all commercial satellite services, in 2009, totalled US $90,500m., of which US $18,000m. was for satellite communications services (US $68,000m. for direct-to-home television; US $2,500m. for satellite radio; and US $2,000m. for commercial remote sensing).

There are both fixed satellite communications service companies and mobile satellite communications service companies. Several of the larger companies are found in this glossary. Fixed satellite service refers to geostationary communications satellites used for television, for broadcast of radio stations, for networks and for telephony and data communications. Mobile satellite service refers to networks of communications satellites intended for use with mobile and portable wireless communications that can operate in air, on land and sea. Satellites for this service can be placed in geostationary, medium-Earth, or low-Earth orbits. Significant players in this area include **Inmarsat** and Globalstar (www.globalstar.com).

Key fixed satellite service companies include: AMOS by Spacecom, Israel (www.amos-spacecom.com); **Antrix Corporation Ltd**, India; APT Satellite, Hong Kong (www.apstar.com); Arabsat, Saudi Arabia (www.arabsat.com);

AsiaSat, Hong Kong (www.asiasat.com); Broadcast Satellite System Corporation, Japan (www.b-sat.co.jp); Embratel, Star One, Brazil, (www.starone.com.br); Eutelsat, Europe; Gascom, Russia (www.gascom.ru); **Intelsat SA**, Luxembourg; Korea Telecom (KT), Republic of Korea (www.kt.com); **Loral Space & Communications**; MEASAT Satellite Systems, Malaysia (www.measat.com); Nilesat, Egypt (www.nilesat.com); Russian Satellite Communications Company, Russia (sibir.statpro.ru); Satmex, Mexico (www.satmex.com); **SES Group**, Luxembourg; Singapore Telecommunications, Singapore (info.singtel.com); SKY Perfect JSAT Group, Japan; Telenor Satellite Broadcasting, Norway (www.telenorsbc.com); Telkom Indonesia, Indonesia (www.telkom.co.id); **Telesat**, Canada; and Thaicom Public Co. Ltd, Thailand (www.thaicom.net).

Sea Launch, Norway, Russia, Ukraine and USA, www.sea-launch.com

Sea Launch is a multinational corporation for commercial space launch that formed in 1995 and began operations in 1999. The company launches the Ukrainian Zenit rocket from the equatorial regions of the Pacific Ocean. It has developed a mobile space launch platform that is transported to the launch site from the port of Long Beach, California in the USA. Sea Launch expanded operational capabilities with the creation of a subsidiary, Land Launch, which conducts commercial launches of Zenit from the Baikonur Cosmodrome. The first successful Land Launch took place in 2008. From 1999 to 2009, Sea Launch and Land Launch conducted 33 successful launches. In 2009, Sea Launch filed for bankruptcy protection. As of 2010, Sea Launch was finalizing its plan of reorganization and was working toward emerging from bankruptcy.

There are four partners associated with Sea Launch: **Boeing Company** of the USA; **RSC Energia** of Russia; SDO Yuzhnoye and Yuzmash of Ukraine (see **Yuzhnoye State Design Office**); and Aker ASA Group of Norway. Boeing owns 40% of Sea Launch and provides for the payload fairing, spacecraft integration and mission operations; RSC Energia owns 25% and provides for the upper stage, launch vehicle integration and mission operations; Yuzhnoye owns 15% and provides the Zenit launch vehicle for Sea Launch, and also vehicle integration support and mission operations; and Aker ASA Group, with a 20% stake, does not provide operational services, though its predecessor, Kvaerner AS, provided construction of the Sea Launch Commander vessel and conversion services for the Odyssey Launch Platform for Sea Launch.

In 2003, Sea Launch, through Boeing, entered into a reciprocal agreement, the Launch Services Alliance, with **Arianespace** and **Mitsubishi Heavy Industries**. The launchers for the Launch Services Alliance are Zenit of Sea Launch, the European Ariane and Japan's H-II. The agreement enables customers to transition seamlessly among the three space launch vehicles.

Secure World Foundation (SWF), USA, www.secureworldfoundation.org

The Secure World Foundation is a private operating foundation dedicated to maintaining the secure and sustainable use of space. SWF engages with academics, policy-makers, industry, scientists and advocates in the space community to support space security; to foster peace and security on Earth; and to facilitate human security through the use of space services for development, disaster relief, humanitarian assistance and environmental protection. SWF was granted observer status to the United Nations Committee on the Peaceful Uses of Outer Space in 2008.

SES Group, Luxembourg, www.ses.com

SES Group is a network of satellite operators located primarily in Europe through SES Astra (www.ses-astra.com), and in the USA through SES World Skies (www.ses-usg.com). The global satellite fleet of SES reaches 99% of the world's population. SES was formed in 1985 as the first private satellite operator in Europe. That same year, SES signed an agreement with **Arianespace** to launch telecommunications satellites. In 1989 ASTRA 1A became the first operational satellite for SES. As of 2010 SES operated a fleet of 39 satellites, with six more planned for operations in the short term. Revenue for 2009 was US $2,300m. SES Astra operates the ASTRA satellite system for SES.

SES World Skies was formed in 2009 from the former Americom Government Services (AGS). AGS, which developed and launched geosynchronous communications satellites from 1973, became part of the SES group in 2001 as SES Americom. In turn, SES acquired a number of companies, most notably New Skies in 2006. This created SES New Skies. In 2009 SES Americom and SES New Skies integrated their satellite operating fleets, creating SES World Skies. SES World Skies provides bandwidth and hosted payload opportunities to **US Government**, intelligence and civilian agencies through a fleet of geosynchronous communications satellites.

SKY Perfect JSAT Corporation, Japan, www.sptvjsat.com

The merger of JSAT Corporation, SKY Perfect Communications and Space Communications Corporation in 2008 resulted in the creation of SKY Perfect JSAT Corporation, the core operating company of the SKY Perfect JSAT Group. SKY TK group operates 12 satellites covering Japan, Asia, Oceania and North America. Mobile satellite services were established in 2008 through JSAT MOBILE Communications. SKY Perfect JSAT expanded its services in the field of satellite communications with the sale of transponders in Asia and joint initiatives with **Intelsat SA**. In 2009, revenue was US $1,500m.

Space Exploration Technologies Corporation (SpaceX), USA, www.spacex.com

Space Exploration Technologies Corporation was established in 2002. Since then, SpaceX has been developing the Falcon family of space launch vehicles for light-lift to heavy-lift capabilities. As of 2010 the light-class Falcon 1 was operational and the heavy-lift Falcon 9 was successfully developed and flight tested.

In addition, SpaceX secured funding from the **National Aeronautics and Space Administration** (NASA), to demonstrate delivery and return of cargo to the International Space Station (ISS). NASA announced in 2008 the selection of the Falcon 9 launch vehicle and Dragon spacecraft, which is under development by SpaceX, to resupply ISS when Space Shuttle operations are terminated (current plans call for termination in 2011). SpaceX signed a US $1,600m. contract for a minimum of 12 flights, with an option to order additional missions for a cumulative total contract value of more than US $3,000m. with NASA's Commercial Crew and Cargo Program.

Space Foundation, USA, www.spacefoundation.org

The Space Foundation was created in 1983 as a non-profit foundation to foster, develop and promote greater understanding of the practical utilization of space. The two main activities of the Space Foundation, include teacher education programmes and space symposia. Space education programmes are aimed at K-12 teachers in all 50 states of the USA. The Space Foundation organizes and hosts the National Space Symposium and the Strategic Space Symposium. The National Space Symposium brings together the space community to discuss strategic issues, and the Strategic Space Symposium focuses on the space-related missions of US Strategic Command. The foundation also publishes an annual report, *The SPACE Report*, which surveys global space activity. The Tauri Group of the USA is the technical lead for *The SPACE Report* (www.taurigroup.com).

Space Generation Advisory Council (SGAC), Austria, www.spacegeneration.org

The Space Generation Advisory Council, founded in 1999 in support of the United Nations Program on Space Applications (see **United Nations**—UN), is a non-governmental organization that aims to represent students and young space professionals to the UN, states and space agencies. SGAC has permanent observer status in the UN Committee on the Peaceful Uses of Outer Space. SGAC focuses on pragmatic space policy advice to policy-makers based on the interests of students and young professionals. The mission of SGAC is to employ youth in advancing humankind through the peaceful uses of space. In addition to policy advice, members carry out a range of projects, including Under African Skies—a grass roots science teaching project in Africa; Yuri's Night—a World Space Party; the Association for the Development of Aerospace Medicine; and a Global Space Education Curriculum project.

Space Policy Institute, USA, www.gwu.edu/~spi

The Space Policy Institute was established in 1987. The institute conducts research on space policy issues, organizes seminars, symposia and conferences on various topics, and offers graduate courses on space policy. It operates as a research and policy programme of the Center for International Science and Technology Policy of the Elliott School of International Affairs at George Washington University. The Space Policy Institute focuses its activities on policy issues related to the space efforts of the USA, and co-operative and competitive interactions in space among the USA and other countries. The Institute is an affiliate of the **International Space University** and the **Aerospace Corporation**.

Space Research and Remote Sensing Organization (SPARRSO), Bangladesh, www.sparrso.gov.bd

The Space Research and Remote Sensing Organization, created in 1980, is the national space research and exploration agency of Bangladesh. SPARRSO acts as an autonomous research and development organization of the government. Space technology applications in Bangladesh started in 1968 through the establishment of the Atomic Energy Centre. In 1972 the Bangladesh Landsat Programme was established to receive data from the **National Aeronautics and Space Administration**'s (NASA) Landsat programme (landsat.gsfc.nasa.gov).

The activities of SPARRSO encompass: space science; remote sensing for agricultural research, disaster monitoring and environmental studies; geographic information systems; advice to the government in matters relating to space technology applications and policy; and dissemination of satellite data. SPARRSO collaborates with NASA, the **Japan Aerospace Exploration Agency**, the **Centre National d'Etudes Spatiales** and the **China National Space Administration**.

Space Research Centre (SRC), Poland, www.cbk.pan.wroc.pl

The Space Research Centre was established in 1977. It is a scientific institute of the Polish Academy of Sciences. SRC carries out basic and applied research in the field of space physics, solar system research and physical and geodesic studies of the planets and Earth. Researchers from SRC have participated as co-investigators in **European Space Agency** (ESA) science projects, which include Ulysses, ISO, SOHO, XMM, Cluster, Double Star, Huygens, Mars Express, Herschel, Planck, XEUS, Integral, Rosetta and BepiColombo. Polish researchers have also had some activities related to the ExoMars rover instruments, and Polish principal investigators are active with the utilization of Envisat data (ESA Earth observation platform). SRC also participates in the ESA Student Space Exploration and Technology Initiative project (www.sseti.net).

Additionally, SRC facilitated an Inter-Ministerial Consultative Group for Space to advise the prime minister of Poland, and played a role in developing co-operative agreements with ESA. In 2007, Poland became an ESA European Co-operating State, following Hungary (2003), Czech Republic (2003) and Romania (2006). Before 2007, Poland concluded co-operation agreements with ESA in 1994 and in 2002.

Space Studies Board (SSB), USA, sites.nationalacademies.org/SSB

The Space Studies Board was established in 1958. SSB provides an independent forum for information and advice on all aspects of space science and applications, and it serves as the focal point within the National Academies for activities on space research. It oversees advisory studies and programme assessments, facilitates international research and promotes communications on space science and science policy among the research community and the **US Government**. The SSB serves as the US National Committee for the International Council for Science **Committee on Space Research**.

Space Systems Development Department of the Naval Research Laboratory, USA, www.nrl.navy.mil

The Space Systems Development Department of the Naval Research Laboratory develops space systems to respond to Navy, Department of Defense, and national mission requirements. NRL pioneered naval research into space, from atmospheric probes with captured V-2 rockets, through direction of the Vanguard project, the first US satellite programme and launch system, to involvement in other projects, such as the Navy's Global Positioning System. The department provides space systems engineering capability from mission concept development, to on-orbit engineering and operational support.

The department is focused on system architecture and development of satellite payloads, with current research including optical systems, spacecraft processors and controllers, signal processing, data management and software development; the design, fabrication, test, integration and launch support for electronic components and systems; the development and implementation of ground stations for communication, data collection, processing and dissemination and spacecraft tracking, telemetry and control systems, as well as tactical communications systems; and space surveillance and satellite navigation technology.

Current programmes include WindSat, a technology demonstration programme designed to provide for wind vector measurements; the Tether Physics and Survivability Experiment and Advanced Tether Experiment, space tether demonstration projects; Global AIS and Data-X International Satellite Constellation, which test small satellite constellation architectures; and the High Temperature Superconductivity Space Experiment.

Space Systems Loral (SS/L), USA, www.ssloral.com

Space Systems Loral is a designer, manufacturer and integrator of geostationary satellites and satellite systems. A subsidiary of Loral Space and Communications, SS/L provides orbital testing, procures insurance and launch services and manages mission operations. SS/L has an international base of commercial and governmental customers whose applications, include fixed satellite communications services, broadband digital communications, wireless telephony, direct-to-home television broadcast, video and radio broadcasting, environmental monitoring and air traffic control. Since 1960, SS/L has manufactured more than 230 satellites. Approximately 25% of all **US Government**-leased commercial transponders are carried on SS/L-built satellites. Revenue in 2009 was US $1,000m.

Space Vehicles Directorate of the Air Force Research Laboratory, USA, www.kirtland.af.mil/afrl_vs

The mission of the Air Force Research Laboratory (AFRL) is to develop, integrate and deliver technologies for warfighting capabilities. Formed in 1997 as the product of an organizational consolidation that integrated previously separate Air Force laboratories (Armstrong, Phillips, Rome and Wright-Patterson) with the Air Force Office of Scientific Research, AFRL consists of a number of directorates. The space-related directorate, Space Vehicles Directorate, comprises 1,000 employees and had an annual budget of US $400m.

The Space Vehicles Directorate provides R&D of space-based command, control, communications, intelligence, surveillance and reconnaissance; of counterspace systems; and of responsive space technologies and capabilities (a successful project of note for responsive space was the Experimental Satellite System-11).

Three distinct departments form the directorate's core operations. The Battlespace Environment Division detects and understands the threats in the aerospace environment to warfighting systems. The Integrated Experiments and Evaluation Division develops, incorporates and demonstrates military space concepts. This section manages and executes space technology tests and experimental projects, which include orbital missions. The Spacecraft Technology Division provides technology for global awareness and control of space.

Spot Image, France, www.spot.com

Spot Image was created in 1982 by the **Centre National d'Etudes Spatiales** (CNES), the Institut Géographique National and several space manufacturers—Matra (part of the Lagardère Group since 1994), Alcatel (now Alcatel-Lucent) and the Swedish Space Corporation—as a public company owned by CNES. In 2008 Spot Image was privatized through the acquisition of all CNES shares by **European Aeronautic Defence and Space Company**

(EADS) Astrium, which owned 81% of Spot Image as of 2010. Other key shareholders include the Swedish Space Corporation and **Telespazio**, at about 14% combined.

Spot Image is a world-wide distributor of geographic information, products and services derived from the SPOT Earth observation satellites. There are plans for follow-on satellites to the SPOT series planned for launch in 2010, 2011 and 2012. Spot Image has established subsidiaries in Australia, Brazil, People's Republic of China, Japan, Singapore and the USA. It distributes optical and radar data acquired by satellites offering low-resolution to high-resolution images.

Spot Image's purpose is to make Earth images available for causes, such as natural and human-made disasters, humanitarian missions, global surveillance of the environment, international security and peacekeeping, education and research. Spot Image is committed to the *International Charter Space and Major Disasters* and the Unosat programme of the **United Nations** (www.unitar.org/unosat), which applies satellite imagery for humanitarian purposes.

SPOT Image, along with **DigitalGlobe** and **GeoEye** of the USA, dominate the commercial remote sensing market. SPOT Image captures approximately 25% of the total market share in sales of data and value-added services globally. Revenue for SPOT Image in 2009 was US $130m.

STARSEM, The Soyuz Company, Russia, www.starsem.com

STARSEM, The Soyuz Company was created in 1996 to provide commercial launch services with the Soyuz family of space launch vehicles. The Soyuz launched Sputnik in 1957, and since then more than 1,750 human and robotic missions have been performed. The European–Russian company brings together organizations—**Arianespace**, **European Aeronautic Defence and Space Company Astrium**, **Russian Federal Space Agency** (Roscosmos) and the **Samara Space Centre** of Russia—for the commercialization of Soyuz. STARSEM offers the Soyuz for satellite telecommunications systems, scientific spacecraft, Earth observations and meteorological platforms.

STARSEM provides for launch vehicle manufacturing, mission preparation at the Baikonur Cosmodrome and in-orbit delivery of payloads. Starting in 2010 Soyuz missions have taken place at Europe's Guiana Space Centre in French Guiana, in co-operation with Arianespace and the **European Space Agency**.

Stichting Ruimte Onderzoek Nederland (SRON), the Netherlands, www.sron.nl

The Stichting Ruimte Onderzoek Nederland (Netherlands Institute for Space Research) was founded in 1983. As part of the Netherlands Organization for Scientific Research, SRON is the national centre for the development and exploitation of satellite instruments for research in astrophysics, Earth sciences and planetary research. SRON acts as the national agency of the Netherlands for space research, and as the national point of contact for **European Space Agency** (ESA) programmes.

SRON has worked on the development of instruments for three satellites: BeppoSAX of the **Agenzia Spaziale Italiana** (ASI); Chandra of the great observatories programme of the **National Aeronautics and Space Administration** (NASA); and the XMM-Newton of the European Space Agency (ESA). Additionally, SRON is involved in the operation of the ESA satellite Integral, which measures gamma radiation. In all of these missions, SRON has been active in the processing and interpretation of the data that the instruments provide.

Surrey Satellite Technology Ltd (SSTL), United Kingdom, www.sstl.co.uk

Surrey Satellite Technology, which started in 1985, is a provider of small satellite missions. SSTL designs, builds and launches small satellites using commercial off-the-shelf satellite technology. SSTL's mission is to provide access to space through the small satellite market for Earth observation and imaging, scientific research, military purposes and technology demonstration.

During the 1970s, a group of aerospace researchers at the University of Surrey in the United Kingdom (UK) developed a satellite using standard consumer technology, known as commercial off-the-shelf components. That first satellite, UoSAT-1 (University of Surrey satellite) was launched in 1981 with the help of the **National Aeronautics and Space Administration** (NASA). Given the operational success of UoSAT-1, UoSAT-2 was launched in 1984. In 1985 the University of Surrey formed Surrey Satellite Technology Ltd to transfer the results of its research into a commercial enterprise.

As of 2010, SSTL launched 34 spacecraft, with seven more under manufacture. In 2006 SSTL acquired SIRA Electro-Optics, which led to capability for developing cameras and visual technology for satellites; in 2008 SSTL established a subsidiary, Surrey Satellite Technology LLC in the USA; in 2009, **European Aeronautic Defence and Space Company Astrium** acquired a 99% shareholding stake in SSTL from the University of Surrey.

SSTL missions, through 2009, included Deimos-1 in 2009 for commercial remote sensing; UK-DMC in 2009 and 2003 as part of the Disaster Monitoring Constellation (see **DMC International Imaging**); RapidEye in 2008 as part of a commercial remote sensing constellation; CFESat in 2007 for technology demonstrations; GIOVE-A in 2005 as a Galileo (the **European Union**'s satellite navigation system) technology demonstration; Beijing-1 in 2005 as part of the Disaster Monitoring Constellation; BLMIT in 2005 as part of the Disaster Monitoring Constellation; TopSat in 2005 as technology demonstration for the UK Ministry of Defence; Nigeria Sat-1 in 2003 as part of the Disaster Monitoring Constellation; BILSAT-1 in 2003 for TÜBİTAK UZAY of Turkey as part of the Disaster Monitoring Constellation; AlSAT-1 in 2002 for Algeria as part of the Disaster Monitoring Constellation; PICOSat in 2001 as a technology demonstration for the US Air Force; Tsinghua-1 in 2000 for Tsinghua University, People's Republic of China in the area of Earth observations; SNAP-1 in 2000 as a technology demonstration; TiungSat-1 for Malaysia in 2000 for Earth observations; Clementine in 1999 for the French military; FASats in 1998 and 1995 for the Chilean Air Force; Thai-Paht for Mahankorn University, Thailand in 1998; CERISE in 1995 for the French military; HealthSat-2 in 1993 for Satellife of the USA (www.healthnet.org); PoSAT-1 for Portugal in 1993 for Earth observations, science and communications; KITSAT-1 in 1992 for the Republic of Korea for Earth observations and communications; S80/T for the **Centre National d'Etudes Spatiales** in 1992; and UoSATs from 1981 to 1999.

Swedish National Space Board (SNSB), Sweden, www.snsb.se

In 1972 the Swedish National Space Board was formed as a governmental agency under the Swedish Ministry of Enterprise, Energy and Communications. SNSB is responsible for national and international activities relating to space research and remote sensing. SNSB has three main tasks: to distribute government grants for space research, technology development and remote sensing; to initiate R&D in space science and remote sensing; and to act as the Swedish contact for international co-operation. In practice, SNSB functions as the national space agency for Sweden.

The Swedish space programme is carried out through international co-operation, in particular through Sweden's membership in the **European Space Agency**. In 1961 Swedish space activities began with a rocket launch, and in 1966 the Swedish Esrange rocket base was established. The main goal of the Swedish space programme is to promote the use of space for public applications encompassing the environment, climate and communications, and to increase the competitiveness of the Swedish space industry (see **Swedish Space Corporation Group**) and scientific institutions. Key areas of competence include magnetosphere and ionosphere research, astronomical studies, remote sensing and observation of Earth's atmosphere and environment and specialized industrial competence, including development of satellites.

Swedish Space Corporation Group (SSC), Sweden, www.ssc.se

The Swedish Space Corporation Group was founded by the Swedish Government in 1972. It is wholly owned by the Swedish state and administered by the Ministry of Enterprise, Energy and Communications. The SSC Group, includes: NanoSpace (www.nanospace.se), which develops micromechanical systems; ECAPS (www.ecaps.se), which develops propulsion systems; LSE Space Engineering & Operations (www.ssc.se, Germany and France), which is a space operations consultancy; Santiago Satellite Station (www.ssc.se); and Universal Space Network (www.uspacenetwork.com). SSC designs, develops and tests various types of space systems—satellites, sounding rockets, space subsystems and experiment equipment for research in microgravity.

T

Telesat, Canada, www.telesat.ca

Telesat is a Canadian satellite communications company founded in 1969. In 1972 it launched the Canadian Anik A1, the world's first communications satellite in geostationary orbit operated by a commercial company. In 2007 **Loral Space & Communications**, with a 64% stake, and the Public Pension Investment Board of Canada acquired Telesat from Bell Canada. At the same time, Telesat merged with Loral Skynet, a subsidiary of Loral Space & Communications. Telesat is one of the largest fixed satellite communications service providers in the world (see **Satellite Communication Service Companies**). Telesat revenue for 2009 was US $700m.

Telespazio, Italy and France, www.telespazio.it

Telespazio was created in 1961 under the auspices of the Italian National Research Council and the Italian Post and Telecommunications Ministry. It is part of the Space Alliance between **Finmeccanica** (67%) and Thales Group (33%), employs 1,900 people, operates a network of space centres, and generated revenue of US $600m. in 2009. Telespazio plays a role in European space programmes such as Constellation of Small Satellites for the Mediterranean Basin Observation, European Geostationary Navigation Overlay System, Galileo, and Global Monitoring for Environment and Security. The space-related focus of Telespazio encompasses the business areas of satellite operations, military satellite communications, Earth observations, navigation and infomobility, and scientific programmes.

In 2009 Telespazio incorporated the MARS Center (www.marscenter.it). The MARS Center was created in 1988 to focus on the design, engineering, construction and operation of experimental platforms, specifically on sounding rockets, the Space Shuttle and the International Space Station (ISS). The experiments on-board ISS range from the integration of payloads to their in-orbit operation, and to the dissemination of the scientific data. Telespazio, through its investment in the MARS Center, is also taking part in the **European Union**'s scientific and technological research project ULISSE, the objective of which is to assimilate scientific data from the ISS with that from other microgravity missions. In addition, Telespazio plays an active role in research for direct exploration missions to near-Earth objects, small celestial bodies at risk of colliding with Earth, and Europe's Space Situational Awareness programmes.

Thales Alenia Space, France and Italy, www.thalesgroup.com

The origins of Thales Alenia Space (which is part of the Thales Group) go back to Alcatel Alenia Space. Alcatel Alenia Space was established in 2005 by the merger of Alcatel Space, owned by Alcatel and **Finmeccanica**, and Alenia Spazio, in which **Telespazio** had a stake. In 2007, Thales Alenia Space was formed when the Thales Group acquired Alcatel Alenia Space. Thales Alenia Space develops and operates satellite systems and orbital infrastructures. It is a joint venture between the Thales Group (67%) and Finmeccanica (33%), and forms a Space Alliance with Telespazio. Thales Alenia Space employs 7,200 people and had revenue of US $3,000m. in 2009.

Thales Alenia Space is involved in programmes and projects in various areas, including in Earth Observations, the Global Monitoring for Environment and Security programme, with Sentinel-1 and Sentinel-3; meteorology as prime contractor for Meteosat satellites for the **European Organisation for the Exploitation of Meteorological Satellites**; and climatology as prime contractor for the Soil Moisture and Ocean Salinity mission. In defence it is the prime contractor for the French Syracuse and the Italian Sicral telecommunications systems; prime contractor for the Italian Earth Observation system Constellation of Small Satellites for the Mediterranean Basin Observation; and main partner of the German telecom programme, Satcom BW, and of Earth observation programmes Helios (France) and SAR-Lupe (Germany). In navigation it is the prime contractor of European Geostationary Navigation Overlay System, the precursor of Galileo, while in science and exploration it is the prime contractor for the European ExoMars mission, and for the Gravity Field and Steady-State Ocean Circulation Explorer satellite. In space infrastructure and transportation it is the provider of International Space Station pressurized volume, notably the Italian Multi-Purpose Logistics Modules and the European Columbus laboratory, and is supplier of satellites for defence and security purposes. In facility operational services the company provides services for the **European Space Agency**, the **Centre National d'Etudes Spatiales** and the **Agenzia Spaziale Italiana**; it provides services for the French, Italian and German ministries of defence; and services for the defence procurement agency of France. In the civil–military satellite export market, Koreasat 5 in South Korea, Star One in Brazil and Yahsat in the United Arab Emirates are developed by Thales Alenia Space.

TÜBİTAK Space Technologies Research Institute (TÜBİTAK UZAY), Turkey, www.uzay.tubitak.gov.tr

TÜBİTAK Space Technologies Research Institute, which was founded in 1985 within the framework of a protocol signed between the Scientific and

Technological Research Council of Turkey (TÜBİTAK) and the Middle East Technical University (UZAY), is a publicly funded research institute. TÜBİTAK UZAY specializes in space technologies, electronics, information technologies and related fields. The institute takes part in R&D projects with a focus on small satellites. In 2010, a Turkish reconnaissance satellite developed by TÜBİTAK UZAY, Rasat, was launched.

Turkey did not have a national space agency as of 2010. Rather, TÜBİTAK UZAY is tasked to co-ordinate the National Space Research Programme of Turkey. The goals of the programme are to establish Turkey's space research and development infrastructure; to strengthen Turkish space industry; to support developments in space sciences, life sciences and Earth system sciences; to develop education programmes; to apply the uses of space technologies and space-based services for society; and to create and disseminate knowledge to support decision-making.

Tunisian Association for Communication and Space Sciences—*see* **Association Tunisienne de la Communication et des Sciences Spatiales (ATUCOM), Tunisia.**

U

United Kingdom Space Agency, United Kingdom,
www.ukspaceagency.bis.gov.uk

In 2010, the United Kingdom Space Agency was established to replace the British National Space Centre, to bring all civil space activities under one single organizational management. The British National Space Centre formed in 1985 from government departments and public bodies in the United Kingdom (UK).

The space strategy of the UK, implemented by the UK Space Agency, established five goals: to win an increasing share of the global market in space systems, services and applications in the race to develop tomorrow's economy; to deliver world-leading space systems for managing our changing planet; to be a partner of choice in global scientific missions to explore the universe; to benefit society by strengthening innovation from space, and stimulate the creation of new products and services for everyday use; and to develop a major channel for skills development and outreach for a high technology future, and improve public and political recognition of the value of space systems as part of the critical national infrastructure.

In turn, the United Kingdom Space Agency has five key mission areas: exploring the solar system; understanding the universe; studying Earth and the Sun; navigation and communications; and Earth observations. The United Kingdom Space Agency is involved with the **European Space Agency** and **European Union** space projects, namely Galileo, Global Monitoring for Environment and Security, and satellite broadband communications systems. The United Kingdom Space Agency also collaborates with other organizations through co-operation on Earth Observation data acquisition and dissemination. Between 2000 and 2005 the British National Space Centre led a project—Micro Satellite Applications in Collaboration, or Mosaic—to develop small satellites. In 2009 total British National Space Centre funding was US $100m.

The United Kingdom Space Agency mission areas are realized through a number of projects, mainly within the context of European space programmes, which involve the participation of the agency. These include: exploring the solar system—Aurora, BepiColombo, Cassini-Huygens, Chandrayaan-1, Deep Impact, ExoMars, Mars Express, Rosetta, SMART-1, Stardust and Venus Express; understanding the universe—AKARI, Gaia, Herschel, Hipparcos, Hubble Space Telescope, Infrared Space Observatory, International

United Nations (UN)

GammaRay Astrophysics Lab, International Ultraviolet Explorer, James Webb Space Telescope, LISA Pathfinder, Planck, Swift and XMM-Newton; study of Earth and Sun—Cluster, Double Star, Hinode, Solar Heliospheric Observatory, STEREO and Ulysses; navigation and communications—EGNOS, Galileo, HOTBIRD and **Inmarsat** satellite series; and Earth observations—Aura, CryoSat-2, Disaster Monitoring Constellation, Envisat, ERS-1 and ERS-2, GOCE, Meteosat Second Generation, MetOp-A, Project for Onboard Autonomy, SMOS and TopSat.

United Launch Alliance (ULA), USA, www.ulalaunch.com

Formed in 2006, United Launch Alliance is a joint venture owned by **Lockheed Martin Space Systems Company** and the **Boeing Company**. It was created to provide space launch services for the **US Government**. ULA offers a variety of launch vehicles and payload accommodation options through the Atlas V, Delta IV and Delta II space launch systems. Government launch customers include the Department of Defense, **National Aeronautics and Space Administration**, the **National Reconnaissance Office** and **National Oceanic and Atmospheric Administration**, among other organizations. ULA employs 4,000 people and has an annual revenue of US $2,000m.

The heavy-lift Atlas V and Delta IV launch vehicles were developed in partnership with the US Air Force Evolved Expendable Launch Vehicle programme to meet the demand for assured access to space for national security purposes. The attempts to commercialize the Atlas V and Delta IV have failed to date, largely due to factors of demand and competition in the commercial space launch market.

Delta II is a medium-class space launch system in service since 1989, which was originally designed and built by McDonnell Douglas. Delta II rockets were later built by Boeing Integrated Defense Systems (McDonnell Douglas was acquired by the Boeing Company in 1997), until Delta II became the responsibility of ULA in 2006. ULA markets Delta II to government customers, and Boeing Launch Services markets Delta II to commercial customers.

United Nations (UN), www.oosa.unvienna.org

The space activities of the United Nations are organized by the UN Office for Outer Space Affairs (UNOOSA). UNOOSA serves as the secretariat for the UN General Assembly's committee dealing exclusively with international cooperation in the peaceful uses of space, the UN Committee on the Peaceful Uses of Outer Space (COPUOS, www.oosa.unvienna.org/oosa/en/COPUOS/copuos.html), which was established by the UN General Assembly in 1959. UNOOSA is responsible for implementing the UN secretary-general's responsibilities under international space law and maintaining the UN Registry of Objects Launched into Outer Space.

UNOOSA activities include: the UN Program on Space Applications, which started in 1971 to conduct international workshops; training courses and pilot projects on topics that entail basic space science; use and applications of global navigation satellite systems; natural resources management and environmental monitoring; satellite communications; and space technology and disaster management. On these topics, the programme on space applications supports a number of regional centres for space science and technology education, and prepares and distributes reports, studies and publications. Additionally, the UN Platform for Space-based Information for Disaster Management and Emergency Response programme established a network of providers of space-based solutions to support disaster management activities. The Secretariat of the International Committee on Global Navigation Satellite Systems (GNSS) promotes the use of GNSS infrastructure on a global basis. Workshops and programmes address the use of space technology for mitigating and adapting to climate change, and for advancing sustainable development. There is also a forum for international and national space activities of member states and international organizations, national research on space debris and nuclear power sources and research in the areas of near Earth objects and space weather. Finally, there are global meetings and conferences, including the Inter-Agency Meeting on Outer Space Activities (the 30th session was held in 2010), and the Exploration and Peaceful Uses of Outer Space (UNISPACE) conferences.

To date, the UN has held three global conferences on UNISPACE. The first, in 1968, reviewed progress in space science, technology and applications, and called for increased international co-operation, with particular regard to the benefit of developing states. The conference recommended the creation of the post of Expert on Space Applications, which led to the creation, in 1971, of the Space Applications Program of the UN. The second UNISPACE reviewed progress of the Space Applications Program to enable developing states to benefit from the uses of space technology. The third conference (UNISPACE III) was held in 1999. This conference was convened as a special session of COPUOS. UNISPACE III formulated a blueprint for the peaceful uses of space in the 21st century—*The Space Millennium: Vienna Declaration on Space and Human Development*—and evaluated the work of the UN in space to enable more effective uses of space technology in the pursuit of economic, social and cultural development and in protecting the environment.

COPUOS was established to review the scope of international co-operation; to devise programmes to be implemented under UN auspices; to encourage continued research and the dissemination of information on space matters; and to study legal problems arising from the exploration and uses of space. There are 69 UN member states in COPUOS, which has two standing subcommittees—the Scientific and Technical Subcommittee and the Legal Subcommittee. The committee and its two subcommittees meet annually to consider questions put before them by the UN General Assembly, reports submitted to them and issues raised by the UN member states, in particular, from the Inter-Agency Meeting on Outer Space Activities.

COPUOS is the primary international forum for the development of laws and principles governing space. The specific five treaties of the Outer Space Treaty Regime include: *Treaty on Principles Governing the Activities of States in the Exploration and Use of Outer Space, including the Moon and Other Celestial Bodies* (*Outer Space Treaty*), which entered into force in 1967; *Agreement on the Rescue of Astronauts, the Return of Astronauts and the Return of Objects Launched into Outer Space* (*Rescue Agreement*), which entered into force in 1968; *Convention on International Liability for Damage Caused by Space Objects* (*Liability Convention*), which entered into force in 1972; *Convention on Registration of Objects Launched into Outer Space* (*Registration Convention*), which entered into force in 1976; *Agreement Governing the Activities of States on the Moon and Other Celestial Bodies* (*Moon Agreement*), which entered into force in 1984. There are no space powers that have ratified the Moon Agreement, though France and India are signatories.

United Space Alliance, USA, www.unitedspacealliance.com

United Space Alliance was initially formed as a joint venture between Rockwell International (Boeing acquired Rockwell in 1996) and Lockheed Martin in 1995, in response to the interest of the **National Aeronautics and Space Administration** (NASA) to consolidate the numerous Space Shuttle programme contracts to one prime contractor in the USA. United Space Alliance and NASA signed the Space Flight Operations Contract in 1996. Since 1996 United Space Alliance has been equally owned by Boeing and Lockheed Martin. As of 2009 United Space Alliance employed 8,800 people and had US $2,000m. in revenue.

Following the Space Shuttle Columbia accident in 2003, management issues regarding communication and safety were identified. Since Columbia, there have existed more open and transparent communication patterns; operational concerns are discussed between NASA and United Space Alliance in a more inclusive manner, safety is considered independently from operations, and United Space Alliance employees are empowered to address technical problems and issues with the Space Shuttle system.

United Space Alliance supported the Space Flight Operations Contract until 2006. Since then, United Space Alliance has operated the Space Shuttle programme under the Space Program Operations Contract. This contract establishes United Space Alliance as NASA's primary industry partner in human space flight operations, including the day-to-day management of the Space Shuttle fleet, and planning, training and operations for the International Space Station (ISS). As a member of the Orion and Ares team for NASA's project Constellation, United Space Alliance was also lending its expertise to the design and development of the next generation of space exploration vehicles for NASA; however, in April 2010, the Barack Obama

Administration called for a reduction in the scope of project Constellation, which would lead to the elimination of the Ares crew launch vehicle and a smaller Orion human spacecraft.

United Space Alliance has also created two subsidiary companies, United Space Alliance Space Operations and Space Flight Operations, to take on additional work, including: ISS contracts with Boeing; the ISS cargo mission contract with Lockheed Martin; the mission support operations contract with Lockheed Martin; the facilities development and operations contract with Lockheed Martin; the integrated mission operations contract with NASA Johnson Space Center; the Constellation Space Suit System with Oceaneering (www.oceaneering.com); Extravehicular Activity Systems with Hamilton Sundstrand (www.hamiltonsundstrand.com); engineering support to NASA Marshall Space Flight Center, Progress-Soyuz cargo operations services with LZ Technology (www.lzt.biz); and exploration ground launch services for NASA.

Universities Space Research Association (USRA), USA, www.usra.edu

Universities Space Research Association is a private, non-profit corporation founded in 1969 under the auspices of the National Academy of Sciences. The membership consists of 105 universities in the USA and abroad that have graduate programmes in sciences and engineering related to space. USRA focuses on space-related technical competencies with the goal of expanding knowledge and developing technology for the benefit of the academic community, space-related industries and the **National Aeronautics and Space Administration**'s (NASA) mission to pioneer space exploration, scientific discovery and aeronautics research.

The purpose of USRA is to enable universities and other research organizations to co-operate with one another, with the **US Government** and with other organizations toward the development of knowledge associated with space science and technology; to acquire, plan, construct and operate laboratories and other facilities; and to formulate policies under contract with the US Government for R&D and education associated with space science and technology.

USRA also operates the Lunar and Planetary Institute (LPI, www.lpi.usra.edu/lpi) under a co-operative agreement with the Science Mission Directorate of NASA. LPI is a research institute that provides support services to NASA and the planetary science community and conducts planetary science research.

US Air Force Space Command (AFSPC), USA, www.afspc.af.mil

The mission of the US Air Force (USAF) is to fly, fight and win in air, space and cyberspace. US Air Force Space Command leads the effort to accomplish this mission in space and cyberspace. AFSPC was created in 1982 to serve as the executive agent for space for the Department of Defense (DoD). From

1982 to 2008, AFSPC was responsible for the development, acquisition and operation of the USAF space and missile systems. In 2009 missile system responsibility shifted to a newly created command, Global Strike, and AFPSC assumed cyberspace responsibilities in addition to retaining the space force mission. 46,000 people, including 21,000 active-duty military and civilians, 13,000 contractor employees and 12,000 reservists and Air National Guard combine to perform AFSPC missions.

AFSPC has two numbered air forces. The 14th Air Force manages the generation and employment of space forces to support US Strategic Command and North American Aerospace Defense Command operational plans and missions. The 24th Air Force assures the warfighter freedom of action by establishing, extending, operating and defending assigned portions of the military networks to provide capabilities in, through and from cyberspace.

The Space and Missile Systems Center (SMC), under AFSPC, designs and acquires USAF and DoD space systems. It oversees launches, completes on-orbit checkouts and then turns systems over to user agencies. SMC supports the Program Executive Office for Space on the Global Positioning, Defense Satellite Communications and MILSTAR systems. SMC also supports the Evolved Expendable Launch Vehicle programme, Defense Meteorological Satellite, Defense Support programmes and the Space-Based Infrared System. The Space and Missile Systems Center manages the research, design, development, acquisition and sustainment of space launch, command and control, missile systems and satellite systems.

In addition, the Space Innovation and Development Center (SIDC) of AFSPC is responsible for integrating space systems into the operational USAF. This centre, originally known as the Space Warfare Center, is a test centre of air, space and cyberspace power to the warfighter. In 1992, a USAF panel on space recommended establishing a dedicated Space Warfare Center to exploit the capabilities of space-based assets. The Space Warfare Center was dedicated in 1993 and was redesignated SIDC in 2006. SIDC works to improve the exploitation of air, space and cyberspace capabilities through war gaming, exercises, experiments and space range development. The SIDC has also played a role in military space education with the administration of the Advanced Space Operations School since 2009.

US Government Actors with a role in space affairs

The US Government allocates US $60,000m. annually for space. In comparison, the entire combined allocation of resources among all the national space agencies of the world amounted to US $22,000m. as of 2009.

Department of Defense (DoD) military space spending, primarily through the US Air Force (USAF) as the executive agent for space for DoD, and the **National Aeronautics and Space Administration** (NASA) constitute 75% of US Government space budgets. DoD space spending is estimated at US $26,000m. and NASA was funded at US $19,000m.

Additional monies are allocated for the space intelligence sector. This entails the development and operations of intelligence-gathering reconnaissance satellites and the analysis of the data acquired from those satellites. This sector includes the **National Reconnaissance Office** (NRO) and the **National Geospatial-Intelligence Agency** (NGA), which together received US $15,000m. in 2010.

All other federal entities comprise the remainder of budgetary allocations. The more significant ones are the **National Oceanic and Atmospheric Administration** (NOAA) at US $1,000m. in 2010, for the operations of weather satellites, and the licensing and oversight of commercial remote sensing; the **Federal Aviation Administration Office of Commercial Space Transportation** (FAA-AST), to license and regulate commercial space launch; Department of Energy (DoE) for propulsion and nuclear power research; and the National Science Foundation (NSF), which funds basic space science research primarily within the academic sector. Total funding for FAA-AST, DoE and NSF (space science research only) was US $500m. in 2010.

Other US Government organizations with a role in space include: the Central Intelligence Agency, which co-manages and co-staffs the NRO, along with the USAF and DoD; Congress, which authorizes policy and appropriates civil, military and intelligence space budgets, oversees the implementation of space programmes and investigates space programme administration and management; the Department of Commerce, which regulates export controls (NOAA and the **US Office of Space Commercialization** are also within the Department of Commerce); the Department of State, which administers international space co-operation agreements and regulates export controls; the Federal Communications Commission, which regulates interstate and international communications by satellite, including the allocation of spectrum use for satellite operators; the National Security Council and the Office of Science and Technology Policy, which assist in the formation of national space policy for the Office of the President and co-ordinate interagency processes for space policy formulation; the Office of Management and Budget, which develops and co-ordinates the federal budget as proposed by the Office of the President, including budgets for all space activities of the US Government; the Office of the President, which formulates national space policy, proposes budgets for space programmes and establishes the priorities for US Government space programmes; and the US Geological Survey, which co-manages the Landsat programme with NASA.

US Office of Space Commercialization, www.space.commerce.gov

The US Office of Space Commercialization was initially established as the Office of Space Commerce in 1988. In 1998, the current name was assumed. The Office of Space Commercialization is the principal unit for space commerce policy activities within the US Department of Commerce. Space commerce refers to businesses using the medium of space to benefit the economy. The mission of this office is to foster the conditions for the economic growth and technological advancement of the US commercial space industry. The office focuses on several sectors of the space commerce industry, including satellite navigation, commercial remote sensing, space transportation and entrepreneurial activities. The office also participates in broad governmental discussions of national space policy and other space-related issues.

V

Virgin Galactic, USA, www.virgingalactic.com

Virgin Galactic, part of the Virgin Group, was founded in 2004 to develop space tourism. The company plans to provide suborbital space flights to the paying public from Spaceport America (www.spaceportamerica.com), along with suborbital space science missions and orbital launches of small satellites. Virgin Galactic owns and operates privately built spaceships under development by Scaled Composites (www.scaled.com), which is owned by Northrop Grumman.

World Space Week Association (WSWA), USA, www.worldspaceweek.org

World Space Week Association, founded in 1981, is a space education organization and a partner of the **United Nations** (UN) in the global co-ordination of World Space Week. This non-governmental, non-profit international organization is based in the USA. World Space Week is under the guidance of the UN Committee on the Peaceful Uses of Outer Space and the UN Office for Outer Space Affairs. The UN General Assembly declared in 1999 that World Space Week would be held annually from 4–10 October. These dates commemorate two events: the launch of Sputnik 1 and the signing of the *Treaty on Principles Governing the Activities of States in the Exploration and Peaceful Uses of Outer Space, including the Moon and Other Celestial Bodies.*

X

X Prize Foundation, USA, www.xprize.org

The X Prize Foundation is a non-profit prize institute that designs and manages public competitions to encourage technological development. A high-profile X Prize was the Ansari X Prize relating to spacecraft development, awarded in 2004. This prize was intended to stimulate R&D for space exploration. The Ansari X Prize was inspired by the Orteig Prize offered in 1919 by French hotelier Raymond Orteig for a nonstop flight across the Atlantic ocean.

In 1996, the X Prize Foundation offered a US $10m. prize to any privately financed team that could build and fly a three-passenger vehicle 100 km into space twice within two weeks. The contest, later called the Ansari X Prize for Suborbital Spaceflight, motivated 26 teams world-wide to invest more than US $100m. in pursuit of the prize. In 2004 the Ansari X Prize was won by Mojave Aerospace Ventures in SpaceShipOne developed by Scaled Composites. Scaled Composites, which is owned by Northrop Grumman as of 2010, is developing a second generation spacecraft under licence to **Virgin Galactic**, for the development of suborbital space tourism.

Y

Yuzhnoye State Design Office (SDO Yuzhnoye), Ukraine, www.yuzhnoye.com

Yuzhnoye State Design Office was established in 1951 as a design department of Yuzhnoye machine-building plant for military ballistic missile development. In 1954, the design department was reorganized as an independent design office. SDO Yuzhnoye is a state design office, or company, focused on space launch vehicle technology. SDO Yuzhnoye includes the Yuzhmash State Enterprise, PO Yuzhmash, which is an aerospace manufacturer (www.yuzhmash.com).

Starting in the 1970s, SDO Yuzhnoye developed solid-rocket motors. More recently, SDO Yuzhnoye and PO Yuzhmash developed the Zenit launch system used by **Sea Launch**. Other space systems during the history of Yuzhnoye and Yuzhmash, include Kosmos launcher, Interkosmos, Cyclone and Dnepr, which is operated as a commercial launch vehicle by the International Space Company Kosmotras (www.kosmotras.ru).

SDO Yuzhnoye consists of design departments on launchers, spacecraft, rocket engines, equipment development, ground units and systems processing and new technologies. SDO Yuzhnoye is also involved in the joint Ukrainian–Brazilian satellite launching project from the Centro de Lançamento de Alcântara in Brazil (Alcântara Launch Centre, www.cla.aer.mil.br).

Z

Zentrum für Angewandte Raumfahrttechnologie und Mikrogravitation (ZARM), Germany, www.zarm.uni-bremen.de

The Zentrum für Angewandte Raumfahrttechnologie und Mikrogravitation (Centre of Applied Space Technology and Microgravity) is a scientific institute at the University of Bremen in Germany. ZARM concentrates on the investigation of fluid mechanics phenomena under microgravity conditions (simulated by use of a drop tower) and questions related to space technology. ZARM's space activities started in 1989 with the small satellite project, BREM-SAT. Since then, the Space Technology Group at ZARM has worked on other satellite projects and on the support of scientific experiments onboard the Space Shuttle. The research efforts of the Space Technology Group focus on guidance, navigation and control of spacecraft ranging from satellites to re-entry vehicles, and from hardware to software for trajectory and attitude control.

Projects at ZARM include: First Look, fundamental physics mission Gravity Probe B; Laser Interferometer Space Antenna Pathfinder; Gaia, a science satellite mission; Thruster Actuation System—experiments in physics or space observatories; Global Positioning System/Inertial Measurement Unit at the Center of Applied Space Technology and Microgravity; Satellite Test of the Equivalence Principle; and Magnetic Torques, which interact with the Earth's magnetic field and create magnetic torque attitude control.

Figures

FIGURES

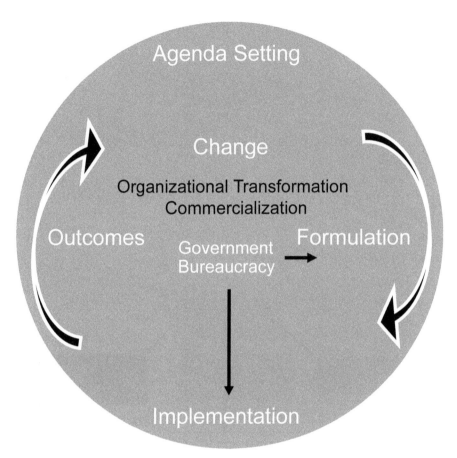

Figure 1.1 Space Politics and Policy
Source: developed by Eligar Sadeh.

FIGURES

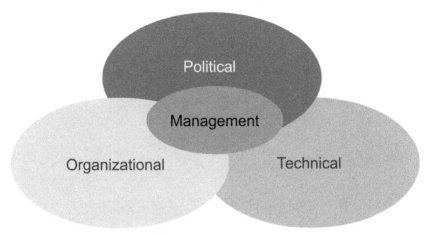

Figure 1.2 Space Policy Implementation

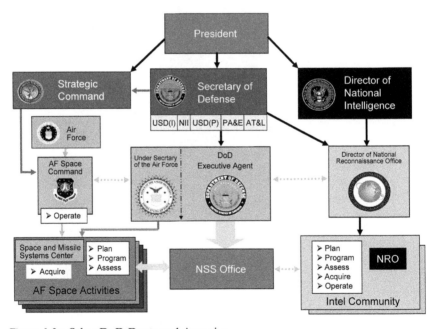

Figure 1.3a Other DoD Depts and Agencies

Figures

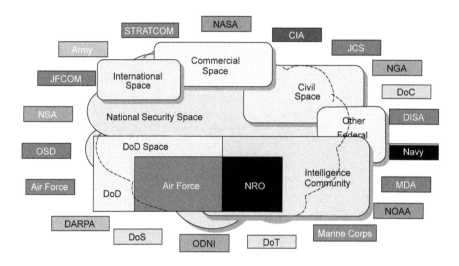

Figure 1.3b National Security Space Community
Source: United States National Security Office.

Notes: All actors shown in figure are part of the United States Government. USD is Undersecretary of Defense; I is USD for Intelligence; P is USD for Policy; AT&L is USD for Acquisition, Technology, and Logistics; NII is Assistant Secretary of Defense for Networks and Information Integration; PA&E is Program Analysis and Evaluation; AF is Air Force; DoD is Department of Defense; NSS is National Security Space; NRO is National Reconnaissance Office; Intel is Intelligence; Depts is Departments; OSD is Office of the Secretary of Defense; NSA is National Security Agency; JFCOM is Joint Forces Command; STRATCOM is Strategic Command; NASA is National Aeronautics and Space Administration; CIA is Central Intelligence Agency; JCS is Joint Chiefs of Staff; NGA is National Geospatial-Intelligence Agency; DoC is Deportment of Commerce; MDA is Missile Defense Agency; NOAA is National Oceanic and Atmospheric Administration; DoT is Department of Transportation; ODNI is Office of the Director of National Intelligence; DoS is Department of State; and DARPA is Defense Advanced Research Projects Agency.

FIGURES

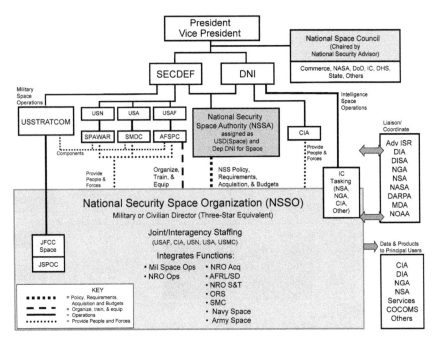

Figure 1.4 Proposed National Security Space Authority

Source: This figure was previously published in Sadeh, Eligar, "Report: Strategic Space and Defense 2008," *Astropolitics* 7: 1 (2009): 5. Original source is: Leadership, Management, and Organization for National Security Space: Report to Congress of the Independent Assessment Panel on the Organization and Management of National Security Space (September 2008).

Notes: All actors shown in figure are part of the United States Government. SECDEF is Secretary of Defense; DNI is Director of National Intelligence; USSTRATCOM is United States Strategic Command; USN is United States Navy; USA is United States Army; USAF is United States Air Force; SPAWAR is Space and Naval Warfare Systems Command; SMDC is Army Space and Missiles Defense Command; AFSPC is Air Force Space Command; USD is Undersecretary of Defense; Dep is Deputy; CIA is Central Intelligence Agency; NSA is National Security Agency, NGA is National Geospatial-Intelligence Agency; Adv is Advanced; ISR is Intelligence, Surveillance, and Reconnaissance; DIA is Defense Intelligence Agency; NASA is National Aeronautics and Space Administration; DoD is Department of Defense; IC is Intelligence Community; DHS is Department of Homeland Security; State refers to Department of State; DARPA is Defense Advanced Research Projects Agency; MDA is Missile Defense Agency; NOAA is National Oceanic and Atmospheric Administration; JFCC is Joint Functional Combatant Command; JSPOC is Joint Space Operations Center; NSS is National Security Space; USMC is United States Marine Corps; Mil is Military; Ops is Operations; NRO is National Reconnaissance Office; Acq is acquisitions; AFRL/SD is Air Force Research Laboratory Space Vehicles Directorate; S&T is Science and Technology; ORS is Operationally Responsive Space; SMC is Air Force Space and Missile Systems Center; and COCOMS are Combatant Commands (now called Unified Combatant Commands).

Figures

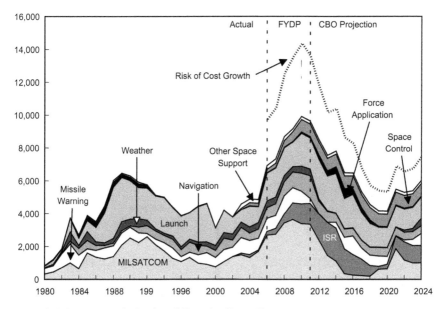

Figure 2.1 Investment in National Security Space Programmes
Source: Congressional Budget Office, "The Long-Term Implications of Current Plans for Investment in Major Unclassified Military Space Programs," (Washington, DC: Congressional Budget Office, 12 September 2005), 3, www.cbo.gov/ftpdocs/66xx/doc6637/09-12-MilitarySpace.pdf (accessed May 2009).

Notes: FYDP is Future Years Defense Program; MILSATCOM is military satellite communications; and ISR is intelligence, surveillance, and reconnaissance. The investment costs shown in this figure comprise research, development, test, evaluation, and procurement associated with major unclassified programs; they exclude general research and development related to space technologies. FYDP assumes that procurement and development of individual satellites is spread over four years. Satellites are assumed to have an average lifetime of eight years. The projection assumes that constellations of satellites are reconstituted as necessary and that block upgrades occur as planned.

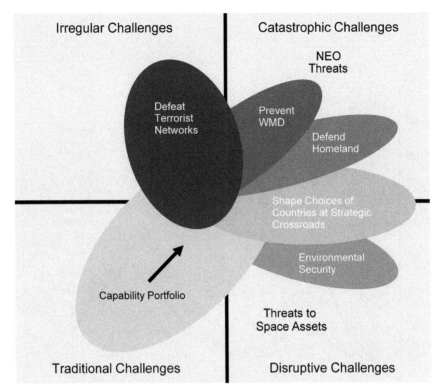

Figure 2.2 Investment in National Security Space Programmes

Source: Developed by Eligar Sadeh; adapted and updated from Quadrennial Defense Review Report, United States Department of Defense, 6 February 2006, 19, www.comw.org/qdr/qdr2006.pdf (accessed May 2009).

Notes: WMD is weapon of mass destruction; and NEO is Near Earth Objects. "Prevent WMD" implies the prevention of proliferation, development, and acquisition of chemical, biological, and nuclear weapons.

'Environmental Security' entails conflict and its prevention, and state authority or control over sovereignty as they are linked to national, regional and global environmental factors. There are several dimensions to environmental security, which include: global environmental dangers imperil national security by undermining natural support systems on which all human activity depends; environmental scarcity is inextricably linked to socioeconomic and sociopolitical instability, which engender conflicts between states; and climate change leads to sustained natural and humanitarian disasters where societal demands exceed the capacity for governments to cope, making climate change a threat multiplier for instability and conflict.

The US Air Force conducted a review of the NEO challenge in 2008; see AF/A8XC Natural Impact Hazard (Asteroid Strike) Interagency Deliberate Planning Exercise After Action Report (Directorate of Strategic Planning, Headquarters, US Air Force, December 2008). Of particular concern are potentially hazardous asteroid NEOs that are 150m in diameter and larger, approach Earth within 7.5m. km and have the potential for impacting the Earth. The issue of NEOs deals with planetary defence involving the detection and possible mitigation of potentially hazardous NEOs. This is

Figure 2.2 continued

central to the question of what role space will play in the advancement of life on Earth and in the future of humanity. Planetary defence is ultimately about providing for the security of Earth similar to avoiding nuclear war (issue of WMD) and global environmental change (which is a disruptive challenge in the figure above). Further, the political evolution of the NEO issue demonstrates problems of authority and governance. In the USA, NASA has taken the lead on this and co-operates with the US Department of Defense for detection, but no one has authority over the problem. The same situation holds true at the international level, though the UN has expressed an interest in the issue; every year the UN Committee on the Peaceful Uses of Outer Space invites Member States, international organizations and other entities to provide the committee with information on research in the field of NEOs; see www.oosa.unvienna.org/oosa/natact/neo/2006.html (accessed May 2009).

Military and other space assets are vulnerable to interference and disruption, by electronic means and kinetic anti-satellite weapons. Such vulnerability is a shared risk that all spacefaring states face. Risks can be mitigated by working with allies and other spacefaring states in the international arena through data-sharing approaches to space situational awareness, realizing collective security regimes for space assets, establishing multilateral space deterrence and satellite security doctrines, and formulating and agreeing to rules of the road or best practices on the expected peaceful behaviour in the space domain. The USA has also adopted a national approach to this issue, which allows for political and operational flexibility in meeting the challenges of space protection. At the same time, counter space capabilities that are often associated with space protection face obstacles to successful implementation. Given the priorities for space situational awareness, communications, positioning, navigation, timing and reconnaissance, the resources are not likely to exist for developing a comprehensive set of capabilities to achieve effective space control and denial. This is further exacerbated by obstacles posited by space acquisition processes that are plagued by cost and scheduling problems.

FIGURES

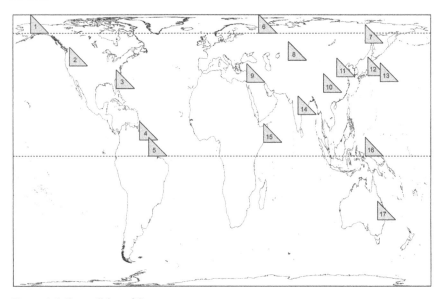

Figure 4.1 Geopolitics of Space

Source: Developed by Eligar Sadeh; adapted and updated from Haverluk, Terrence W., *Geopolitics From the Ground Up: Geography and Global Politics* (Hoboken, New Jersey: John Wiley and Sons, 2007).

Notes: Since 1957, satellites and payloads have been boosted above the atmosphere from spaceports around the world. 1 is United States Alaska Spaceport, Kodiak Island, Latitude 67.50 North (N); 2 is United States Vandenberg Air Force Base, Latitude 34.40 N; 3 is United States Cape Canaveral Air Force Station and Kennedy Space Center, Latitude 28.50 N; 4 is Kourou, French Guiana, Latitude 5.20 S; 5 is Brazil Alcantara Launch Center, Latitude 2.30 S; 6 is Russia Plesetsk Cosmodrome, Latitude 62.80 N;7 is Russia Svobodny Cosmodrome, Latitude 51.40 N; 8 is Russia Baikonur Cosmodrome (today located in Kazakhstan), Latitude 45.60 N; 9 is Israel Palmachim Air Base, Latitude 31.50 N; 10 is China Xichang Space Launch Center, Latitude 28.25 N; 11 is China Jiuquan Space Launch Center, Latitude 40.60 N; 12 is Japan Kagoshima, Latitude 31.20 N; 13 is Japan Tanegashima Island, Latitude 30.40 N; 14 is India Sriharikota Island, Latitude 13.90 N; 15 is Italy San Marco Range off the Kenya coast, Latitude 2.90 S; 16 is Sea Launch (launches from a Pacific Ocean platform near the equator in international waters); and 17 is Australia Woomera, Latitude 31.10 S. Sea Launch is a multinational corporation involving companies from the United States (Boeing), Russia (RSC Energia), Ukraine (SDO Yuzhnoye/ PO Yuzhmash), and Norway (Aker ASA).

An issue of geopolitics is the struggle for the control of key terrestrial and space 'chokepoints' that facilitate access to space. For example, locations at certain latitudes are more important than others. The best place to launch space payloads into geostationary Earth orbit is on the equator because of the benefits of the centrifugal force of the Earth. Kourou, in French Guiana, the European spaceport, provides a 17% fuel efficiency advantage over Cape Canaveral. Likewise, for the benefits of equatorial launch, the Brazilian spaceport at Alcantara is highly sought after by China, Russia and the USA. The launch direction will be east, because that is the direction of Earth's rotation; therefore, it is better to launch from a coastal location. In case of a malfunction, the debris will fall harmlessly in the ocean. More than 60 degrees N latitude is important

Figure 4.1 continued

because orbital perturbations degrade the stability of most orbits, except at this latitude. Satellites launched at this latitude will maintain an orbital inclination greater than 60 degrees (63.4 is ideal) and will maintain greater orbital stability over time, greatly reducing the expenditure of fuel, increasing satellite life expectancy and effectiveness. This is why Russia built the Plesetsk Cosmodrome in 1957, and the reason for the USA Alaska Spaceport, which serves commercial launches and the US Air Force launches for polar orbit.

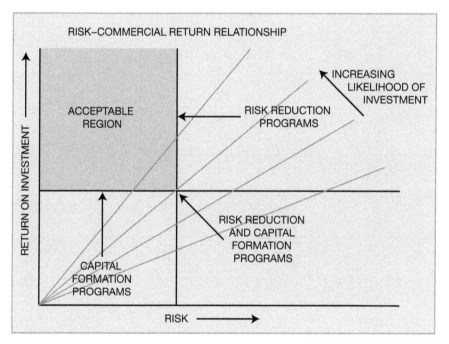

Figure 5.1 Risk to Commercial Return Relationship
Source: Developed by Eligar Sadeh.
Notes: The government, whose motivation is to act in the public interest and create public value, undertakes the risk of research and development and formulates appropriate political and legal regimes that facilitate public-private relationships. One major barrier to commercial space projects is the inability of companies to raise the private capital needed for implementation of business plans. Important factors related to this barrier are the lack of credible near-term revenue markets for many commercial space activities, and political and legal issues. In the final sum, the private sector cannot be confident that it will get an acceptable return-on-investment. This sector often looks to the government to share the technological and capital risks.

Tables

Tables

Table 2.1 Space-Enabled Reconnaissance Strike Complex

Operation and duration	System for Guidance	Munition used	Percent
Kuwaiti Theatre of Operations, 1991 Desert Storm: 37 Days 1 Mbps/5K	Unguided Laser EO	245,000 20,450	92 8
Serbia, 1999 Allied Forces: 78 Days 24.5 Mbps/5K	Unguided Laser EO GPS	16,000 7,000 700	66 31 3
Afghanistan, 2001–02 Enduring Freedom: 90 Days 68.2 Mbps/5K	Unguided Laser EO GPS	9,000 6,000 7,000	41 27 32
Iraq, 2003 Iraqi Freedom: 29 Days 51.1 Mbps/5K	Unguided Laser EO and GPS	9,251 19,948	32 68

Source: Data on precision-guided munitions and communications are derived from Central Air Forces, 'Operation Iraqi Freedom: By the Numbers', Prince Sultan Air Base, Saudi Arabia, 30 April 2003; Benjamin S. Lambeth, *The Transformation of American Airpower* (Ithaca, NY: Cornell University Press, 2000); and Eliot A. Cohen, Gulf War Air Power Survey, Summary Report (Washington, DC: Government Printing Office, 1993). For viewpoints that de-emphasize the role of technological factors in modern warfare, see: Stephen Biddle, 'Victory Misunderstood: What the Gulf War tells us about the Future of Conflict', *International Security* 21: 2 (Fall 1996): 139–79; Darrell Press, 'The Myth of Air Power in the Persian Gulf War and the Future of Warfare', *International Security* 26: 2 (Fall 2001): 5–44; and Stephen Biddle, *Military Power: Explaining Victory and Defeat in Modern Battle* (Princeton, NJ: Princeton University Press, 2004)

Notes: Mbps is megabits (10^6 bits) per second of data transfer; data transfer rates are for satellite communications speeds into theatre per 5,000 deployed troops (battalion-sized units); EO is electro-optical; and GPS is Global Positioning System.

Table 2.2 Force Enhancement Space Systems

Environmental Monitoring	Communication	PNT	ITWAA	ISR
LEO	*GEO*	*MEO*	*GEO* *MEO*	*Various Orbits*
DMSP	DSCS II	GPS	DSP	IMINT
NPOESS	DSCS III	GPS II	GPS	SIGINT
(cancelled)	UFO	GPS IIR	SBIRS	*FIA (cancelled)*
	Milstar	GPS IIR-M	STSS	IOSA
	GBS	GPS IIF	*ISP III*	*Commercial Space Radar (cancelled)*
	Iridium	*GPS III*		
	WGS			
	Commercial			
	AEHF			
	MUOS			
	PMSCS			
	TSAT (cancelled)			

Source: Compiled by Peter L. Hays. Future systems are in italics.

Notes: LEO is low Earth orbit; DMSP is Defense Meteorological Satellite Program; NPOESS is National Polar-Orbiting Operational Environmental Satellite System; GEO is Geostationary Earth orbit; DSCS is Defense Satellite Communications System; UFO is Ultra-High Frequency Follow-On satellite system; Milstar was originally an acronym for Military Strategic and Tactical Relay satellite; GBS is Global Broadcast Service; WGS is Wideband Global System; AEHF is Advanced Extremely High Frequency system; MUOS is Mobile User Objective System; PMSCS is Polar Military Satellite Communications System; TSAT is Transformational Satellite Communications System (the plans for TSAT were cancelled in 2009); PNT is Position-Navigation-Timing; MEO is medium Earth orbit; GPS is Global Positioning System; ITWAA is Integrated Tactical Warning and Attack Assessment; DSP is Defense Support Program; SBIRS is Space-Based Infrared System; STSS is Space Tracking and Surveillance System; ISR is Intelligence, Surveillance and Reconnaissance; IMINT is Imaging Intelligence satellites; SIGINT Signal Intelligence satellites; FIA is Future Imagery Architecture; and IOSA is Integrated Overhead SIGINT Architecture.

Satellites in LEO can operate from less than 160km to several hundred kilometres altitude and complete each orbit in approximately 90 minutes. Polar LEO is ideal for many reconnaissance satellites and weather applications because it overlies all parts of the Earth several times each day as the Earth rotates, and also it can be aligned in Sun Synchronous Orbits that arrive overhead the same location at the same time each day. PNT Satellites (such as the US GPS) in MEO are located at approximately 20,000km altitude and complete an orbit every 12 hours. GEO is located approximately 36,000km above the equator, a location where the satellite orbital velocity matches Earth's rate of rotation and the satellite appears to remain motionless above the same spot. This is a very valuable attribute for communications and SIGINT satellites.

Table 2.3 Attributes of Military Space Doctrines

	Primary Value and Functions of Military Space Forces	*Space System Characteristics and Employment Strategies*	*Conflict Missions of Space Forces*	*Appropriate Military Organization for Operations and Advocacy*
Sanctuary	Enhance Strategic Stability Facilitate Arms Control	Limited Numbers Fragile Systems Vulnerable Orbits Optimized for NTMV	Limited	NRO
Survivability	Enhance Strategic Stability Facilitate Arms Control Force Enhancement	Terrestrial Backups Distributed Architectures Autonomous Control	Force Enhancement Degrade Gracefully	Major Command Unified Command
Control	Control Space Assured Force Enhancement	Hardening On-Orbit Spares Crosslinks Manoeuvre Less Vulnerable Orbits Stealth	Control Space Assured Force Enhancement Surveillance Offensive/ Defensive Counterspace	Unified Command Space Force
High Ground	Control Space Assured Force Enhancement Decisive Impact on Terrestrial Conflict BMD	Attack Warning Sensors Deception- Disruption- Denial- Degradation- Destruction Reconstitution Capability Active Defense Convoy	Control Space Assured Force Enhancement Surveillance Offensive/ Defensive Counterspace Force Application: Space-to-Space Space-to-Earth BMD	Space Force

Source: Compiled by Peter L. Hays
Notes: NTMV is National Technical Means of Verification; NRO is National Reconnaissance Office; and BMD is Ballistic Missile Defense.

Table 4.1 Moderate State Civil Satellites

Launch	State	Satellite	Type	Launch State	Launch Vehicle	Term
2006	Korea, Repub.	Mugunghwa 5	COM	Ukraine	Zenit-3SL	
2006	Korea, Repub.	Arirang-2	IMG	Russia	Rokot	
2006	Kazakhstan	Kazsat	COM	Russia	Proton-K/DM-2M	
2006	Israel	EROS B	IMG	Russia	Start-1	
2006	Taiwan	COSMIC 1	SCI	USA	Minotaur	
2006	Taiwan	COSMIC 2	SCI	USA	Minotaur	
2006	Taiwan	COSMIC 3	SCI	USA	Minotaur	
2006	Taiwan	COSMIC 4	SCI	USA	Minotaur	
2006	Taiwan	COSMIC 5	SCI	USA	Minotaur	
2006	Taiwan	COSMIC 6	SCI	USA	Minotaur	
2005	Iran	Sinah-1	IMG	Russia	Kosmos 11K65M	
2004	Ukraine	Sich-1M	IMG	Russia	Tsiklon-3	2006
2004	Ukraine	MK-1TS Mikron	IMG	Russia	Tsiklon-3	2005
2004	Spain	Nanosat 1	TECH	France	Ariane 5G+	
2004	Saudi Arabia	SaudiComsat-1	COM	Russia	Dnepr	
2004	Saudi Arabia	SaudiComsat-2	COM	Russia	Dnepr	
2004	Saudi Arabia	SaudiSat-2	TECH	Russia	Dnepr	
2004	Taiwan	Formosat-2	SCI	USA	Taurus 3210	
2003	Israel	Amos 2	COM	France	Soyuz-FG	
2003	Nigeria	NigeriaSat-1	IMG	Russia	Kosmos 11K65M	
2003	Korea, Repub.	STSat-1	SCI	Russia	Kosmos 11K65M	2004
2003	Turkey	BILSAT-1	IMG	Russia	Kosmos 11K65M	
2003	Canada	SCISAT-1	SCI	USA	Pegasus XL	
2003	Canada	MOST	SCI	Russia	Rokot	
2003	Czech Republic	Mimosa	SCI	Russia	Rokot	
2002	Saudi Arabia	SaudiOSCAR	COM	Russia	Dnepr	
2002	Australia	Fedsat	SCI	Japan	H-IIA 202	
2002	Algeria	AlSat-1	IMG	Russia	Kosmos 11K65M	
2001	Morocco	Maroc-Tubsat	IMG	Russia	Zenit-2	
2001	Pakistan	Badr B	IMG	Russia	Zenit-2	
2001	Sweden	Odin	AST	Russia	Start-1	
2001	Monaco	Turksat 2A	COM	France	Ariane 44P	
2000	Malaysia	Tiungsat-1	IMG	Russia	Dnepr	
2000	Brazil	Brasilsat B4	COM	France	Ariane 44LP	
2000	Egypt	Nilesat 102	COM	France	Ariane 44LP	
2000	Spain	Hispasat 1C	COM	USA	Atlas IIAS	
1999	Korea, Repub.	Arirang-1	IMG	USA	Taurus 2110	
1999	Brazil	SACI-1	SCI	China	Chang Zheng 4B	1999
1999	Korea, Repub.	Mugunghwa 3	COM	France	Ariane 42P	
1999	Korea, Repub.	Uribyol-3	IMG	India	PSLV	2003

(*Continued on next page*)

Table 4.1 (continued)

Launch	State	Satellite	Type	Launch State	Launch Vehicle	Term
1999	Denmark	Orsted	SCI	USA	Delta 7920-10	
1999	Taiwan	Formosat-1	SCI	USA	Athena-1	2004
1998	Sweden	Astrid-2	SCI	Russia	Kosmos 11K65M	1999
1998	Argentina	SAC-A	IMG	USA	Space Shuttle	1999
1998	Brazil	SCD-2	IMG	USA	Pegasus H	
1998	Australia	WESTPAC	GEO	Russia	Zenit-2	1998
1998	Brazil	Brasilsat B3	COM	France	Ariane 44LP	
1997	Spain	Minisat-01	SCI	USA	Pegasus XL	
1996	Argentina	SAC-B	SCI	USA	Pegasus XL	1996
1996	Argentina	Victor	IMG	Russia	Molniya 8K78M	1998
1996	Czech Republic	Magion-5	SCI	Russia	Molniya 8K78M	2002
1996	Turkey	Turksat 1C	COM	France	Ariane 44L	
1996	Korea, Repub.	Mugunghwa 2	COM	USA	Delta 7925	2000
1995	Canada	Radarsat	COM	USA	Delta 7920-10	
1995	Ukraine	Sich-1	IMG	Russia	Tsiklon-3	2001
1995	Korea, Repub.	Mugunghwa 1	COM	USA	Delta 7925	2006
1995	Czech Republic	Magion-4	SCI	Russia	Molniya 8K78M	1997
1995	Israel	'Ofeq-3	IMG	Israel	Shaviyt 1	1999
1995	Brazil	Brasilsat B2	COM	France	Ariane 44LP	
1995	Sweden	Astrid	SCI	Russia	Kosmos 11K65M	1995
1994	Mexico	Solidaridad 2 Satmex 4	COM	France	Ariane 44L	
1994	Mexico	Brasilsat B1	COM	France	Ariane 44LP	
1994	Turkey	Turksat 1B	COM	France	Ariane 44LP	2005
1993	Korea, Repub.	Uribyol 2 KO-25	COM	France	Ariane 40	2002
1993	Brazil	SCD-1	COM	USA	Pegasus	
1992	Canada	CTA	TECH	USA	Space Shuttle	1992
1992	Sweden	Freja	SCI	China	Chang Zheng 2C	1996
1992	Korea, Repub.	Kitsat KO-23	COM	France	Ariane 42P	2001
1991	Czech Republic	Magion-3	SCI	Soviet Union	Tsiklon-3	1992
1990	Pakistan	BADR	COM	China	Chang Zheng 2E	1990
1990	Israel	'Ofeq-2	TECH	Israel	Shaviyt	1990
1989	Czech Republic	Magion-2	SCI	Soviet Union	Tsiklon-3	1990
1988	Israel	'Ofeq-1	TECH	Israel	Shaviyt	1988
1986	Brazil	Brasilsat 2	COM	France	Ariane 3	2004
1986	Sweden	Viking	SCI	France	Ariane 1	1987
1985	Mexico	Morelos 2 (Satmex 2)	COM	USA	Space Shuttle	2004

(*Continued on next page*)

Table 4.1 (continued)

Launch	State	Satellite	Type	Launch State	Launch Vehicle	Term
1985	Mexico	Morelos 1 (Satmex 1)	COM	USA	Space Shuttle	1994
1985	Brazil	Brasilsat 1	COM	France	Ariane 3	2002
1978	Czech Republic	Magion	SCI	Soviet Union	Kosmos 11K65M	1981
1976	Canada	Hermes	COM	USA	Delta 2914	1979
1974	Spain	Intasat	SCI	USA	Delta 2310	1976
1974	Netherlands	ANS	SCI	USA	Scout D-1	1977
1971	Canada	Isis 2	SCI	USA	Thor Delta E1	1984
1969	Canada	Isis 1	SCI	USA	Thor Delta E1	1984
1967	Australia	WRESAT	SCI	USA	SPARTA	1967
1965	Canada	Alouette 2	SCI	USA	Thor SLV-2 Agena	1975
1962	Canada	Alouette 1	SCI	USA	Thor Agena B	1971

Note: Table excludes: USA, USSR/Russia, China, ESA/ESRO/EU, Japan, India, United Kingdom, France, Germany, Italy, INMARSAT and INTELSAT.
Source: "World Civil Satellites 1957–2006," *Space Security Index Project, Other Resources*, http://spacesecurity.org/publications.htm.

Table 4.2 Moderate State Military Satellites

Satellite	Launch Vehicle	Launch State	Function	Orbit	Launch Year
Chile					
Fasat-Bravo	Zenit-2	Russia	Imaging	LEO	1998
Germany					
SAR-Lupe 1	Kosmos-11K65M	Russia	Imaging Radar	LEO	2006
Israel					
Ofeq-5	Shavit 1	Israel	Imaging	LEO	2002
Italy					
Sicral	Ariane 44L	France	Communications	GEO	2001
Korea, Repub.					
Koreasat 5	Zenit-3SL	France	Communications	GEO	2006
Spain					
XTAR-EUR	Ariane 5ECA	France	Communications	HEO	2005
Spainsat	Ariane 5ECA	France	Communications	GEO	2006

Note: Table excludes the United States, Russia, China, France, United Kingdom, and Japan.
Source: "Active Military Satellites 2006," *Space Security Index Project, Other Resources*, http://spacesecurity.org/publications.htm.

TABLES

Table 4.3 Moderate State Space Programmes

State	Space Programme	Own Satellite	Satellite Services	Space Use	Partnership	Space Launch
Algeria	O	Y	I/S	I/S	Y	
Argentina	O	Y	I/T/S	I/S	Y	S
Australia	O	Y	I/T/M	S	Y	S
Austria	O		I/S/T/O	I/S/T/S	Y	
Azerbaijan	O		I	S	Y	
Brazil	O	Y	I/T/M/S/O	T/S	Y	S
Canada	O	Y	I/T/M/S	T/S	Y	S
Chile	P	Y	I/M/T	S	Y	
Cuba	O		I/T/O/S	S	Y	
Czech Repub.	O	Y	I/T	S		
Egypt	P	Y	I/T/M		Y	
Finland	O	Y	I/T/S	S	Y	
Germany	O	Y	I/T/M	T/I/S	Y	S
Indonesia	O	Y	T/I	S	Y	
Iran	O	Y	I/M	S	Y	S
Ireland				S		
Israel	O	Y	I/T/M		Y	
Italy	O	Y	T/M	S	Y	S
Kazakhstan	O	Y	T		Y	
Korea, Democratic People's Repub.	O	Y	I/M		Y	S
Korea, Repub.	O	Y	M/T/I/S	T/S	Y	S
Lebanon			T/I	T/I	Y	
Luxembourg	Y	Y	I/S	S	Y	
Malaysia	Y	Y	T/I	T/I	Y	
Mexico		Y	T	T	Y	S
Mongolia	O		I	Y/S	Y	
Morocco	O		T/I	T/I	Y	
Netherlands	O	Y	I/T/M/S	S		
Nigeria	O	Y	T/I	T/I/S	Y	
Norway	O	Y	I/T/S	T/I/S	Y	
Pakistan	O	Y	T/S/M	I/S	Y	S
Philippines			N	T	Y	
Peru	O		T/I		Y	
Poland	O		I	S	Y	
Portugal	P	Y	I		Y	
Saudi Arabia	P	Y	T/S/I		Y	S
Singapore			N	T		
Slovakia	Y		I	N/S		
Slovenia			N	N/S	Y	
South Africa	P		N	I/T/S	Y	S
Spain	O	Y	M/T/I	I	Y	
Sweden	O	Y	T	T/I/S	Y	S
Syria	O	Y	I		Y	S

(*Continued on next page*)

Table 4.3 (continued)

State	Space Programme	Own Satellite	Satellite Services	Space Use	Partnership	Space Launch
Taiwan	Y	Y	M/T	T	Y	
Thailand	O	Y	T/M/I	T/S		
Tunisia	Y		T/I		Y	
Turkey	Y	Y	T	T/S	Y	
UAE	P		T	S	Y	
Ukraine	O	Y	I/M/T	S	Y	O
Uruguay	O		T		Y	
Vietnam	O		T/I		Y	

Notes: Space Programs: P is policies, and O is organization; Satellite Use: I is imaging, T is telecommunications, S is science/research, M is military, and O is Other; Space Launch: S is sub-orbital and O is orbital. European countries' space program partnerships exclude activities through the European Union and European Space Agency.

Sources: *Space Security 2008* (Spacesecurity.org, 2008; ISBN 978-1-895722-70-3); *Country Profiles, Secure World Foundation*, http://www.secureworldfoundation.org/index.php?id=58&pid=3&page=Country_profiles; *Country Profiles, Nuclear Threat Initiative*, http://www.nti.org/e_research/profiles; *Current and Future Space Security, Center for Nonproliferation Studies*, http://cns.miis.edu/research/space/index.htm; Space, GlobalSecurity.org, http://www.globalsecurity.org/space/index.html.
Country specific sources:
Finland: http://www.avaruus.info/en/finland
Germany: http://www.dlr.de/en/desktopdefault.aspx/tabid-38
Mexico: http://www.astronautix.com/country/mexico.htm
Mongolia: http://www.aprsaf.org/data/p_news/mong_mainpro.pdf
Netherlands: http://www.nivr.nl/index.php?ac=lang&lang=en
Norway: http://www.spacecentre.no/English/Satellites
Saudi Arabia: http://www.astronautix.com/country/saurabia.htm
South Africa: http://www.info.gov.za/speeches/2009/09030909451002.htm
UAE: http://www.spaceref.com/news/viewpr.html?pid=24520
Ukraine: http://www.nkau.gov.ua/NSAU/nkau.nsf/indexE%21openform

TABLES

Table 6.1 Functions of Selected Bodies in the United Nations

UN	OOSA	ITU	WMO	UNESCO	ICAO
Members					
192 states	69 states	191 states 700 sector members & associates	188 states	193 states 6 associate states	188 states
Charter Date					
Charter of the United Nations 1945	General Assembly Resolution 1472 (XIV) 1959	International Telegraph Convention 1865 Specialized Agency UN 1947	WMO Convention 1950	UNESCO Constitution 1945	International Civil Aviation Convention 1944
Function					
Preserves global peace through international co-operation and collective security; provide the means to help resolve international conflicts and formulate policies on world issues	Reviews the scope of international co-operation in peaceful uses of space; devises programmes in this field under United Nations auspices; encourages continued research and dissemination of information; studies legal problems	Global information and communication focal point for governments and the private sector; helps global communications through three core sectors: radio-communication, standardization and development	Provides world leadership in expertise and international co-operation in weather, climate, hydrology and water resources, and related environmental issues	Contributes to peace and security by promoting collaboration through education, science and culture to further justice, the rule of law, human rights and fundamental freedoms	Develops international civil aviation in a safe and orderly manner; establishes international air transport services on the basis of equality of opportunity and sound economic operations
Space Activities					
Oversees the space activities of the UN General Assembly (OOSA/COPUOS) and Specialized Agencies (WMO, ITU, ICAO, UNESCO), among other UN organs that deal with space issues	Five Space Treaties; many legal principles governing space activities; registry of launchings; UNISPACE Conferences; GNSS; UN-SPIDER	Co-ordination and recording procedures for space systems and Earth stations; capture, processing and publication of data; examination of frequency assignment notices; management of the procedures for space assignment	Co-ordinate environmental satellite matters within WMO; develop Global Observing System; promote satellite data use for weather, water, climate and other applications	Natural resource management; environmental planning; environmental monitoring; distance learning; natural disaster reduction and reaction; ethical principles for space policy	Some aspects of aviation overlap space functions; future involvement in international space issues

Source: Compiled by Henry R. Hertzfeld

Notes: EU is European Union; ESA is European Space Agency; and APSCO is Asia-Pacific Space Cooperation Organization.

Table 6.2 United Nations Treaties and Resolutions of Space Law

Agreement	Year	Ratifications	Signatures
Treaty on Principles Governing the Activities of States in the Exploration and Use of Outer Space, including the Moon and other Celestial Bodies	1967	98	27
Agreement on the Rescue of Astronauts, the Return of Astronauts and the Return of Objects Launched into Outer Space	1968	91	25
Convention on International Liability for Damage Caused by Space Objects	1972	87	25
Convention on Registration of Objects Launched into Outer Space	1975	51	4
Agreement Governing the Activities of States on the Moon and Other Celestial Bodies	1979	13	4

Declaration or Legal Principle	Year	GA Res
Declaration of Legal Principles Governing the Activities of States in the Exploration and Uses of Outer Space	1963	1962 (XVIII)
Principles Governing the Use by States of Artificial Earth Satellites for International Direct Television Broadcasting	1982	37/92
Principles Relating to Remote Sensing of the Earth from Outer Space	1986	41/65
Principles Relevant to the Use of Nuclear Power Sources in Outer Space	1992	47/68
Declaration on International Cooperation in the Exploration and Use of Outer Space for the Benefit and in the Interest of All States, Taking into Particular Account the Needs of Developing Countries	1996	51/22
Application of the concept of the 'launching State'	2005	59/115
Recommendations on Enhancing the Practice of States and International Intergovernmental Organizations in Registering Space Objects	2008	62/101

Source: Compiled by Henry R. Hertzfeld
Notes: IADC is Inter-Agency Space Debris Coordination Committee; UNIDROIT is International Institute for the Unification of Private Law; OECD is Organisation for Economic Co-operation and Development; and CEOS is Committee on Earth Observation Satellites.

Table 6.3 Regional Space Organizations

	EU	*ESA*	*APSCO*
Members	27 member states	17 member states	9 member states
Charter Date	Treaty on European Union Maastricht 1993	Convention of the European Space Agency 1975	Convention of the Asia-Pacific Space Cooperation Organization 2005
Function	Peace, prosperity and freedom for Europe through economic and political partnerships	Promotes co-operation among European states in space research and technology and their space applications, with a view to their being used for scientific purposes and for operational space applications systems	Promotes the peaceful use of outer space in the Asia-Pacific region through attention to space science, technology, education, training and co-operative research; multilateral space co-operation to enable the Asia-Pacific countries to benefit from each others' strengths and help address the technological and financial challenges to their space causes

(*Continued on next page*)

Tables

Table 6.3 (continued)

	EU	ESA	APSCO
Contributions to Space Activities	Develops an effective space policy that will allow the EU to take global leadership in selected strategic policy areas	Elaborates and implements a long-term European space policy; elaborates and implements activities and programmes in the space field; co-ordinates the European space programme and national programmes, and integrates the latter progressively and as completely as possible into the European space programme, in particular with regard to the development of application satellites; elaborates and implements the industrial policy appropriate to its programme and recommends a coherent industrial policy to the Member States	Promotes and strengthens the development of collaborative space programmes among its Member States by establishing the basis for co-operation in peaceful applications of space science and technology; takes effective actions to assist the Member States in such areas as space technological research and development, applications and training by elaborating and implementing space development policies; promotes co-operation, joint development, and shares achievements among the Member States in space technology and its applications, as well as in space science research by tapping the co-operative potential of the region; enhances co-operation among relevant enterprises and institutions of the Member States and promotes the industrialization of space technology and its applications; contributes to the peaceful uses of space by international co-operative activities in space technology and its applications

335

Source: Compiled by Henry R. Hertzfeld.
Notes: IADC is Inter-Agency Space Debris Coordination Committee; UNIDROIT is International Institute for the Unification of Private Law; OECD is Organization for Economic Cooperation and Development; and CEOS is Committee on Earth Observation Satellites.

Table 6.4 Global Space Organizations

	IADC	UNIDROIT	OECD	CEOS	Space & Major Disasters
Members	11 members	61 members	30 members	26 members (space agencies) 20 agencies (associated national and international organizations)	10 members
Charter Date	Terms of Reference for the IADC 1993	UNIDROIT Statute 1940 Convention on International Interests in Mobile Equipment 2001	Convention on the Organisation for Economic Co-operation and Development 1961	CEOS Terms of Reference 1984	International Charter, 'Space and Major Disasters' 2000
Function	Co-ordinates world-wide activities related to the issues of human-made and natural debris in space	Studies needs and methods for modernizing, harmonizing and co-ordinating private and commercial law among states	Brings together the governments of countries committed to democracy and the market economy to support sustainable economic growth; assists other countries' economic development; contributes to growth in world trade	Co-ordinates mechanism for international civil space missions designed to observe and study the Earth	Aims to provide a unified system of space data acquisition and delivery to those affected by natural or human-made disasters, through authorized users

(Continued on next page)

Table 6.4 (continued)

	IADC	UNIDROIT	OECD	CEOS	Space & Major Disasters
Space Activities	Exchanges information on space debris research activities between member space agencies; facilitates opportunities for co-operation in space debris research; reviews the progress of ongoing co-operative activities; identifies debris mitigation options	Protocol to Convention to facilitate commercial transactions in communications satellites to help develop space commerce; space agreement still being debated	Development of space economy; assists space agencies and governments by providing evidence-based analysis on the space infrastructure (evaluating data and socioeconomic indicators), so that the potential of space for the larger economy is more fully realized	Optimizes the benefits of Earth observations through co-operation of its participants in mission planning; develops compatible data products, formats, services, applications and policies; serves as a focal point for international co-ordination of Earth observation activities; exchanges policy and technical information to encourage compatibility of observation and data exchange systems	Supplies data and critical information for the anticipation and management of potential crises to states that are exposed to an imminent risk, or are victims, of natural or technological disasters; participates in the organization of emergency assistance or subsequent operations by means of these data and of the information and services from the use of space facilities

Source: Compiled by Henry R. Hertzfeld.
Notes: EU is European Union; ESA is European Space Agency; and APSCO is Asia-Pacific Space Cooperation Organization.

Tables

Table 8.1 Government Sector Interventions in Space

		International Global	International Regional	National	Sub-National
Function	**Regulatory Communication**	ITU Intelsat Inmarsat ITSO	EU Eutelsat Arabsat	FCC Telesat Canada	
	Science	ISS	ESA	NASA CNSA	State, regional and private science centres
	Social Scientific	ITSO meteorological	Regional meteorological	NOAA Rosaviakosmos (Russian Federal Space Agency) ISRO	University space research institutes Commercial users
	Remote Sensing	GEOSS	GMES	Landsat Spot Image Envisat GeoEye DigitalGlobe Other commercial providers from India, Israel and Russia	Academic researchers Commercial users
	Transportation	Future – international flight control	ESA Arianespace EU	National spaceports	State and regional spaceports
	Quasi-Public		International Launch Services Sea Launch Eurokot	RSC Energia Khrunichev State Space Centre Lockheed Martin EADS Great Wall Industry Khrunichev State Research and Production CenterKB Yuzhnoye PO Yuzhmash	Commercial users Government customers Academic users
	Navigation		EU Galileo	US GPS Russian system Chinese system Japanese system Indian system	Academic Commercial users

Source: Compiled by Roger B. Handberg.
Notes: ITU is International Telecommunications Union; Intelsat is a private company today; Inmarsat is a private company today (originally it was the International Maritime Satellite Organization); ITSO is International Telecommunications Satellite Organization; ISS is International Space Station; GEOSS is Global Earth Observation System of Systems; EU is European Union; Eutelsat was originally set up in 1977 as an inter-

TABLES

governmental organization to develop and operate a satellite-based telecommunications infrastructure for Europe (today, it is a private company); ESA is European Space Agency; GMES is Global Monitoring for Environment and Security (the European Commission, acting on behalf of the European Union, is responsible for the overall initiative, setting requirements, and managing the program); FCC is Federal Communications Commission of the United States; NASA is National Aeronautics and Space Administration; CNSA is China National Space Administration; NOAA is National Oceanic and Atmospheric Administration; ISRO is Indian Space Research Organization; and EADS is European Aeronautic Defence and Space Company.

Documentation

DOCUMENTATION 6.1. INTERNATIONAL COOPERATION IN THE PEACEFUL USES OF OUTER SPACE

Resolution adopted by the United Nations General Assembly: International cooperation in the peaceful uses of outer space (selected text).

The General Assembly,

Recognizing the common interest of mankind as a whole in furthering the peaceful use of outer space,

Believing that the exploration and use of outer space should be only for the betterment of mankind and to the benefit of States irrespective of the stage of their economic or scientific development,

Desiring to avoid the extension of present national rivalries into this new field,

Recognizing the great importance of international cooperation in the exploration of outer space being undertaken by the international scientific community,

Believing also that the United Nations should promote international cooperation in the peaceful uses of outer space,

1. Establishes a Committee on the Peaceful Uses of Outer Space, consisting of Albania, Argentina, Australia, Austria, Belgium, Brazil, Bulgaria, Canada, Czechoslovakia, France, Hungary, India, Iran, Italy, Japan, Lebanon, Mexico, Poland, Romania, Sweden, the Union of Soviet Socialist Republics, the United Arab Republic, the United Kingdom of Great Britain and Northern Ireland and the United States of America, whose members will serve for the years 1960 and 1961, and requests the Committee:
   a) To review, as appropriate, the area of international co-operation, and to study practical and feasible means for giving effect to programs in the peaceful uses of outer space which could appropriately be undertaken under United Nations auspices, including, inter alia:
      i) Assistance for the continuation on a permanent basis of the research on outer space carried on within the framework of the International Geophysical Year;
      ii) Organization of the mutual exchange and dissemination of information on outer space research;

iii) Encouragement of national research programs for the study of outer space, and the rendering of all possible assistance and help towards their realization;
(b) To study the nature of legal problems, which may arise from the exploration of outer space;
2. Requests the Committee to submit reports on it activities to the subsequent sessions of the General Assembly.

Source: www.unoosa.org/oosa/en/SpaceLaw/gares/html/gares_14_1472.html (accessed May 2009)
Note: Adopted by the United Nations General Assembly on 12 December 1959.

DOCUMENTATION 6.2. EUROPEAN SPACE AGENCY

Convention of establishment of a European Space Agency (selected text).

The European Space Agency's purpose shall be to provide for and to promote, for exclusively peaceful purposes, cooperation among European States in space research and technology and their space applications, with a view to their being used for scientific purposes and for operational space applications systems:

By elaborating and implementing a long-term European space policy, by recommending space objectives to the Member States, and by concerting the policies of the Member States with respect to other national and international organisations and institutions;

By elaborating and implementing activities and programmes in the space field;

By coordinating the European space programme and national programmes, and by integrating the latter progressively and as completely as possible into the European space programme, in particular as regards the development of applications satellites;

By elaborating and implementing the industrial policy appropriate to its programme and by recommending a coherent industrial policy to the Member States.

Source: www.esa.int/convention (accessed May 2009)
Notes: The European Space Agency (ESA) functioned de facto from 31 May 1975. The ESA Convention entered into force on 30 October 1980. The ESA has 18 Member States: Austria, Belgium, Czech Republic, Denmark, Finland, France, Germany, Greece, Ireland, Italy, Luxembourg, The Netherlands, Norway, Portugal, Spain, Sweden, Switzerland and the United Kingdom. Canada takes part in ESA projects under a co-operation agreement. Hungary, Romania and Poland are participating in the Plan for European Cooperating States, while other countries are in negotiations to join this Plan. Of note is that not all member countries of the European Union (EU) are members of the ESA, and not all ESA Member States are members of the EU. The ESA is an entirely independent organization, although it

maintains close ties with the EU through a Framework Agreement. The two organizations share a joint European Strategy for Space and have together developed European Space Policy.

DOCUMENTATION 6.3. EUROPEAN SPACE POLICY

European Space Policy: Strategic Mission (selected text).

The development of a truly European Space Policy is a strategic choice for Europe, if it does not want to become irrelevant. Space systems are strategic assets demonstrating independence and the readiness to assume global responsibilities. Initially developed as defence or scientific projects, they now also provide commercial infrastructures on which important sectors of the economy depend and which are relevant in the daily life of citizens. However, the space sector is confronted with high technology and financial risks and requires strategic investment decisions.

Europe needs an effective space policy to enable it to exert global leadership in selected policy areas in accordance with European interests and values. To fulfill such roles, the European Union increasingly relies on autonomous decision making, based on space-based information and communication systems. Independent access to space capabilities is therefore a strategic asset for Europe.

To respond to the challenges described above, the strategic mission of a European space policy will be based on the peaceful exploitation of Outer Space by all states and will seek:

To develop and exploit space applications serving Europe's public policy objectives and the needs of European enterprises and citizens, including in the field of environment, development, and global climate change;

To meet Europe's security and defence needs as regards space;

To ensure a strong and competitive space industry, which fosters innovation, growth, and the development and delivery of sustainable, high quality, and cost-effective services;

To contribute to the knowledge-based society by investing strongly in space-based science, and playing a significant role in the international exploration endeavour;

To secure unrestricted access to new and critical technologies, systems and capabilities in order to ensure independent European space applications.

To achieve this strategic mission will require the European Union, European Space Agency, and their Member States to improve the efficiency and effectiveness of their space activities by taking significant new steps in:

Establishing a European Space Programme and the coordination of national and European level space activities, with a user-led focus;

Increasing synergy between defence and civil space programmes and technologies, having regard to institutional competencies; and

Developing a joint international relations strategy in space.

Source: Commission of the European Communities, Communication from the Commission to the Council and the European Parliament: European Space Policy (Brussels, Belgium: 2007)

DOCUMENTATION 6.4. ASIA-PACIFIC SPACE COOPERATION ORGANIZATION

Convention on Asia-Pacific Space Cooperation Organization (selected text).

To promote and strengthen the development of collaborative space programs among its Member States by establishing the basis for cooperation in peaceful applications of space science and technology;

To take effective actions to assist the Member States in such areas as space technological research and development, applications, and training by elaborating and implementing space development policies;

To promote cooperation, joint development, and to share achievements among the Member States in space technology, and its applications as well as in space science research by tapping the cooperative potential of the region;

To enhance cooperation among relevant enterprises and institutions of the Member States and to promote the industrialization of space technology and its applications;

To contribute to the peaceful uses of outer space in the international cooperative activities in space technology and its applications.

Source: Asia-Pacific Space Cooperation Organization, Convention of the Asia-Pacific Space Cooperation Organization (APSCO) (Beijing, China: 2005)

DOCUMENTATION 6.5. INTERNATIONAL TELECOMMUNICATION UNION

Preamble of the International Telecommunication Union Constitution (selected text).

While fully recognizing the sovereign right of each State to regulate its telecommunication and having regard to the growing importance of telecommunication for the preservation of peace and the economic and social development of all States, the States Parties to this Constitution, as the basic instrument of the International Telecommunication Union, and to the Convention of the International Telecommunication Union [...] which complements it, with the object of facilitating peaceful relations, international cooperation among peoples and economic and social development by means of efficient telecommunication services [...]

Source: www.itu.int/net/about/basic-texts/constitution/preamble.aspx (accessed May 2009)
Notes: The International Telecommunication Union is based in Geneva, Switzerland, and its membership includes 191 Member States and more than 700 Sector Members and Associates.

Documentation

DOCUMENTATION 6.6. WORLD METEOROLOGICAL ORGANIZATION

World Meteorological Organization Space Programme (selected text).

The WMO Programme coordinates environmental satellite matters and activities throughout all WMO Programmes and provides guidance on the potential of remote-sensing techniques in meteorology, hydrology and related disciplines and applications.

It aims at continuously improving the provision of data, products and services from operational and R&D satellites contributing to the Global Observing System (GOS), as well as facilitating and promoting the wider availability and meaningful use of these data, products and services around the globe.

Space Programme activities are developed around the following four cornerstones:

- Satellite observation requirements from WMO and co-sponsored programmes;
- Developint the space-based component of the GOS;
- Enhancing the availability of satellite data, products and services;
- Enhancing users' capability to take advantage of satellite applications.

Source: www.wmo.int/pages/prog/sat/Goalsandobjectives.html (accessed May 2009)

DOCUMENTATION 6.7. COMMITTEE ON EARTH OBSERVATION SATELLITES

Committee on Earth Observation Satellites Terms of Reference (selected text).

Remote sensing from space has evolved from an early period of limited satellite programs to a point where distinctions among existing missions result from the technology employed, rather than from the disciplines served in systems operations. In the future, a number of international and national spaceborne Earth observation systems will operate simultaneously and support both interdisciplinary and international activities.

The organization of international cooperation in spaceborne Earth observation systems also is evolving, from mission-specific reviews to the interdisciplinary coordination of multimission programs. Beginning with the first Multilateral Meeting on Remote Sensing held in Ottawa on May 8–9, 1980, which was attended by agency representatives from Canada, the European Space Agency, France, India, Japan, and the United States, current and potential operators of Earth observation systems have met several times to discuss the means by which mutually beneficial cooperation and coordination could be achieved in both the near and longer term [...]

This framework of initial discussion and cooperation has enhanced the utility of spaceborne Earth observation data to users worldwide, has encouraged the coordination of program plans among spaceborne Earth observation system operators, and has fostered international receptivity to and acceptance of spaceborne Earth observation system activities and applications.

Consequently, the assembled representatives of international and national spaceborne Earth observation systems affirmed the following:

AWARE of the overlap of spaceborne Earth observation mission objectives and of the interdisciplinary applications of remotely sensed data;

RECOGNIZING the advantages of ongoing communication and cooperation among spaceborne Earth observation system operators; and

DESIRING to promote the international growth and potential benefits of spaceborne observation of the Earth, Members have affirmed the value of the activities described above and have agreed to coordinate informally their current and planned systems for Earth observation from space through the organization of a Committee on Earth Observation Satellites (CEOS).

Cooperation in the development and management of remote sensing and associated data management programs can be of benefit to operators of spaceborne Earth observation systems, and to users of Earth observation data. Redundancy among systems and the utility of data can be optimized through the appropriate coordination of complementary and compatible space and ground segments, data management practices and products, and Earth observation systems research and development [...]

CEOS has three primary objectives:

- To optimize the benefits of spaceborne Earth observations through cooperation of its Members in mission planning, and in the development of compatible data products, formats, services, applications and policies;
- To aid both its Members and the international user community by ... serving as the focal point for international coordination of space-related Earth observations activities, including those related to global change;
- To exchange policy and technical information to encourage complementarity and compatibility among spaceborne Earth observations systems currently in service or development, and the data received from them; issues of common interest across the spectrum of Earth observations satellite missions will be addressed.

Source: www.ceos.org/images/PDFs/CEOS%20Terms%20of%20Reference.pdf (accessed May 2009)

Notes: Terms of Reference were adopted on 25 September 1984 and amended on 18 November 1993.

DOCUMENTATION 6.8. GLOBAL EARTH OBSERVATION SYSTEM OF SYSTEMS

The Global Earth Observation System of Systems (GEOSS) 10-Year Implementation Plan (selected text).

Understanding the Earth system—its weather, climate, oceans, atmosphere, water, land, geodynamics, natural resources, ecosystems, and natural and human-induced hazards—is crucial to advancing human health, safety and welfare, alleviating human suffering including poverty, protecting the global environment, reducing disaster losses, and achieving sustainable development. Observations of the Earth system constitute critical input for advancing this understanding.

Interested countries and organizations have collaborated to develop this Plan to ensure comprehensive and sustained Earth observations. It builds on and adds value to existing Earth observation systems by coordinating their efforts, addressing critical gaps, supporting their interoperability, sharing information, reaching a common understanding of user requirements and improving delivery of information to users […]

The purpose of GEOSS is to achieve comprehensive, coordinated, and sustained observations of the Earth system, in order to improve monitoring of the state of the Earth, increase understanding of Earth processes, and enhance prediction of the behavior of the Earth system. GEOSS will meet the need for timely, quality long-term global information as a basis for sound decision making, and will enhance delivery of benefits to society in the following initial areas:

- Reducing loss of life and property from natural and human-induced disasters;
- Understanding environmental factors affecting human health and well-being;
- Improving management of energy resources;
- Understanding, assessing, predicting, mitigating, and adapting to climate variability and change;
- Improving water resource management through better understanding of the water cycle;
- Improving weather information, forecasting, and warning;
- Improving the management and protection of terrestrial, coastal, and marine ecosystems;
- Supporting sustainable agriculture and combating desertification;
- Understanding, monitoring, and conserving biodiversity.

GEOSS is a step toward addressing the challenges articulated by the United Nations Millennium Declaration and the 2002 World Summit on Sustainable Development, including the achievement of the Millennium Development Goals. GEOSS will also further the implementation of international environmental treaty obligations.

Source: earthobservations.org/docs/10-Year%20Implementation%20Plan.pdf (accessed May 2009)
Note: Plan was adopted 16 February 2005.

DOCUMENTATION 6.9. SPACE AND MAJOR DISASTERS

Charter on Cooperation to Achieve the Coordinated Use of Space Facilities in the Event of Natural or Technological Disasters (selected text).

Recognising the potential applications of space technologies in the management of disasters caused by natural phenomena or technological accidents, and in particular Earth observation, telecommunications, meteorology and positioning technologies;

Recognising the development of initiatives concerning the use of space facilities for managing natural or technological disasters;

Recognizing the interest shown by rescue and civil protection, defence and security bodies and the need to respond to that interest by making space facilities more easily accessible;

Desirous to strengthen international cooperation in this humanitarian undertaking;

Having regard to United Nations Resolution 41/65 of 1986 on remote sensing of the Earth from space;

Believing that by combining their resources and efforts, they can improve the use of available space facilities and increase the efficiency of services that may be provided to crisis victims and to the bodies called upon to help them;

Hereby Agree As Follows: [...]

In promoting cooperation between space agencies and space system operators in the use of space facilities as a contribution to the management of crises arising from natural or technological disasters, the Charter seeks to pursue the following objectives:

- supply during periods of crisis, to States or communities whose population, activities or property are exposed to an imminent risk, or are already victims, of natural or technological disasters, data providing a basis for critical information for the anticipation and management of potential crises;
- participation, by means of this data and the information and services resulting from the exploitation of space facilities, in the organisation of emergency assistance or reconstruction and subsequent operations.

Source: www.disasterscharter.org/charter_e.html (accessed May 2009)
Notes: Adopted on 25 April 2000.

DOCUMENTATION 6.10. INTER-AGENCY SPACE DEBRIS COORDINATION COMMITTEE

Inter-Agency Space Debris Coordination Committee (IADC) Terms of Reference (selected text).

1. Purpose: The primary purpose of the IADC is to exchange information on space debris research activities between member space agencies, to facilitate

Documentation

opportunities for cooperation in space debris research, to review the progress of ongoing cooperative activities and to identify debris mitigation options.
2. Rationale: The members share a number of common interests in space debris research which may be developed into a variety of cooperative research activities. Such ventures are likely to increase in frequency and scope in the future. It is highly desirable to exchange information on current research activities so as to identify future cooperative activities. Therefore, the IADC is established to identify, plan, and assist in the implementation of joint cooperative activities that are of mutual interest and benefit.
3. Scope: The IADC will:
   a. review all ongoing cooperative space debris research activities between member organizations;
   b. recommend new opportunities for cooperation;
   c. serve as the primary means for exchanging information and plans concerning orbit debris research activities;
   d. identify and evaluate options for debris mitigation.

Any specific cooperative activities endorsed by the IADC will be implemented through arrangements negotiated between member organisations.
Members should exchange data resulting from national orbital debris programs as appropriate. Data and information exchanged through the IADC will normally be exchanged without restrictions as to use or disclosure. In the event that technical data is exchanged which is considered to be proprietary, and for which protection is desired, the data shall be marked with a notice indicating the use and disclosure restrictions, and the recipient agrees to abide by the terms of such notices.

Source: www.iadc-online.org/index.cgi?item=torp (accessed May 2009)
Notes: Terms of Reference status as of 4 October 2006; Members of the IADC are: Italian Space Agency, British National Space Centre, Centre National d'Etudes Spatiales, China National Space Administration, German Aerospace Centre, European Space Agency, Indian Space Research Organisation, Japan Aerospace Exploration Agency, National Aeronautics and Space Administration, National Space Agency of the Ukraine, and Russian Federal Space Agency.

DOCUMENTATION 6.11. WASSENAAR ARRANGEMENT

Wassenaar Arrangement on Export Controls for Conventional Arms and Dual-Use Goods and Technologies: Guidelines and Procedures, including the Initial Elements (selected text).

As originally established in the Initial Elements adopted by the Plenary of 11–12 July 1996, and as exceptionally amended by the Plenary of 6–7 December 2001.

1. The Wassenaar Arrangement has been established in order to contribute to regional and international security and stability, by promoting transparency and greater responsibility in transfers of conventional arms and dual-use goods and technologies, thus preventing destabilising accumulations. Participating States will seek, through their national policies, to ensure that transfers of these items do not contribute to the development or enhancement of military capabilities which undermine these goals, and are not diverted to support such capabilities.

2. It will complement and reinforce, without duplication, the existing control regimes for weapons of mass destruction and their delivery systems, as well as other internationally recognised measures designed to promote transparency and greater responsibility, by focusing on the threats to international and regional peace and security which may arise from transfers of armaments and sensitive dual-use goods and technologies where the risks are judged greatest.

3. This Arrangement is also intended to enhance co-operation to prevent the acquisition of armaments and sensitive dual-use items for military end-uses, if the situation in a region or the behaviour of a state is, or becomes, a cause for serious concern to the Participating States.

This Arrangement will not be directed against any state or group of states and will not impede bona fide civil transactions. Nor will it interfere with the rights of states to acquire legitimate means with which to defend themselves pursuant to Article 51 of the Charter of the United Nations.

In line with the paragraphs above, Participating States will continue to prevent the acquisition of conventional arms and dual-use goods and technologies by terrorist groups and organisations, as well as by individual terrorists. Such efforts are an integral part of the global fight against terrorism.

Source: www.wassenaar.org/2003Plenary/initial_elements2003.htm (accessed May 2009)

eBooks – at www.eBookstore.tandf.co.uk

A library at your fingertips!

eBooks are electronic versions of printed books. You can store them on your PC/laptop or browse them online.

They have advantages for anyone needing rapid access to a wide variety of published, copyright information.

eBooks can help your research by enabling you to bookmark chapters, annotate text and use instant searches to find specific words or phrases. Several eBook files would fit on even a small laptop or PDA.

NEW: Save money by eSubscribing: cheap, online access to any eBook for as long as you need it.

Annual subscription packages

We now offer special low-cost bulk subscriptions to packages of eBooks in certain subject areas. These are available to libraries or to individuals.

For more information please contact webmaster.ebooks@tandf.co.uk

We're continually developing the eBook concept, so keep up to date by visiting the website.

www.eBookstore.tandf.co.uk

EUROPA WORLD
www.europaworld.com

Librarians, to register for your 30-day **FREE** trial visit:
http://tinyurl.com/ewfreetrial

EUROPA WORLD AND THE EUROPA REGIONAL SURVEYS OF THE WORLD ONLINE

An interactive online library for all the countries and territories of each of the world regions

Europa World enables you to subscribe to *The Europa World Year Book* online together with as many of the nine *Europa Regional Surveys of the World* online as you choose, in one simple annual subscription. The *Europa Regional Surveys of the World* complement and expand upon the information in *The Europa World Year Book* with in-depth, expert analysis at regional, sub-regional and country level.

Key Features:

- Impartial coverage of issues of regional importance from acknowledged experts
- A vast range of up-to-date economic, political and statistical data
- Book and periodical bibliographies – direct you to further research
- Extensive directory of research institutes specializing in the region
- Ability to search by content type across regions
- Thousands of click-through web links to external sites.

A CHOICE Outstanding Resource

30-DAY FREE TRIALS AVAILABLE   30-DAY FREE TRIALS AVAILABLE   30-DAY FREE TRIALS

Alternatively, to register for your free 30-day trial or for more information please contact your sales representative:

UK and Rest of World customers:
Tel: +44 (0)20 701 76062
Fax: +44 (0)20 701 76699
Email: **online.sales@tandf.co.uk**

US, Canadian and South American customers:
Tel: 1-888-318-2367
Fax: 212-244-1563
Email: **e-reference@taylorandfrancis.com**

CPSIA information can be obtained
at www.ICGtesting.com
Printed in the USA
JSHW021512221219
3113JS00001BA/41